Fixed Point Theory
— in —
p-Vector Spaces

Fixed Point Theory
── in ──
p-Vector Spaces

George Xianzhi Yuan

Chongqing University of Technology, Chongqing, China
& Sun Yat-sen University, Guangzhou, China

World Scientific

NEW JERSEY · LONDON · SINGAPORE · BEIJING · SHANGHAI · TAIPEI · CHENNAI

Published by

World Scientific Publishing Co. Pte. Ltd.
5 Toh Tuck Link, Singapore 596224
USA office: 27 Warren Street, Suite 401-402, Hackensack, NJ 07601
UK office: 57 Shelton Street, Covent Garden, London WC2H 9HE

Library of Congress Control Number: 2024056898

British Library Cataloguing-in-Publication Data
A catalogue record for this book is available from the British Library.

FIXED POINT THEORY IN P-VECTOR SPACES

ISBN 978-981-12-7787-0 (hardcover)
ISBN 978-981-12-7788-7 (ebook for institutions)
ISBN 978-981-12-7789-4 (ebook for individuals)

For any available supplementary material, please visit
https://www.worldscientific.com/worldscibooks/10.1142/13456#t=suppl

Desk Editors: Kannan Krishnan/Gabriel Rawlinson

Typeset by Stallion Press
Email: enquiries@stallionpress.com

To all members of my family

Preface

This monograph explores the latest developments in fixed point theory under a unified framework of the "best approximation approach" in p-vector spaces, a core component of nonlinear analysis in mathematics, where $p \in (0, 1]$ (the same applies for p in the following unless specified otherwise). This book presents some key aspects of a new fixed point theory, organized into the following four parts.

First, a general framework is presented to establish the Schauder fixed point theorem in p-vector spaces for both single-valued continuous and upper semicontinuous set-valued mappings, which affirmatively answers the Schauder conjecture in Haunsdorff topological vector spaces (or, in general, p-vector spaces). This conjecture was raised by Schauder [237] in the 1930s, we solve it by using the method of nonlinear analysis with the help of the embedding theorem of compact p-convex sets in locally p-convex spaces, which was established by Kalton [119] in 1977, with $p \in (0, 1)$.

Second, the best approximation approach for classes of semiclosed 1-set contractive single-valued and set-valued mappings in p-vector and locally p-convex spaces is developed in a unified way.

Third, we present the general principles of nonlinear analysis, including the variational principle established by Ekeland [75] in 1972, the Leray–Schauder alternative, the Takahashi minimization theorem, and the Oettli–Théra theorem. For this, we apply the Caristi fixed point theorem, which was established by Caristi [39] in 1976 as a fundamental and powerful tool.

Fourth, fixed point theorems for non-self mappings, in particular, for single-valued and set-valued nonexpansive mappings in locally complete convex spaces, are established as applications of the Caristi fixed point theorem.

The book consists of nine chapters. Chapter 1 gives a brief introduction to the book's goals and the topics covered. Chapter 2 covers the basics of p-vector spaces, where $p \in (0, 1]$. Note that p is defined the same in the following unless specified otherwise. Chapter 3 establishes fixed point theorems under the p-normed spaces from finite-dimensional p-normed spaces to general infinite-dimensional p-normed spaces. In particular, a new concept called the "p-normal structure" in p-normed spaces is introduced, which is used to establish fixed point theorems of nonexpansive mappings in p-normed spaces. Also, an original new tool, called the Dugundji-type extension theorem (Lemma 3.1.4), has been established for p-vector spaces, where $p \in (0, 1]$. Chapter 4 establishes the fixed point theory for single-valued mappings in p-vector spaces, which affirmatively answers the Schauder conjecture [237]. According to this conjecture, a single-valued continuous self-mapping defined on a nonempty compact convex subset of topological vector spaces has at least one fixed point. In addition, we apply the Caristi fixed point theorem to present the general principles of nonlinear analysis, including those given by Ekeland, Leray–Schauder, Takahashi, and Oettli–Théra. Chapter 5 focuses on the development of new fixed point theorems for set-valued mappings in locally p-convex spaces and p-vector spaces; in particular, fixed point theorems of upper semi-continuous set-valued mappings defined on s-convex subsets in Hausdorff p-topological vector spaces provide a positive answer to Schauder's open question. Chapter 6 establishes the general best approximation results for both single-valued and set-valued mappings in locally p-convex spaces. Chapter 7 focuses on the development of a new fixed point theory and a new principle of nonlinear analysis for single-valued mappings in p-vector spaces, while Chapter 8 focuses on the development of the same for set-valued mappings in locally p-convex spaces. In particular, fixed point theorems for non-self mappings of set-valued nonexpansive mappings in locally complete convex spaces are established as applications of the Caristi fixed point theorem [39]. Chapter 9 consists of some notes and remarks on the topics presented in Chapters 1–8, and it also provides a brief discussion of the topics related to fixed point theory not touched upon in this book, with some essentially information or supplementary materials in helping readers better understanding related issues discussed by each chapter.

Specifically, the book focuses on the development of a general new fixed point theory for both single- and set-valued mappings under the framework of both p-vector and locally convex spaces for $p \in (0, 1]$ by including topological vector spaces and locally convex spaces as special classes, respectively. It affirmatively answers the Schauder conjecture under the general setting of p-vector spaces and locally p-convex spaces. The book establishes best approximation results for upper semicontinuous and 1-set contractive set-valued mappings, which are used as tools to establish new fixed point theorems for non-self set-valued mappings with either inward or outward set conditions in various situations. These results improve or unify the corresponding results in the existing literature for nonlinear analysis. They also allow for the establishment of a fundamental general theory for the development of fixed point theorems in topological vector spaces following the Schauder conjecture raised in the 1930s for nonlinear analysis in mathematics. In addition, this book also demonstrates the power of the fixed point theorem by showing the equivalence among the Ekeland variational principle, the Takahashi minimization theorem, the Oettli–Théra theorem, the Caristi–Kirk-type fixed point theorem, and other related principles in nonlinear functional analysis. We summarize the other highlights of this book as follows.

This book reports for the first time new developments in fixed point theory and its associated principle for nonlinear analysis by applying the best approximation approach for the classes of semi-closed 1-set contractive single-valued and set-valued mappings in p-vector spaces for $p \in (0, 1]$ (with p remaining the same in the following), which include topological vector spaces as a special class.

Secondly, general new fixed point theorems for single-valued continuous and upper semicontinuous set-valued mappings in topological vector spaces or p-vector spaces are systematically established, which provides an affirmative answer to the Schauder conjecture under the setting of p-vector spaces.

Thirdly, the best approximation result for upper semicontinuous and 1-set contractive set-valued is established, which acts as a useful tool to establish fixed points of non-self set-valued mappings with either inward or outward set conditions.

Fourth, the book introduces a new concept called the "p-normal structure" for p-normed spaces. This new notion has the potential to serve as a key tool for the development of a new geometry theory for p-normed spaces in general. In addition, a new Dugundji-type extension theorem/lemma (i.e., Lemma 3.1.4) and selection theorem (i.e., Lemma 5.3.4) for upper semicontinuous set-valued mappings have been established for the first time and originally under the framework of locally p-convex spaces, which are expected to prove useful as tools/methods for nonlinear analysis in functional analysis and topology.

Of course, the biggest shortcoming of this book is the lack of discussion on topics such as partial differential equations, optimization, and game theory and disciplines such as social sciences and economics, as well as other related areas that use the fixed point approach for nonlinear analysis. We refer interested readers to some of the classical monographs, such as Agarwal *et al.* [1], Agarwal and O'Regan [3], Aubin and Ekeland [10], Bauschke and Combettes [14], Borwein and Lewis [25], Brézis [26], Browder [36], Chang [46], Cobzaş [61], Ekeland [77], Granas and Dugundji [102], Isac [118], Jachymski *et al.* [117], Kalton *et al.* [122], Lax [153], Park [189–195], Rockafellar [226], Rockafellar and Wets [227], Rothe [231], Takahashi [255], Zălinescu [296], Zeidler [297], Zhang [299], and the comprehensive references therein.

Overall, this book offers an easily accessible approach to establishing a new theory in the development of fixed point theorems and results, making it suitable for senior-level undergraduate students majoring in mathematics, physical sciences, social sciences, and other related disciplines. We do hope that this monograph will become a staple textbook for both undergraduate and postgraduate students, while also serving as a reference for researchers in the field of fixed point theory in nonlinear functional analysis. In addition, it is easily accessible to general readers in mathematics and related disciplines.

The author is thankful to his colleagues at home and abroad for their support during the preparation of this book and to his family members for their understanding, warm encouragement, and constant patience. Of course, his sincere thanks also go to World Scientific for their strong support, particularly to Rebecca Fu and Britney Jiang at the Singapore and Shanghai offices, Gabriel Rawlinson at the London office, and their associated technical and editorial teams led by Kannan Krishnan (Book Editor, Academic Consulting and Editorial Services (ACES) Pvt. Ltd., World Scientific Group Company). Without their support, the publication of this monograph would not have been possible in such a short time.

In addition, the author thanks Professor S. S. Chang (Shi-Sheng Zhang), Professor K. K. Tan, Professor Hong Ma, Professor Jian Yu, Professor Jinlu Li, Professor B. Sims, Professor Bevan Thompson, Professor Y. J. Cho, Professor S. Park, Professor Jian-Zhong Xiao, Professor Mohamed Ennassik, Professor Mohamed Aziz Taoudi, Professor M. Nashed, Professor D. O'Regan, and Professor Bruce Smith for their constant encouragement for more than two decades; and my thanks also go to Professor H. Bauschke, Professor Hong-Kun Xu, Professor Tiexin Guo, Professor Xiao-Long Qin, Professor Ganshan Yang, Professor Xian Wu, Professor Nanjing Huang, Professor Dewen Xiong, and Professor Shi-Qing Zhang, as well as my colleagues and friends across China, Australia, Canada, the UK, the US, and elsewhere.

Part of the original research presented in this book was carried out with the support of the National Natural Science Foundation of China [grant numbers 71971031 and U1811462].

Finally, the author welcomes any suggestions and comments from the readers. He can be reached anytime via email at "george_yuan99@yahoo.com".

George Xianzhi Yuan
Chongqing University of Technology,
Chongqing, China and
Sun Yat-sen University,
Guangzhou, China

About the Author

George Xianzhi Yuan is a professor at Chongqing University of Technology (Chongqing, China). He also serves as a guest or visiting professor at several other universities, including Sun Yat-sen University (Guangzhou) and East China University of Science and Technology (Shanghai). He is the editor-in-chief of the *International Journal of Financial Engineering* (IJFE) and a member of the editorial boards of several academic journals, both domestic and international. Dr. Yuan has also been a distinguished professor at various universities and research institutions, including Tongji University (Shanghai), Soochow University (Suzhou), Nankai University (Tianjin), and the School of Management of the former Graduate School of the Chinese Academy of Sciences (Beijing). In addition, he has also served as the chief scientist or business R&D manager for several financial technology (fintech) and digital technology companies in the industry.

Dr. Yuan has over 30 years of work and study experience both domestically and internationally, including in the United States, Canada, Australia, and China. He has published over 160 academic research papers in Science Citation Index and Social Sciences Citation Index academic journals, as well as multiple monographs in nonlinear analysis and applications in mathematical finance and two textbooks on fintech.

Dr. Yuan has made significant internationally recognized contributions by combining theory and practice in nonlinear analysis, particularly in its related fields such as fixed point theory, mathematical finance, mathematical economics, game theory, and their applications. He has also participated in and led several major national-level projects and has mentored numerous leading academics and industry elites. Dr. Yuan's research in nonlinear analysis and its related applications has been highly praised by internationally renowned scholars such as Professor Ken Arrow (Stanford University), who won the Nobel Prize in 1972, and Professor John Nash (Princeton University), who won the Nobel Prize in 1994.

Dr. Yuan holds bachelor's and master's degrees from Sichuan University (Chengdu, China), a PhD in mathematics and a master's degree in statistics from Dalhousie University (Halifax, Canada), and a master's degree in financial engineering from the University of Toronto (Toronto, Canada). He gained postdoctoral work experience in mathematics from The University of Queensland (Brisbane, Australia).

Dr. Yuan has authored several monographs and textbooks, including *The Study of Minimax Inequalities and Applications to Economies and Variational Inequalities.* Memoirs of the American Mathematical Society, Volume 132, *Amer. Math. Soc.*, 1998; *KKM Theory and Applications in Nonlinear Analysis.* Marcel Dekker, New York, 1999; *The CME Vulnerability: The Impact of Negative Oil Futures Trading.* World Scientific Publishing, Singapore, 2020; *The Textbook of Fintech in Practice.* Tsinghua University Press, Beijing, 2023; and *Introduction to Machine Learning Algorithms for Fintech and Big Data Risk Control.* China Science Press, Beijing, 2023.

Contents

Chapter 1

Introduction

The goal of this chapter is to give a brief overview to the contents of this book and its structure. This chapter consists of two short sections.

Section 1.1 contains a brief introduction to the contents covered in this book. Section 1.2 explains the structure of the material presented in the book, which provides an introduction to its main content and an outline of how the contents are organized and applied.

1.1 Overview

It is known that the class of p-seminorm spaces for $p \in (0,1]$ is an important generalization of normed spaces with rich topological and geometrical structures. Its study has received considerable attention, e.g., see Alghamdi *et al.* [5], Balachandran [13], Bayoumi [15], Bayoumi *et al.* [16], Bernuées and Pena [21], Ding [65], Ennassik and Taoudi [79], Ennassik *et al.* [78], Gal and Goldstein [93], Gholizadeh *et al.* [94], Jarchow [116], Kalton [119–121], Kalton *et al.* [122], Machrafi and Oubbi [162], Park [189–200], Qiu [215–220], Qiu and Rolewicz [221], Rolewicz [228], Silva *et al.* [249], Simons [247], Tabor *et al.* [253], Tan [257], Wang [263], Xiao and Lu [268], Xiao and Zhu [270], Yuan [286–291], Zeidler [297], Zhong [300], and Zhong *et al.* [301, 302].

However, to the best of our knowledge, the corresponding fixed point theories, which are basic tools, and associated results in the field of nonlinear functional analysis have not been well developed. Therefore, the goal of this book is to develop some important tools in the nonlinear analysis of 1-set contractive singe-valued and set-valued mappings under the framework of p-vector and locally p-convex spaces, with emphasis on the development of "fixed point theory in p-vector spaces", where $p \in (0, 1]$, by including topological vector spaces and locally convex spaces as special classes of p-vector and locally p-convex spaces with $p = 1$.

In particular, we first develop the **general fixed point theorems for continuous single-valued mappings**, which affirmatively answer the Schauder conjecture raised in 1930 under the general framework of p-vector spaces, and then **general fixed point theorems for upper semicontinuous set-valued mappings** under Hausdorff topological vectors (or p-vector spaces) are also established without requiring the metric on underlying spaces in Chapter 5 for $p \in (0, 1]$ by applying the functional analysis method. Then, the best approximation results for upper semicontinuous and 1-set contractive mappings are given with boundary conditions, which are used as tools to establish fixed points for non-self single-valued or set-valued mappings with either inward or outward set conditions. Finally, we derive existence results for solutions of Birkhoff–Kellogg problems (proposed by Birkhoff and Kellog [23] in 1922), the general principle of nonlinear alternatives by including the Leray–Schauder alternative [154], and related results as special classes. The results given in this book not only include the corresponding results in the existing literature as special cases but are also expected to serve as useful tools for studying nonlinear problems arising in social sciences, engineering, applied mathematics, and related topics and areas.

Before reviewing the studies on best approximation and related nonlinear analysis tools within the framework of p-vector spaces, we would first like to share with readers a few remarks on these aspects. Although most results in nonlinear analysis are normally highly associated with convexity hypotheses in local convex topological vector spaces, it is surprising that p-vector spaces – despite generally lacking a local convex structure compared to locally convex spaces – exhibit some elegant properties. These properties naturally arise with various convenient approximations and better (i.e., larger) structures for the so-called "convexities" of p-convex subsets. They also play a critical role in describing the Birkhoff–Kellogg problems [23] and related nonlinear problems, including fixed point problems, in topological vector spaces based on the behavior of p-vector spaces for $p \in (0, 1]$. Here, we recall that, since the first Birkhoff–Kellogg problem was introduced and its associated theorem was proved by Birkhoff and Kellogg [23] in 1922 while discussing the existence of solutions for the equation $x = \lambda F(x)$, where λ is a real parameter and F is a general nonlinear non-self mapping defined on an open convex subset U of a topological vector space E, the focus has shifted to finding a general form of the Birkhoff–Kellogg problem in the so-called invariant direction for nonlinear set-valued mappings F, i.e., to find $x_0 \in \overline{U}$ and $\lambda > 0$ such that $\lambda x_0 \in F(x_0)$, here $\overline{U}$ is the closure of U.

Since Birkhoff and Kellogg developed their theorem, the study of Birkhoff–Kellogg problems has received significant attention from scholars. For example, in 1934, one of the fundamental results in nonlinear functional analysis, famously called the Leray–Schauder alternative, was established via the topological degree by Leray and Schauder [154]. Thereafter, certain other types of Leray–Schauder alternatives were proved using techniques different from the topological degree. For example, see the works by Granas and Dugundji [102] and Furi and Pera [92] in the Banach space setting and applications to the boundary value problems for ordinary differential equations in noncompact problems. These include a general class of mappings for nonlinear alternatives of the Leray–Schauder-type in normal topological spaces and some Birkhoff–Kellogg-type theorems for general class mappings in topological vector spaces by Agarwal *et al.* [1], Agarwal and O'Regan [2,3], Park [186–189], and O'Regan [180], who used the Leray–Schauder-type coincidence theory to establish the Birkhoff–Kellogg problem and Furi–Pera-type result for a general class of set-valued mappings. In this book, based on the application of best approximation as a tool for general 1-set contractive set-valued mappings, we develop the general principle for the existence of solutions for Birkhoff–Kellogg problems. We also establish related nonlinear alternatives, which then allows us to analyze the general existence of Leray–Schauder-type and related fixed point theorems for non-self mappings in general p-vector spaces for $p \in (0, 1]$. These new results given in this book not only include the corresponding results in the existing literature as special cases but are also expected to be useful for the study of nonlinear problems arising from theory and practice.

Now, we give a brief introduction to the best approximation theorem related to the development of nonlinear analysis, which is a powerful tool in mathematics.

We know that best approximation is naturally related to fixed points for non-self mappings, which are tightly linked with the classical Leray–Schauder alternative based on the continuation theorem developed by Leray and Schauder [154]. This theorem is a remarkable result in nonlinear analysis. In addition, there exist several continuation theorems, which have many applications in the study of nonlinear functional equations (see O'Regan and Precup [182]). Historically, it seems that the continuation theorem is based on the idea of obtaining a solution to a given equation, starting with one of the solutions of a simpler equation. The essential part of this theorem is the "Leray–Schauder boundary condition". It seems that the "continuation method" was initiated by Poincare [211] and Bernstein [20]. Certainly, Leray and Schauder [154] were the first to give an abstract formulation of the "continuation principle" in 1934 using the topological degree theory; see also Granas and Dugundji [102], Isac [118], Rothe [230, 231], and Zeidler [297]. However, in this book, we examine how the best approximation method could be used for the study of fixed point theorems in p-vector space for $p \in (0, 1]$, which, as a basic tool, helps us develop the principle of nonlinear alternatives, the Leray–Schauder alternative, and the fixed point theorems of Rothe, Petryshyn, and Altman types for non-self mappings; see Almezel *et al.* [6], O'regan [181], and Ruzhansky *et al.* [234] on the development of fixed point theory and applications in general, as well as Qiu [220] and Zhong *et al.* [302] for the study of the Ekeland variational principle and its relation to fixed point theory and applications. Of course, for a list of comprehensive references, recent developments in the general framework of fixed point theory, and its study and applications, we refer interested readers to Park [189–200], Xu [272–275], Xu and Muglia [277], Yuan [286–291], Yuan and Xiao [294, 295], and the related references therein.

It is well known that best approximation is an important aspect in the study of nonlinear problems related to the solvability of partial differential equations, dynamic systems, optimization, mathematical programs, and operation research. In particular, it is an approach that is widely accepted and used in the study of nonlinear problems in optimization, complementarity problems, and variational inequalities problems. This approach is strongly based on what is today called Fan's best approximation theorem. Developed by Fan [84–86] in 1969, this theorem serves as a powerful tool in nonlinear analysis; see the book by Singh *et al.* [248] for a related discussion and study on the fixed point theory and best approximation with the KKM-map principle. Among the related tools are the Rothe type and the principle of the Leray–Schauder alternative in topological vector spaces (in short, TVSs) and local topological vector spaces (in short, LCSs), which are comprehensively studied by Chang *et al.* [49], Chang *et al.* [53–55], Carbone and Conti [38], Ennassik and Taoudi [79], Ennassik *et al.* [78], Isac [118], Granas and Dugundji [102], Kirk and Shahzad [137], Liu [160], Park [189], Rothe [230,231], Shahzad [238,240], Xu [272], Yuan [283–291], Zeidler [297], and the references therein.

On the other hand, the celebrated KKM principle [142], established in 1929, is based on the well-known Sperner combinatorial lemma and was first applied to derive a simple proof of the Brouwer fixed point theorem. Later, it became clear that these three theorems are mutually equivalent. They were even regarded as a sort of mathematical trinity by Park [189]. Since Fan [82–85] extended the classical KKM theorem to infinite-dimensional spaces in 1961, there have been a number of generalizations and applications in numerous areas of nonlinear analysis, as well as fixed point theorems in TVS and LCS as developed by Browder [30–36] and the related references therein. Among them, the Schauder fixed point theorem [237] in normed spaces is a powerful tool in dealing with nonlinear problems in analysis. Most notably, it has played a major role in the development of fixed point theory and related nonlinear analysis and the mathematical theory of partial and ordinary differential equations. A generalization of the Schauder theorem from normed space to general topological vector spaces is an old conjecture in fixed point theory, which is explained in Problem 54 in *The Scottish Book* by Mauldin [167]. It is stated as follows:

Schauder Conjecture: "Does every nonempty compact convex set in a topological vector space have the fixed point property? Or, in its analytic statement: Does a continuous function defined on a compact convex subset of a topological vector space to itself have a fixed point?"

In a recent breakthrough, this conjecture has been solved was announced by French mathematician Cauty, starting with his initial work [41] in 2001. But this was followed by several subsequent works in 2005 [42], 2007 [43], and 2010 [44], which, based on his initial work, filled in the gaps. Therefore, he could affirmatively solve the Schauder conjecture completely under the framework of linear topological metric spaces by introducing a new concept called "algebraic absolute neighborhood retracts" (algebraic ANRs), which comes under the category of algebraic topology in mathematics.

However, in 2021, two mathematicians, Ennassik and Taoudi, from Morocco in their work [79] provided a smart and remarkable alternative method to affirmatively solve the Schauder conjecture for general linear topological spaces and p-vector spaces, where $p \in (0, 1]$. They applied a functional analysis method that was supported by the embedding theorem of compact p-convex subsets in locally p-convex spaces, established by another remarkable mathematician, Kalton [119], in 1977. Actually, in Chapter 5, we establish new general **fixed point theorems for upper semicontinuous set-valued mappings in both locally p-convex and p-vector spaces, which provide a unified positive answer to the Schauder conjecture**.

In this book, we explain how the general fixed point theory in p-vector spaces could be developed by following the analysis method used by Ennassik and Taoudi in their work [79], which is widely used as a primary tool in functional analysis, a core component in mathematics.

Finally, we would like to point out here that on the path to solving the Schauder conjecture, spanning the past 90 years or so, many scholars around the world have extensively worked on this question and the related fixed point theory. Here, we only mention a few of them, as follows: Ennassik and Taoudi [79] and Ennassik *et al.* [78] used the p-seminorm method (analysis method) under p-vector spaces; significant contributions were made by Cauty [41–44], as well as many others, under the general framework of p-vector spaces for even non-self set-valued mappings for $p \in (0, 1]$. Other contributors include Askoura and Godet-Thobie [8], Ben-El-Mechaiekh and Mechaiekh [17], Browder [30–36], Chang [47], Chang *et al.* [49], Chang and Yen [55], Chen [56], Dobrowolski [67], Fan [81–86], Gholizadeh *et al.* [94], Goebel and Kirk [96], Górniewicz [100], Górniewicz *et al.* [101], Granas and Dugundji [102], Isac [118], Jachymski [117], Kirk and Sims [138], Ko and Tsai [143], Kryszewski [147], Li [157, 158], Li *et al.* [156], Liu [160], Nhu [175], Okon [178],

Park [188–200], Reich [222], Leray and Schauder [154], Shahzad [240], Shih and Tan [243], Smart [250], Tarafdar and Yuan [259], Weber [264, 265], Xiao and Lu [268], Xiao and Zhu [270], Xu [278], Xu *et al.* [279], Yuan [283–291], and related references therein.

The goal of this book is to establish general new tools for nonlinear analysis under the framework of general p-vector spaces and locally p-convex spaces for 1-set contractive mappings, where $p \in (0, 1]$. Secondly, we do wish that these new results, such as best approximation; theorems of the Birkhoff–Kellogg type; nonlinear alternatives; fixed point theorems for non-self set-valued with boundary conditions; and Rothe-type, Petryshyn-type, Altman-type, Leray–Schauder-type, and other related nonlinear problems, will play important roles for the nonlinear analysis of p-seminorm spaces.

In addition, our discussion and results on the fixed point theorem for the development of new fixed point theorems for both single-valued and upper semicontinuous set-valued mappings in both locally p-convex spaces and p-vector spaces defined on s-convex subsets in Hausdorff p-topological vector spaces are presented systematically for the first time. Most results are new; they provide a positive answer for the Schauder conjecture under the general setting of p-vector spaces (which are not be locally convex; see Kalton [119–121], Kalton *et al.* [122], Jarchow [116], Roloewicz [228], Fan [81–85], Singh *et al.* [248], and related references).

Hopefully, this book will serve as a staple textbook for undergraduate and postgraduate students, a reference book for researchers in the field of fixed point theory in nonlinear functional analysis, and an easily accessible text for general readers in mathematics and related disciplines.

1.2 Structure of the Book

In this section, we briefly explain the organization of the material covered in this book, which offers an introduction to its main contents and an outline of how the contents are applied.

First, the book starts with an introduction to the basics of p-vector spaces, which will be used to develop nonlinear analysis by applying the best approximation approach for the classes of semiclosed 1-set contractive set-valued mappings in p-vector spaces for $p \in (0, 1]$, which include nonexpansive mappings, Caristi type fixed point theorem and related principles in nonlinear analysis.

Second, the general fixed point theorems of upper semicontinuous set-valued mappings in Hausdorff topological vector spaces and p-vector spaces provide an affirmative answer to the Schauder conjecture [237] under the setting of p-vector spaces.

Third, the best approximation result for upper semicontinuous and 1-set contractive set-valued mappings is established, which is used as a helpful tool to establish fixed points of non-self set-valued mappings with either inward or outward set conditions.

Fourth, this book provides an easily accessible approach to establishing a new theory, making its contents easy to understand for senior-level undergraduates majoring in mathematics, physical sciences, and even social sciences.

Chapter 2

The Basics of p-Vector Spaces

The goal of this chapter is to present a basic discussion of p-vector spaces, where $p \in (0, 1]$. The chapter consists of the following three sections.

The first section introduces some key concepts of p-vector spaces; the second section discusses some properties of p-vector spaces; and the final section presents a discussion on the embedding theorem for compact p-convex sets for $p \in (0, 1)$ given by Kalton [119] in 1977 in topological vector spaces, which allows us to prove the Schauder fixed point theorem for single-valued continuous mappings defined on the compact convex sets in p-vector spaces (where $p \in (0, 1]$), which include topological vector spaces as a special class.

Throughout this chapter and the remainder of this book, for the convenience of our discussion, if X is a vector space, then θ denotes the zero element (or, alternatively, the origin) and the numerical zero in space X, respectively. (X, T_X) and (Y, T_Y) are assumed to be two topological spaces with topology structures T_X and T_Y, respectively, which are in short denoted by X and Y whenever there is no risk of confusion. All p-vector spaces are assumed to be Hausdorff, and $p \in (0, 1]$ unless specified otherwise.

We denote by $\mathbb{N}$ the set of all positive integers, i.e., $\mathbb{N} = \{1, 2, \ldots, \}$. For a given set X, 2^X denotes the family of all subsets of X. We denote by $\mathbb{R}^n$ finite n-dimensional Euclidean spaces, where n is a positive integer. The letter $\mathbb{K}$ is used to denote the field of either real numbers or complex numbers unless specified otherwise. For the purposes of this book, we also use the symbols $\mathbb{R}$ and $\mathbb{C}$ to represent the real field and the complex field, respectively. Of course, more notions and mathematical symbols will be introduced in chapters or sections later when they are required and used.

2.1 Some Concepts of p-Vector Spaces

The goal of this section is to introduce some key concepts of p-vector spaces by including general topological vector spaces as a special class in functional analysis, where $p \in (0, 1]$. We now recall some notions and definitions for p-convex topological vector spaces which will be used in the chapter (see Balachandran [13], Bayoumi [15], Jarchow [116], Kalton [119–121], Qiu and Rolewicz [221], Rolewicz [228], Gholizadeh *et al.* [94], Ennassik *et al.* [78], Ennassik and Taoudi [79], Xiao and Lu [268], Xiao and Zhu [270], Yuan [286–291], and the references therein).

Definition 2.1.1. Let $p \in (0, 1]$. A set A in a vector space E is said to be p-convex if for any $x, y \in A$, we have $sx + ty \in A$, whenever $0 \leq s, t \leq 1$ with $s^p + t^p = 1$. The set A is said to be absolutely p-convex if for any $x, y \in A$, we have $sx + ty \in A$, whenever $|s|^p + |t|^p \leq 1$. In the case of $p = 1$, the concept of (absolutely) 1-convexity is simply the usual (absolutely) convexity defined in vector spaces.

Definition 2.1.2. Let $p \in (0, 1]$. If A is a subset of a topological vector space E, the p-convex hull of A and its closed p-convex, denoted by $co_p(A)$ and $\overline{co}_p(A)$, respectively, are the smallest p-convex set containing A and the smallest closed p-convex set containing A in E, respectively.

Definition 2.1.3. Let $p \in (0, 1]$, A be p-convex, and $x_1, \ldots, x_n \in A$, $t_i \geq 0$ with $\sum_1^n t_i^p = 1$. Then, $\sum_1^n t_i x_i$ is called a p-convex combination of $\{x_i\}$ for $i = 1, 2, \ldots, n$. If $\sum_1^n |t_i|^p \leq 1$, $\sum_1^n t_i x_i$ is called an absolutely p-convex combination. It is easy to see that $\sum_1^n t_i x_i \in A$ for a p-convex set A.

Definition 2.1.4. A subset A of a vector space E is called balanced (or circled) if $\lambda A \subset A$ holds for all scalars λ satisfying $|\lambda| \leq 1$. We say that A is absorbing if for each $x \in X$, there is a real number $\rho_x > 0$ such that $\lambda x \in A$ for all $\lambda > 0$ with $|\lambda| \leq \rho_x$.

Definition 2.1.5. Let E be a vector space and $\mathbb{R}^+$ be a non-negative part of a real line $\mathbb{R}$. Then, a mapping $P : E \longrightarrow \mathbb{R}^+$ is said to be a p-seminorm if it satisfies the following requirements for $p \in (0, 1]$:

  (i) $P(x) \geq 0$ for all $x \in E$;
 (ii) $P(\lambda x) = |\lambda|^p P(x)$ for all $x \in E$ and $\lambda \in R$;
(iii) $P(x + y) \leq P(x) + P(y)$ for all $x, y \in E$.

We recall that a p-seminorm P is called a p-norm if $x = 0$ whenever $P(x) = 0$. A vector space E with a specific p-norm is called a p-normed space. A linear space E on which there is a p-norm is called a p-normed space and denoted by $(E, \| \cdot \|_p)$. If $p = 1$, then it is a usual normed space. A p-normed space is also a metric linear space with a translation-invariant metric $d_p(x, y) := \|x - y\|_p$ for $x, y \in E$.

We now introduce the following concept of Minkowski p-functional (or, alternatively, p-gauge), which plays a key role in the study of p-vector spaces, where $p \in (0, 1]$ (see Balachandra [13], Jarchow [116], and the classical references therein).

Definition 2.1.6. Let A be an absorbing subset of a vector space E. For $x \in E$ and $p \in (0, 1]$, set $P_A := \inf\{\alpha > 0 : x \in \alpha^{\frac{1}{p}} A\}$. Then, the non-negative real-valued function P_A is called the p-gauge (or simply gauge if $p = 1$). The p-gauge of A is also known as the Minkowski p-functional for $p \in (0, 1]$.

The concept of Minkowski p-functionals (p-gauges) allows us to construct a corresponding p-seminorm for each given absolutely p-convex (and thus also absorbing) subset in vector spaces. Indeed, by Proposition 4.1.10 of Balachandra [13], we have the following proposition.

Proposition 2.1.7. *Let A be an absorbing subset of E and $p \in (0, 1]$. Then, the p-gauge P_A has the following properties:*

  (i) $P_A(0) = 0$;
 (ii) $P_A(\lambda x) = |\lambda|^p P_A(x)$ *if* $\lambda \geq 0$;
(iii) $P_A(\lambda x) = |\lambda|^p P_A(x)$ *for all* $\lambda \in R$, *provided A is balanced* (*circled*);
(iv) $P_A(x + y) \leq P_A(x) + P_A(y)$ *for all* $x, y \in A$, *provided A is p-convex.*

In particular, P_A is a p-seminorm if A is absolutely p-convex (and also absorbing).

Remark 2.1.8. It is worthwhile to note that a zero-neighborhood in a topological vector space E being an absolutely θ-neighborhood is also absorbing (by Lemma 2.1.16 of Balachandran [13] or Proposition 2.2.3 of Jarchow [116]). This leads us to the following definition for a topological p-vector space (in short, p-vector space) by using the concept of Minkowski p-functionals for $p \in (0, 1]$.

Definition 2.1.9. A vector space E is said to be a topological p-vector space (p-vector space) if the base of the origin in E is generated by a family of Minkowski p-functionals (p-gauges), where $p \in (0, 1]$.

By incorporating Proposition 2.1.7, it seems that the following is a natural way of obtaining the definition for locally p-convex spaces, where $p \in (0, 1]$.

Definition 2.1.10. A topological p-vector space E is said to be locally p-convex if the origin element θ in E has a fundamental set of absolutely p-convex θ-neighborhoods. This topology can be determined by p-seminorms, which are defined in the obvious way (see Bayoumi [15, p. 52], Jarchow [116], or Rolewicz [228]). When $p = 1$, a locally p-convex space E is reduced to a usual locally convex space.

We know that the topology of every Hausdorff locally bounded topological vector space is given by some p-norm (e.g., see Jarchow [116]). Secondly, the space $L^p(\mu)$ is a p-normed space based on the complete measure space $(\Omega, \mathcal{M}, \mu)$, with the p-norm given by

$$\|f(t)\|_p = \int_\Omega |f(t)|^p d\mu, \text{ for } f \in L^p(\mu),$$

where Ω is a nonempty set, $\mathcal{M}$ is a σ-algebra in Ω, $\mu : \mathcal{M} \to [\theta, \infty)$ is a positive measure, and

$$L^p(\mu) = \left\{ f | f : \Omega \to \mathbb{K} \text{ is measurable and } \int_\Omega |f(t)|^p d\mu < \infty \right\}.$$

If μ is the Lebesgue measure on $[0, 1]$, then it is customary to write $L^p[0, 1]$ instead of $L^p(\mu)$. If μ is the counting measure on

$\Omega = \{1, 2, \ldots, n\}$ or $\Omega = \{1, 2, \ldots, \}$, then the corresponding spaces are denoted by $l^p(n)$ and l^p, respectively. Here, we also note that $l^p(n)$ is an n-dimensional space and l^p is a complete separable space. The class of p-normed spaces for $p \in (0, 1]$ is an important generalization of the class of usual normed spaces of p-normed spaces with $p = 1$.

It is well known that most of the fixed point theorems are concerned with some convex sets. But the unit ball with center θ in a p-normed space for $p \in (0, 1)$ is not a convex set. We know that every open ball in l^p for $p \in (0, 1)$ does not contain an open convex subset; and there is no open convex subset in $L^P[0, 1]$ for $p \in (0, 1)$, except for $L^p[0, 1]$ itself, and indeed its dual space only contains the zero element (that is, the only continuous functional is zero) (see Kalton *et al.* [122], Rudin [232, pp. 36–37] and related references). Here, we also point out that the existence of fixed points for nonconvex sets is very useful both in theory and in applications. Even in usual normed spaces, the existence of fixed points for operators on nonconvex sets is often considered, and much attention has been given to these problems. For example, Klee [140] established some fixed point theorems in Hausdorff topological vector spaces without local convexity under some symmetry conditions. Bayoumi [15] also established the generalized Brouwer fixed point theorem and the generalized Kakutani fixed point theorem for a p-convex subset in p-convex Fréchet spaces. Zeidler [297] gives a very comprehensive discussion on the development of fixed point theory in functional analysis with applications in mathematics and various related fields.

We now introduce some fundamental definitions for a given set in vector spaces. In mathematics, the linear span (also called the linear hull or, simply, *span*) of a subset S of vectors (from a vector space E), denoted by $span(S)$ or $lin(S)$, is defined as the set of all linear combinations of the vectors in S. For example, two linearly independent vectors span a plane. The linear span can be characterized either as the intersection of all linear subspaces that contain S or as the smallest subspace containing S. The linear span of a set of vectors is therefore a vector space itself.

To express that a vector space E is a linear span of a subset S, one commonly uses one of the following phrases: S spans E, S is a spanning set of E, E is spanned/generated by S, or S is a generator or generator set of E.

Definition 2.1.11 (linear hull (span)). Given a vector space E over a field $\mathbb{K}$ (which may be either a real field $\mathbb{R}$ or a complex field $\mathbb{C}$), the span of a subset S of vectors (not necessarily finite) is defined as the intersection W of all subspaces of E that contain S. W is referred to as the subspace spanned by S or by the vectors in S. Conversely, S is called a spanning set of W, and we say that S spans W.

Alternatively, the linear hull (span) of S may be defined as the set of all finite linear combinations of elements (vectors) of S, which follows from the above definition as follows:

$$lin(S)(\text{or } span(S)) := \left\{ \sum_{i=1}^{n} \lambda_i v_i : n \in \mathbb{N}, v_i \in S, \lambda_i \in \mathbb{K} \right\}.$$

By following Axler [11, p. 18], we recall that a subset U of E is called a subspace of a vector space E if U is also a vector space (using the same addition and scalar multiplication as on E). Thus, the conditions for U of E is a subspace of E if and only if U satisfies the following three conditions: (1) additive identity, i.e., $o \in U$; (2) closed under addition, i.e., $u, w \in U$ implies $u + w \in U$; and (3) closed under scalar multiplication, i.e., $\alpha \in \mathbb{K}$ (a real field $\mathbb{R}$ or a complex field $\mathbb{C}$), $u \in U$, $\alpha u \in U$.

We now have the following proposition, which provides another equivalence of definitions for the linear hull (span) of S in vector spaces, i.e., the "span" is the smallest containing subspace, as shown in the following.

Proposition 2.1.12. *The set of all linear combinations of a subset S of a vector space E over $\mathbb{K}$ (either a real field or a complex field) is the smallest linear subspace of E containing S.*

Proof. Suppose $v_1, \ldots, v_m$ is a list of vectors in E (see Axler [11, p. 29]). We first show that $span(v_1, \ldots, v_m)$ is a subspace of E. The additive identity is in $span(v_1, \ldots, v_m)$ because $0 = 0v_1 + \cdots + 0v_m$. Also, $span(v_1, \ldots, v_m)$ is closed under addition because

$$(a_1 v_1 + \cdots + a_m v_m) + (c_1 v_1 + \cdots + c_m v_m)$$
$$= (a_1 + c_1)v_1 + \cdots + (a_m + c_m)v_m.$$

Furthermore, $span(v_1, \ldots, v_m)$ is closed under scalar multiplication because

$$\lambda(a_1 v_1 + \cdots + a_m v_m) = \lambda a_1 v_1 + \cdots + \lambda a_m v_m.$$

Thus, $span(v_1, \ldots, v_m)$ is a subspace of E.

Note that each v_j is a linear combination of $v_1, \ldots, v_m$, and thus $span(v_1, \ldots, v_m)$ contains each v_j. Conversely, because subspaces are closed under scalar multiplication and addition, every subspace of V containing each v_j contains $span(v_1, \ldots, v_m)$. Thus, $span(v_1, \ldots, v_m)$ is the smallest subspace of E containing all the vectors $v_1, \ldots, v_m$. This completes the proof. □

In mathematics, the affine hull (or affine span) of a subset S in an Euclidean space $\mathbb{R}^n$ is the smallest affine set containing S, or, equivalently, the intersection of all affine sets containing S. Here, an affine set may be defined as the translation of a vector subspace.

Definition 2.1.13 (affine hull). The affine hull (denoted by aff(S)) of S is the set of all affine combinations of elements of S, that is,

$$\mathrm{aff}(S) = \left\{ \sum_{i=1}^{n} \alpha_i x_i : n \in \mathbb{N}, x_i \in S, \alpha_i \in \mathbb{R}, \sum_{i=1}^{n} \alpha_i = 1 \right\}.$$

For a vector space E over a field $\mathbb{K}$ (either a real or a complex field), we know that the span of a subset A of vectors (not necessarily finite) can also be defined as the intersection W of all subspaces of E that contain A. W is referred to as the subspace spanned by W or by the vectors in A. Conversely, A is called a spanning set of W, and we say that A spans W, also denoted by $span(A)$ for W.

For a given p-normed space E, we use $B(x, r)$ to denote the open ball of E with center $x \in E$ and radius $r > 0$ for $p = 1$, and B_p to denote the closed unit ball with center θ in $l^p(n)$ for $n \in \mathbb{N}$, i.e., $B_p := \{x \in l^p(n), \|x\|_p \leq 1\}$.

Here, if there is an $r > 0$ such that $B(x, r) \cap span(A) \subset A$, then x is said to be an interior point with respect to the relative topology of $span(A)$ and denoted by $x \in ri(A)$ (or $ri(A)$ is denoted by $int(A)$ or $rint(A)$ as we will see below).

In functional analysis, a branch of mathematics, the **algebraic interior** (or radial kernel) of a subset in a vector space is a refinement of the often used concept of interior points, for which we recall the following.

By following Zălinescu [296], we first recall that a subset $A \subset E$ of a vector space E is said to be **radial** at a given point $a_0 \in A$ if for every point $x \in E$, there exists a real $t_x > 0$ such that for all $t \in [0, t_x]$, we have $a_0 + tx \in A$. Geometrically, this means A is radial at a_0 if for every $x \in E$, there is some (non-degenerate) line segment (depending on x) emanating from a_0 in the direction of x that lies entirely in A. By the definition, indeed every radial set is a star domain (although not conversely), and all points at which a set is radial are called **(algebraic) internal points**. On the other hand, in topology, the interior of a subset A of a topological space E is the union of all subsets of A that are open in E. A point in the interior of A called an interior point of A and denoted by $int(A)$. Thus, **an internal point** should not to be confused with the topological concept of **an interior point**.

Definition 2.1.14. Assume that A is a subset of a vector space E. Then, the **algebraic interior** (or radial kernel) of A with respect to E is the set of all points at which A is a radial set. Here, we recall that a point $a_0 \in A$ is called an **(algebraic) internal point** of A, and A is said to be radial at a_0 if for every $x \in E$, there exists a real number $t_x > 0$ such that for every $t \in [0, t_x]$, $t \in [0, t_x]$, $a_0 + tx \in A$. This last condition can also be written as $a_0 + [0, t_x]x \subset A$, where the set $a_0 + [0, t_x]x := \{a_0 + tx : t \in [0, t_x]\}$, which is the line segment (or closed interval) starting at a_0 and ending at $a_0 + t_x x$. Thus, the **algebraic interior** of A (with respect to X) is the set of all such points, or, in other words, it is the subset of points contained in a given set with respect to which they are the **radial points** of the set.

By the definition above, it is clear that the points at which a set is radial are internal points, and the set of all points at which $A \subset E$ is radial is equal to the algebraic interior. In addition, let M be a vector subspace of E and $A \subset E$. Then, we may in fact state a general definition for algebraic interior of A with respect to M as follows:

$$aint_M A := \{a \in E : \text{ for all } m \in M, \exists t_m > 0$$

$$\text{such that } a + [0, t_m] \cdot m \subset A\},$$

where $aint_M A \subset A$ always holds and if $aint_M A \neq \emptyset$, then $M \subset$ aff$(A - A)$. Here, aff$(A - A)$ is the affine hull of $A - A$ (i.e., the $span(A - A)$).

Definition 2.1.15 (algebraic interior (core)). In the special case where $M := E$, the set $aint_E A$ above is simply the set of all (algebraic) internal points and is called the **algebraic interior** or **core** of A, denoted by $core(A)$ or A^i.

Here, we would like to emphasize that the concept of the algebraic interior for a given set A in topological vector space E is different from another concept called "topological (relatively) interior point" (denoted by $int(A)$), as shown by the following facts (for more details, see the comment given in Remark 2.1.17 below.).

Let A be a subset of a topological vector space E, and denote by $int(A)$ the topological (relatively) interior points of A in E. Then, we have that (see Berge [19], Bauschke and Combettes [14, Proposition 6.12 in pp. 116], or Ben-EI-Mechaiekh and Mechaiekh [17, pp. 185]):

(1) $int(A) \subset core(A)$;
(2) if A is a nonempty convex subset of a finite-dimensional space E, then $int(A) = core(A)$;
(3) if A is a convex with nonempty (topological) interior, then $int(A) = core(A)$; and
(4) if A is a closed convex and E is a complete metric space, then $int(A) = core(A)$ (if $int(A) \neq \emptyset$).

We note that every absorbing subset is radial at the origin $a_0 = \theta$, and if the vector space is real, then the converse also holds. That is, a subset of a real vector space is absorbing if and only if it is radial at the origin (some authors use the term "radial" as a synonym for "absorbing").

First, noting that with a relative algebraic interior in vector spaces, we have that if $M = $ aff$(A - A)$, then the set $aint_M A$, denoted by $^i A := aint_{(\text{aff}(A-A))} A$, is called the relative algebraic interior of A. This name stems from the fact that $a \in A^i$ if and only if aff$(A) = E$ and $a \in {}^i A$, where aff$(A) = E$ if and only if aff$(A - A) = E$.

Secondly, with a relative interior in topology spaces, if A is a subset of a topological vector space E, then the relative interior of A respect to aff(A) (denoted by $rint(A)$, or $ri(A)$) is the set $rint(A) := int_{\text{aff}(A)}A$. That is, it is the topological interior of A in aff(A), which is the smallest affine linear subspace of E containing A. In addition, we recall that for the relative interior of A in E (denoted also by $ri(A)$), then we have that $ri(A) = core(A) \neq \emptyset$ if A is a nonempty closed subset of a finite-dimensional space E (by Proposition 6.12 of Bauschke and Comettes [14, p. 116] or Ben-EI-Mechaiek and Mechaiek [17, p. 185]).

Definition 2.1.16. Let C be a subset of a vector space E. We recall that an **internal point** to the subset C is a point $x \in C$ such that each straight line through E which lies in the affine hull aff(C) of C contains x as an (algebraic) interior point.

Remark 2.1.17. Here, we share the following three remarks.

First, we would like to point out that an internal point is different from the topological concept of an interior point. Indeed, by denoting the set of all internal points of C as the core of C, denoted by $core(C)$ (also called the core of C) in a vector space E, if E is a topological vector space, then for the topological interior sets of C (denoted by $int(C)$), we have $int(C) \subset core(C)$ in general. But we also have the equality $int(C) = core(C)$ in a number of situations: (1) when C is a nonempty convex subset of E and $dim(E)$ is finite; (2) when C is a convex subset of E and $int(C) \neq \emptyset$.

Second, the core of a non-degenerate convex subset C of a finite-dimensional space E is always nonempty in aff(C) (the affine hull of C). Here, we recall that aff(C) is the smallest linear variety containing C; it is precisely described by aff$(C) := \{x = \Sigma_{i=1}^{n}\lambda_i x_i \in E$, for $x_1, \ldots, x_n \in C, n \in \mathbb{N}^+, \lambda_i \in \mathbb{R}$, with $\Sigma_{i=1}^{n}\lambda_i = 1\}$.

For a given p-vector space E, where $p \in (0, 1]$, we denote the interior, closure, and boundary of a given subset A in a p-normed space E by $A^0, \overline{A}$, and $\partial(A)$, respectively; and the origin (zero) element is denoted by θ for E if there is no confusion.

Definition 2.1.18. The point x is said to be **frontal** with respect to C (with respect to $\overline{x}$) if there exists some $\overline{x} \in C$ such that the open line segment $(\overline{x}, x) = \{\overline{x} + t(x - \overline{x}), t \in (0, 1)\}$ is contained in

C and the open half-ray $\{\overline{x} + t(x - \overline{x}) : t > 1\}$ does not meet C; and the point x is also said to be a **frontal point** of C (with respect to $\overline{x}$). We note that if C is bounded, each frontal point x is simply a boundary point of C (commonly denoted by ∂C or $\partial(C)$, unless specified otherwise, in the mathematical literature).

Now, for the sake of simplicity, we give following general discussion in a topological vector space E (i.e., a special case of a p-vector space with $p = 1$ by using the classic concept of Minkowski functionals (i.e., the case of Minkowski p-functionals with $p = 1$) to describe the frontal points (also known as frontiers or boundary) of a given subset C with respect to the origin element θ in E. We first recall the following definition for gauge functions (Minkowski functionals) of a convex set C containing the zero element θ in a real vector space E.

Definition 2.1.19. The gauge function (the Minkowski functional) of a convex set C containing θ (the zero element) in a real vector space E is the extended function $J_c : E \to \mathbb{R} \cup \{+\infty\}$ given by

$$J_c(x) = \begin{cases} \inf\{t > 0 : x \in tC\} & \text{if } \{t > 0 : x \in tC\} \neq \emptyset, \\ +\infty & \text{otherwise.} \end{cases} \qquad (2.1)$$

Clearly, $C \subset \{x \in E : J_c(x) \le 1\}$. Thus, $C \subset dom(J_c)$, the effective domain of J_c consisting of all points $x \in E$ with $J_c < \infty$. The gauge of a convex set C is a non-negative sublinear functional with $J_c(0) = 0$.

In addition, we recall that a convex set S in a vector space E is said to be semi-bounded (with respect to the origin element θ) if every half-line originating from θ contains at least one point not belonging to C. We know that in $\mathbb{R}^n$, $n \in \mathbb{N}$, a bounded set is semi-bounded, but in general, the converse is not true.

Now, if C is a convex subset of a vector space E with the origin element $\theta \in C$, we have the following general facts to describe the boundary C by the Minkowski functional J_c.

Proposition 2.1.20. *Let C be a convex subset containing the origin element θ in a real vector space E and $J_c : E \to \mathbb{R} \cup \{+\infty\}$ be a Minkowski functional, as defined above. Then, we have the following general facts:*

(i) θ is an internal point of C if and only if $dom(J_c) = E$;
(ii) C is semi-bounded with respect to θ if and only if $J_c(x) > 0$ for all $x \in E \setminus \{0\}$;
(iii) if $\partial_a C$ (or simply by $\partial(C)$) denotes the set of all frontal points (also known as frontiers or boundary) of C with respect to θ, then $C \cup \partial_a C = \{x \in E : J_c(x) \le 1\}$; and
(iv) if C has a nonempty interior in E, a normed space, and θ is internal to C, then J_c is continuous on E.

Indeed, by using topological language in mathematics, the boundary of a subset S in a topological vector space E is the set of points in the closure of S, not belonging to the (topological) interior of S. For the convenience of our discussion, we use $\partial(S)$ or $bd(S)$ to denote the boundary of a subset S in E, and we also prefer to use both notions of "boundary" or "frontier" whenever there is no risk of confusion or unless specified otherwise in this book.

In addition, for a given p-vector space E, where $p \in (0, 1]$, by following Definition 2.1.6 above, the corresponding p-gauge function (Minkowski p-functional) (denoted by) J_c^p can be defined by following Definition 2.1.19 above. Then, a similar result by comparing with Proposition 2.1.20 above for a p-convex subset C containing origin element θ could be obtained. we leave this to interested readers as an exercise due to space constraints in this book.

Now, let (X, T_X) and (Y, T_Y) be two topological spaces with topology structures T_X and T_Y, respectively. We recall that a function $f : X \to Y$ is said to be bijective if and only if it is invertible; that is, a function $f : X \to Y$ is bijective if and only if there is a function $g : Y \to X$, the inverse of f, such that each of the two ways of composing the two functions produces an identity function: $g(f(x)) = x$ for each $x \in X$, and $f(g(y)) = y$ for each $y \in Y$. We now recall the following definition (see also Rudin [232, p. 8 and p. 17]).

Definition 2.1.21. Let (X, T_X) and (Y, T_Y) be two topological spaces and the mapping $f : X \to Y$ be a bijection. Then, f is said to be homeomorphism if the bijection $f : X \to Y$ is continuous, and its inverse $f^{-1} : Y \to X$ is also continuous, with respect to the given topologies. If two spaces X and Y are such that there exists a homeomorphism between them, then we say that X and Y are homeomorphic or topologically equivalent. In addition, we can verify that the relation of being homeomorphic is an equivalence relation.

We remark that the relation "**homeomorphic**" between two topological spaces is one of the most fundamental relations in topology because two topological spaces that are homeomorphic are indistinguishable from a topological point of view: they are topologically equivalent.

An alternative definition is that a bijection $f : X \to Y$ is said to be a "**homeomorphism**" if and only if both f and its inverse mapping f^{-1} map open sets to open sets. Thus, if (X, T_X) and (Y, T_Y) are homeomorphic, then not only are the elements of X and Y in one-to-one correspondence but so also are their open sets. We can thus regard (Y, T_Y) as being essentially the same space as (X, T_X) as far as its purely topological properties are concerned: (X, T_X) and (Y, T_Y) are merely two different ways of presenting the same space. We now also recall the following definition (see also Rudin's book [232, pp. 7–8]).

Definition 2.1.22. Let (X, T_X) and (Y, T_Y) be two topological vector spaces. A linear mapping $T : X \to Y$ is said to be an isomorphism if T is one-to-one and onto, and the two vector spaces X and Y are said to be isomorphic.

By the definition, we note that in mathematics language, an isomorphism is a structure-preserving mapping between two structures of the same type that can be reversed by an inverse mapping. Two mathematical structures are isomorphic if an isomorphism exists between them.

For a positive integer $n \in \mathbb{N}$, the real field $\mathbb{R}$, and the complex field $\mathbb{C}$, perhaps the simplest finite-dimensional Banach spaces are $\mathbb{R}^n$ and $\mathbb{C}^n$, the standard n-dimensional vector spaces over $\mathbb{R}$ and $\mathbb{C}$, respectively, normed by the mean of the usual Euclidean metric as follows.

If, for example, for each $z_i \in \mathbb{C}$, $i = 1, 2, \ldots, n$, $z = (z_1, \ldots, z_n)$ is a vector in $\mathbb{C}^n$, then

$$\|z\| = (|z_1|^2 + \cdots + |z_n|^2)^{\frac{1}{2}}.$$

Other norms can be defined on $\mathbb{C}^n$, for example,

$$\|z\| := |z_1| + \cdots + |z_n|, \text{ or } \|z\| := \max\{|z_i| : 1 \le i \le n\}.$$

These norms correspond, of course, to different metrics on $\mathbb{C}^n$ (when $n > 1$), but one can see very easily that they all induce the same topology on $\mathbb{C}^n$. For example, we note that if X is a topological vector space over $\mathbb{C}$ (the complex field) with n dimensions (denoted by $dimX = n$), then every basis of X induces an isomorphism of X onto $\mathbb{C}$. In fact, we can draw stronger conclusions, as shown by the results discussed in the following.

2.2 Some Properties of *p*-Vector Spaces

The goal of this section is to introduce and discuss some properties for p-vector spaces, which will be used for the development of a new fixed point theory in p-vector spaces for $p \in (0, 1]$ in this book.

Now, we discuss some properties of p-convex subsets in p-convex topological vector spaces (see also Balachandran [13], Bayoumi [15], Jarchow [116], Kalton [119], Rolewicz [228], Gholizadeh *et al.* [94], Ennassik *et al.* [78], Ennassik and Taoudi [79], Xiao and Lu [268], Xiao and Zhu [270], and the references therein).

We recall that a p-seminorm P is called an p-norm if $x = 0$ whenever $P(x) = 0$. A topological vector space with a specific p-norm is called a p-normed space. Of course, if $p = 1$, X is a usual normed space. By Lemma 3.2.5 of Balachandra [13], the following proposition gives a necessary and sufficient condition for a p-seminorm to be continuous.

Proposition 2.2.1. *Let X be a topological vector space; P is a p-seminorm on X, and $V := \{x \in X : P(x) < 1\}$. Then, P is continuous if and only if $0 \in int(V)$, where $int(V)$ is the interior of V.*

Now, given a p-seminorm P, the p-seminorm topology determined by P (in short, the p-topology) is the class of unions of open balls $B(x, \epsilon) := \{y \in X : P(y - x) < \epsilon\}$ for $x \in X$ and $\epsilon > 0$.

By Proposition 4.1.12 of Balachandra [13], we also have the following proposition.

Proposition 2.2.2. *Let A be a subset of a vector space X, which is absolutely p-convex for $p \in (0,1]$ and absorbing. Then, we have the following:*

(i) *The p-gauge P_A is a p-seminorm such that if $B_1 := \{x \in X : P_A(x) < 1\}$ and $\overline{B_1} = \{x \in X : P_A(x) \leq 1\}$, then $B_1 \subset A \subset \overline{B_1}$; in particular, $\ker P_A \subset A$, where $\ker P_A := \{x \in X : P_A(x) = 0\}$.*
(ii) *$A = B_1$ or $\overline{B_1}$ according to whether A is open or closed in the P_A-topology.*

Remark 2.2.3. Let X be a topological vector space, let U be an open absolutely p-convex neighborhood of the origin, and let ϵ be any given positive number. If $y \in \epsilon^{\frac{1}{p}} U$, then $y = \epsilon^{\frac{1}{p}} u$ for some $u \in U$ and $P_U(y) = P_U(\epsilon^{\frac{1}{p}} u) = \epsilon P_U(u) \leq \epsilon$ (as $u \in U$ implies that $P_U(u) \leq 1$). Thus, P_U is continuous at *zero*; therefore, P_U is continuous everywhere. Moreover, we have $U = \{x \in X : P_U(x) < 1\}$.

Indeed, since U is open and the scalar multiplication is continuous, we have that for any $x \in U$, there exists $0 < t < 1$ such that $x \in t^{\frac{1}{p}} U$, and so $P_U(x) \leq t < 1$. This shows that $U \subset \{x \in X : P_U(x) < 1\}$. The conclusion follows by Proposition 2.2.2 above.

The following result is a very important and useful result, which allows us to make the approximation for convex subsets containing zero in topological vector spaces by p-convex subsets in locally p-convex spaces (see Lemma 2.1 of Ennassik and Taoudi [78], Remark 2.1 of Qiu and Rolewicz [221], or Lemma 2.1 of Yuan [285, 286]).

Lemma 2.2.4. *Let A be a subset of a vector space X. Then, we have that:*

(i) *if A is r-convex, with $r \in (0,1)$, then $\alpha x \in A$ for any $x \in A$ and any $\alpha \in (0,1]$;*
(ii) *if A is convex and $0 \in A$, then A is s-convex for any $s \in (0,1]$; and*
(iii) *if A is r-convex for some $r \in (0,1)$, then A is s-convex for any $s \in (0,r]$.*

Proof.

(i) As $r \leq 1$, by the fact that "for all $x \in A$ and all $\alpha \in [2^{(n+1)(1-\frac{1}{p})}, 2^{n(1-\frac{1}{p})}]$, we have $\alpha x \in A$" is true for all integer $n \geq 0$. Taking into account the fact that $(0,1] = \cup_{n\geq 0}[2^{(n+1)(1-\frac{1}{p})}, 2^{n(1-\frac{1}{p})}]$, the result is thus obtained.

(ii) Assume that A is a convex subset of X with $0 \in A$, and take a real number $s \in (0,1]$. we show that A is s-convex. Indeed, let $x, y \in A$ and $\alpha, \beta > 0$, with $\alpha^s + \beta^s = 1$. Since A is convex, $\frac{\alpha}{\alpha+\beta}x + \frac{\beta}{\alpha+\beta}y \in A$. Keeping in mind that $0 < \alpha+\beta < \alpha^s+\beta^s = 1$, it follows that $\alpha x + \beta y = (\alpha+\beta)(\frac{\alpha}{\alpha+\beta}x + \frac{\beta}{\alpha+\beta}y) + (1-\alpha-\beta)0 \in A$.

(iii) Now, assume that A is r-convex for some $p \in (0,1)$, and pick up any real $s \in (0,p]$. We show that A is s-convex. To see this, let $x, y \in A$ and $\alpha, \beta > 0$ such that $\alpha^s + \beta^s = 1$. First, note that $0 < \alpha^{\frac{p-s}{p}} \leq 1$ and $0 < \beta^{\frac{p-s}{p}} \leq 1$, which imply that $\alpha^{\frac{p-s}{p}}x \in A$ and $\beta^{\frac{p-s}{p}} \in A$. By the p-convexity of A and the equality $(\alpha^{\frac{s}{p}})^p + (\beta^{\frac{s}{p}})^p = 1$, it follows that $\alpha x + \beta y = \alpha^{\frac{s}{p}}(\alpha^{\frac{p-s}{p}}x) + \beta^{\frac{s}{p}}(\beta^{\frac{p-s}{p}}y) \in A$. This competes the sketch of the proof. $\square$

Remark 2.2.5. We would like to point out that the results (i) and (iii) of Lemma 2.2.4 do not hold for $r = 1$. Indeed, any singleton $\{x\} \subset X$ is convex in topological vector spaces; however, if $x \neq 0$, then it is not p-convex for any $p \in (0,1)$.

We also need the following Proposition 2.2.6, which is proposition 6.7.2 of Jarchow [116], given here without its proof due to space constraints.

Proposition 2.2.6. *Let K be compact in a topological vector X and $p \in (0,1]$. Then, the closure $\overline{C}_p(K)$ of the p-convex hull and the closure $\overline{AC}_p(K)$ of absolutely p-convex hull of K are compact if and only if $\overline{C}_p(K)$ and $\overline{AC}_p(K)$ are complete, respectively (thus they both would be compact in its completion space by applying Theorem 3.3.3 of Jarchow [114]).*

Before we close this section, we would like to point out that the structure of p-convexity when $p \in (0,1)$ is really different from what we normally have for the concept of "convexity" in topological vector spaces. In particular, perhaps the following fact is one of the key reasons for us to use better (*p*-convex) structures in *p*-vector spaces to approximate the corresponding structures of the convexity in topological vector spaces (i.e., the *p*-vector space when $p = 1$). We have the following fact, which indicates that each *p*-convex subset is "bigger" than the convex subset in topological vector spaces for $p \in (0, 1)$ (for more on this, see Xiao and Zhu [270, p. 1740]).

Lemma 2.2.7. *Let x be a point of a p-vector space X, where we assume $p \in (0, 1)$. Then, the p-convex hull and the closure of $\{x\}$ are given by*

$$C_p(\{x\}) = \begin{cases} \{tx : t \in (0, 1]\} & \text{if } x \neq 0, \\ \{0\} & \text{if } x = 0; \end{cases} \tag{2.2}$$

and

$$\overline{C_p(\{x\})} = \begin{cases} \{tx : t \in [0, 1]\} & \text{if } x \neq 0, \\ \{0\} & \text{if } x = 0. \end{cases} \tag{2.3}$$

By Lemma 2.2.7, we know that if x is a given point in *p*-vector space X, when $p = 1$, $\overline{C_1(\{x\})} = C_1(\{x\}) = \{x\}$. This shows the significant difference in the structures of the *p*-convexities between $p = 1$ and $p \neq 1$! Thus, we would like to point out that under the general setting of *p*-vector spaces for $p \in (0, 1)$, a set-valued mapping with nonempty *p*-convex values does not include a single-valued mapping as its special class, i.e., a single-valued mapping is not a special case of a general set-valued mapping with nonempty *p*-convex values in *p*-vector spaces; therefore, the fixed point theories for single-valued and set-valued mappings are required to be developed simultaneously, as done in this book later.

In particular, as an application of Lemma 2.2.7, we have the following fact, which says that each set-valued mapping with nonempty *p*-closed values always has a zero element as its trivial fixed point for $p \in (0, 1)$ (but not for the case of $p = 1$).

Lemma 2.2.8. *Let U be a nonempty subset of a p-vector space X for $p \in (0, 1)$ with the origin element $\theta \in U$, and assume that a set-valued mapping $T : U \to 2^X$ is with nonempty closed p-convex values. Then, T has at least one fixed point in U, which is the element zero, i.e., $\theta \in \cap_{x \in U} T(x) \neq \emptyset$.*

Remark 2.2.9. We note that the results (i) and (iii) of Lemma 2.2.4 do not hold for $p = 1$. Indeed, any singleton $\{x\} \subset X$ is convex in topological vector spaces; however, if $x \neq 0$, then it is not p-convex for any $p \in (0, 1)$.

We now state following fact, which is a special case of Lemma 2.4 of Xiao and Zhu [270] (see also Proposition 2.1.20 above for a general case when set C is not bounded).

Lemma 2.2.10. *Let C be a bounded closed p-convex subset of p-seminorm X with the origin element $\theta \in intC$, where $p \in (0, 1)$. For every $x \in X$, define an operator by $r(x) := \dfrac{x}{\max\{1, (P_C(x))^{\frac{1}{p}}\}}$, where P_C is the Minkowski p-functional of C. Then, C is a retract of X and $r : X \to C$ is continuous such that:*

(1) *if $x \in C$, then $r(x) = x$;*
(2) *if $x \notin C$, then $r(x) \in \partial C$ (see its definition as follows);*
(3) *if $x \notin C$, then the Minkowski p-functional $P_C(x) > 1$.*

Proof. All originate from the definition of Minkowski p-functionals. See also Lemma 2.4, letting $s = p$, from Xiao and Zhu [270], as well as Proposition 2.2.4 and Remark above. Thus, the proof is compete. $\square$

Remark 2.2.11. As discussed in Remark 2.2.3, Lemma 2.2.10 still holds if "the bounded closed p-convex subset C of the p-normed space $(X, \| \cdot \|_p)$" is replaced by "X is a p-seminorm vector space and C is a bounded closed absorbing p-convex subset with the origin element $\theta \in intC$ of X" (see Proposition 2.1.20 above).

Before we close this section, we would like to point out that the structure of p-convexity when $p \in (0, 1)$ is really different from what we normally have for the concept of "convexities" used in topological vector spaces. In particular, perhaps the following fact is one of the reasons for us to use better (p-convex) structures in p-vector spaces to approximate the corresponding structures of the convexities used in topological vector spaces (i.e., the p-vector space when $p = 1$).

Remark 2.2.12. By following the discussion presented in Proposition 2.2.4 and Remark thereafter, each given (open) p-convex subset U in a p-vector space X with the zero element $\theta \in int(U)$ always corresponds to a p-seminorm P_U, which is indeed the Minkowski p-functional of U in X, and P_U is continuous in E. In particular, we know that a topological vector space is locally p-convex if the origin element θ of X has a fundamental set (denoted by) $\mathfrak{U}$, which is a family of absolutely p-convex θ-neighborhoods (each denoted by U). Moreover, this topology can be determined by the p-seminorm P_U, which are indeed the family $\{P_U\}_{U \in \mathfrak{U}}$, where P_U is simply the Minkowski p-functional for each $U \in \mathfrak{U}$ in X, where $p \in (0,1]$.

Once again, throughout this book, without loss of generality and unless specified otherwise, for a given p-vector space X, where $p \in (0,1]$, we always denote by $\mathfrak{U}$ the base of the p-vector space X's topology, which is the family of its θ-neighborhoods. For each $U \in \mathfrak{U}$, its corresponding p-seminorm P_U is the Minkowski p-functional of U in X. For a given point $x \in X$ and a subset $C \subset X$, we denote by $d_{p_u}(x, C) := \inf\{P_u(x - y) : y \in C\}$ the distance of x and C by the seminorm P_U, where P_U is the Minkowski p-functional for each $U \in \mathfrak{U}$ in X.

2.3 Kalton's Embedding Theorem for Compact p-Convex Sets in Topological Vector Spaces

The goal of this section is to discuss Kalton's embedding theorem for compact p-convex sets (where $p \in (0,1)$) in topological vector spaces. This allows us to prove the Schauder fixed point theorem for single-valued continuous mappings defined on the compact convex sets in p-vector spaces (but here, $p \in (0,1]$), which include topological vector spaces as a special class.

The embedding theorem established by Kalton [119] in 1977 states that every compact p-convex subset of a topological vector space can be linearly embedded in a locally p-convex topological vector space for $p \in (0,1)$.

We now begin with some important facts, which will be used in this section (and this book) later, as well as some notions and concepts used by Berge [19].

Definition 2.3.1. Let A be a subset of a topological vector space X, and let y be another topological vector space. By following Kalton [119], we shall say that A can be linearly embedded in Y if there is a linear mapping $T : lin(A) \to Y$ (not necessarily continuous) whose restriction to A is a homeomorphism.

Until recently, it was not known whether every compact convex subset of a topological vector space could be linearly embedded into a locally convex space. We know that Roberts [224] constructed a nonempty compact convex subset of $L_p = L_p(0, 1)$ for $p \in (0, 1)$, which has no extreme points and hence cannot be linearly embedded into a locally convex space.

In this section by following Kalton [114] in 1977, we discuss a similar problem for p-convex sets, where $p \in (0, 1)$. In view of the example given by Roberts [224], it is perhaps somewhat surprising that we are able to show that a compact p-convex set can be linearly embedded into a locally p-convex space and always has p-extreme points for $p \in (0, 1)$.

Throughout this section, we shall assume that all vector spaces are over the real field and that all topologies are Hausdorff. We recall that a subset C of a vector space is p-convex if whenever $x, y \in C$ and $a, b \in \mathbb{R}$, with $a, b \in [0, 1]$ and $a^p + b^p = 1$, then $ax + by \in C$. The set C is said to be absolutely p-convex if it is p-convex and $x \in C$ implies that $-x \in C$. A p-extreme point of a subset C is any point $x \in C$ such that whenever $x = ay_1 + by_2$, with $y_1, y_2 \in C$, and $a, b \in [0, 1]$, with $a^p + b^p = 1$, then $x = y_1$ or $x = y_2$. We denote by $\partial_p C$ the set of all p-extreme points of C. If C is any set, we denote by $\Gamma_p(C)$ and $\Delta_p(C)$ the smallest p-convex and absolutely p-convex sets containing C, respectively. We note that in a topological vector space, if $p \in (0, 1)$, a closed p-convex set always contains the origin element θ.

Let K be a compact Hausdorff topological space, and let $\mathcal{C}(\mathcal{K})$ be the Banach space of all real-valued continuous functions on K. Let $\mathcal{M}(\mathcal{K}) = \mathcal{C}(\mathcal{K})^*$ be the dual of $\mathcal{C}(\mathcal{K})$, i.e., the space of all regular Borel measures on K, with the usual dual norm denoted by $\|\cdot\|_1$, and we denote by w^* the weak*-topology on $\mathcal{M}(\mathcal{K})$ induced by $\mathcal{C}(\mathcal{K})$. For $x \in K$, we denote by δ_x or $\delta(x) \in \mathcal{M}(\mathcal{K})$ the unit mass concentrated at x, and let $\delta(K) = \{\delta(x) : x \in K\}$.

Now, suppose $p \in (0, 1)$, and let $\mathcal{M}_p(K)$ be the subspace of $\mathcal{M}(\mathcal{K})$ of all μ of the form

$$\mu = \sum_{n=1}^{\infty} a_n \delta(x_n),$$

where $(x_n : n \in \mathbb{N})$ is a sequence of distinct points K and

$$\|\mu\|_p = \sum_{n=1}^{\infty} \||a_n|^p < \infty.$$

Let $U_p = \{\mu : \|\mu\|_p \leq 1\}$ and $U^+ = \{\mu \geq 0 : \|u\|_p \leq 1\}$. Observe that $\|\cdot\|_p$ is a p-norm on $\mathcal{M}(\mathcal{K})$ (see the discussion given by Waelbroeck [262, p. 3].

We shall define the topology θ_p on $\mathcal{M}_p(K)$ to be the finest vector topology on $\mathcal{M}_p(K)$ which agrees with the w^*-topology on each set nU_p for $n \in \mathbb{N}$. We can give an explicit basic set of neighborhoods for θ_p (see also more discussion given by Wiweger [266]), namely the sets of the form

$$\cup_{n=1}^{\infty} \sum_{k=1}^{n} kU_p \cap W_k,$$

where $\{W_k : k \in \mathbb{N}\}$ is a sequence of w^*-neighborhoods of origin point θ. Since each U_p is p-convex and the w^*-topology is locally convex, we have the following statement.

Lemma 2.3.2. θ_p *is a locally p-convex topology on $\mathcal{M}_p(K)$.*

In the following lemma, we combine two results which have essentially the same proof.

Lemma 2.3.3. (i) U_p *and* U_p^+ *are θ_p-compact.* (ii) *Suppose* $T : \mathcal{M}_p(K) \to X$ *is a linear mapping into a topological vector space satisfying (a) T is continuous for the weak*-topology on $\delta(K)$ and (b) whenever $\mu_n \in U_p$ and $\|\mu_n\|_1 \to 0$, then $t(\mu_n) \to 0$; then T is continuous for the topology θ_p.*

Proof. The operator T in (ii) will be continuous if its restriction to U_p is continuous for the w^*-topology. Using this and the observation that the unit ball of $\mathcal{M}(\mathcal{K})$ is w^*-compact and contains U_p, we see that if either (i) or (ii) is false, we can construct a net $\{\mu_\alpha\}$ in U_p such that $\mu_\alpha \to \mu$ in w^*-topology, and either (1) $\mu \notin U_p$ or (2) $\mu \in U_p$, and there is a neighborhood V of 0 in X such that $T(\mu - \mu_\alpha) \notin V$ for all α.

In either case, by replacing μ_α with a subnet, we may assume that when each μ_α is written in the form

$$\mu_\alpha = \sum_{n=1}^{\infty} a_{\alpha,n} \delta(x_{\alpha,n}),$$

where $(x_{\alpha,n}, n \in \mathbb{N})$ is a sequence of distinct points of K, and $|a_{\alpha,n}| \geq |a_{\alpha,n}|$ for $n\mathbb{N}$, then the limits $\lim_\alpha a_{\alpha,n} (= a_n$, say$)$ and $\lim_\alpha x_{\alpha,n} (= x_n)$ exist. To see this, consider the net $(a_{\alpha,n}; x_{\alpha,b})$ in the compact space $[-1, 1]^{\mathbb{N}} \times K^{\mathbb{N}}$.

Now, $n|a_{\alpha,n}|^p \leq \sigma_{k=1}^n |a_{\alpha,k}|^p \leq 1$, so that $|a_{\alpha,n}| \leq n^{-1/p}$, and hence

$$\left\| \mu_\alpha - \sum_{k=1}^{n} a_{\alpha,n} \delta(x_{\alpha,k}) \right\|_1 = \sum_{k=n+1}^{\infty} |a_{\alpha,k}|$$

$$\leq (n+1)^{1-(1/p)} \sum_{k=n+1}^{\infty} |a_{\alpha,k}|^p$$

$$\leq (n+1)^{1-(1/p)}.$$

By the lower semicontinuity of $\|\cdot\|_1$ with respect to w^*-topology, we have

$$\left\| \mu - \sum_{k=1}^{n} a_k \delta(x_k) \right\|_1 \leq (n+1)^{1-(1/p)},$$

and hence $\sum a_k \delta(x_k) = \mu$ in $\|\cdot\|_1$. However, we clearly have $\sum |a_k|^p \leq 1$, and hence (after combining terms where $x_k = x_l$, with $k, l \in \mathbb{N}$), it is clear that $\mu \in U_p$, contradicting (1).

For (2), pick a symmetric neighborhood W of 0 in X such that $W + W + W \subset V$. Then, there exists $n \in \mathbb{N}$ such that $\|\mu\|_p \leq 1$, and $\|\mu\|_1 \leq (n+1)^{1-(1/p)}$ implies $T\mu \in W$. Since T is continuous for the w^*-topology on $\delta(K)$, we may choose α such that

$$T\left(\sum_{k=1}^{n} a_{\alpha,k} \delta(x_{\alpha,k}) - \sum_{k=1}^{n} \alpha_k \delta(x_k) \right) \in W.$$

Then, we have that

$$T(\mu_\alpha - \mu) = T\left(\sum_{k=n+1}^{\infty} a_{\alpha,k}\delta(x_{\alpha,k})\right)$$

$$+ T\left(\sum_{k=1}^{n} a_{\alpha,k}\delta(x_{\alpha,k}) - \sum_{k=1}^{n} a_k(x_k)\right) - T\left(\sum_{k=n+1}^{\infty} a_k\delta(x_k)\right),$$

$$\in W + W + W \subset V,$$

contradicting (2). This completes the proof. $\qquad\square$

Notably, it is clear that θ_p is the finest topology agreeing with the w^*-topology on U_p (e.g., see Proposition 5 of Waelbroeck [262, p. 48]), we now also have the following result.

Lemma 2.3.4. *Let X be a topological vector space, and suppose $x(t) \in X$ for $0 \le t \le 1$. Suppose that the set $\Delta_p(A)$ is relatively compact, where*

$$A = \{(t - s)^{-(1/p)}(x(t) - x(s)) : 0 \le s < t \le 1\}.$$

Then, $x(t) = x(0)$ for $0 \le t \le 1$.

Proof. Let E be the space of real functions on $[0, 1]$ of the form

$$\psi = \sum_{i=1}^{n} c_i \chi_i, \tag{2.4}$$

where χ_i for $i = 1, \ldots, n$ are characteristic functions of disjoint intervals. We may define a linear mapping $T : E \to X$ so that

$$T(\chi_{(s,t)}) = T(\chi_{[s,t)}) = T(\chi_{(s,t]}) = T(\chi_{[s,t]}) = x(t) - x(s).$$

If $\psi \in E$ is given by (2.4) and

$$\int_0^1 \psi(t)^p dt \le 1,$$

then

$$T(\psi) = \sum_{i=1}^{n} c_i(t_i - s_i)^{1/p}[(t_i - s_i)]^{-(1/p)}(x(t_i) - x(s_i))],$$

where $s_i < t_i$ are the endpoints of the interval whose characteristic function is $\mathcal{X}_i$. As

$$\int_0^1 |\psi(t)|^p dt = \sum_{i=1}^n |c_i|^p (t_i - s_i),$$

we have $T(\psi) \in \Delta_p(A)$, and so T extends uniquely to a compact operator $T : L_p \to X$. Hence, $T = 0$, and so $x(t) = x(0)$ for $0 \leq t \leq 1$ (see, e.g., Corollary 2.3 of Kalton [121]). This completes the proof. $\square$

Lemma 2.3.5. *Let K be a compact subset of a topological vector space X, and suppose $\Delta_p(K)$ is relatively compact. Then, the mapping $T : (\mathcal{M}_p(K), \theta_p) \to X$, defined by*

$$T\left(\sum_{n=1}^\infty a_n \delta(x_n) \right) = \sum_{n=1}^\infty a_n x_n,$$

is continuous. (Here, we note that the series necessarily converges since $\Delta_p(X)$ is bounded.)

Proof. We use Lemma 2.3.3(ii). Clearly, (a) is satisfied by T. To prove (b), suppose the contrary that there is a sequence $\mu_m \in U_p$ such that $\|\mu_m\|_p \leq 1$ and $\|\mu_m\|_1 \to 0$, but that for some neighborhood V of 0 in X, we have $T(\mu_m) \notin V$. Let

$$\mu_m = \sum_{n=1}^\infty a_{m,n} \delta(x_{m,n}),$$

where $(x_{m,n} : n \in \mathbb{N})$ is a sequence of distinct points of K. We define $y_m(t), 0 \leq t \leq 1$ as follows:

$$y_m(t) = \begin{cases} 0, & \text{if } 0 \leq t < |a_{m,1}|^p; \\ \sum_{n=1}^k a_{m,n} x_{m,n}, & \text{if } \sum_{n=1}^k |a_{m,n}|^p \leq t < \sum_{n=1}^{k+1} |a_{m,n}|^p; \\ \sum_{n=1}^\infty a_{m,n} x_{m,n}, & \text{if } \sum_{n=1}^\infty |a_{m,n}|^p \leq t \leq 1. \end{cases} \tag{2.5}$$

We shall show that if $1 \geq t > s \geq 0$ and $t - s \geq 2\|\mu_m\|_1^p$, then $(t-s)^{-(1/p)}(y_m(t) - y_m(s)) \in (\frac{3}{2})^{(1/p)} \Delta_p(K)$. This will be trivially true if either $t < |a_{m,1}|^p$ or $s \geq \sum_{n=1}^{\infty} |a_{m,n}|^p$. Hence, we assume $t \geq |a_{m,1}|^p$ and $s < \sum_{n=1}^{\infty} |a_{m,n}|^p$. Then by (2.5), we have

$$y_m(t) - y_m(s) = \sum_{n=l+1}^{\infty} a_{m,n} x_{m,n},$$

where $0 \leq l < \infty$ and $1 \leq k < \infty$, and

$$t - s \geq \sum_{n=l+2}^{k} |a_{m,n}|^p \geq \sum_{n=l+1}^{k} |a_{m,n}|^p - \|\mu_m\|_1^p$$

$$\geq \sum_{n=l+1}^{k} |a_{m,n}|^p - \frac{1}{2}(t - s).$$

Hence, $t - s \geq \frac{2}{3} \sum_{n=l+1}^{k} |a_{m,n}|^p$, and so

$$(t - s)^{-(1/p)}(y_m(t) - y_m(s)) = \sum_{n=l+1}^{k} a_{m,n}(t - s)^{-(1/p)} x_{m,n} \in \lambda \Delta_p(K),$$

where

$$\lambda^p = (t - s)^{-1} \sum_{n=l+1}^{k} |a_{m,n}|^p \leq \frac{3}{2}.$$

Now, considering $(y_m : m \in \mathbb{N})$ as a sequence in the compact space of all $\Delta_p(K)$-valued functions on $[0, 1]$ with pointwise convergence, we may find a cluster point $y(t)$. Then, since $\|\mu_m\|_1 \to 0$, we will have

$$(t - s)^{-(1/p)}(y(t) - y(s)) \in \left(\frac{3}{2}\right)^{(1/p)} \Delta_p(K),$$

whenever $0 < s < t \leq 1$. Hence, by the preceding lemma, $y(t) = y(0) = 0$ for all $t \in [0, 1]$. However, $y(1)$ is a cluster point of the sequence $T(\mu_m)$ and $T(\mu_m) \notin V$; thus, we have arrived at a contradiction. This completes the proof. $\square$

Theorem 2.3.6 (Kalton's embedding theorem [119]). *Let K be a compact p-convex subset of a topological vector space X, where $p \in (0, 1)$. Then, K can be linearly embedded in a locally p-convex topological vector space.*

Proof. Clearly, $\Delta_p(K) := \{ax - by : 0 \le a, b \le 1, a^p + b^p \le 1, x, y \in K\}$ is compact, and hence we may construct the continuous operator $T : (\mathcal{M}_p(K), \theta_p) \to X$, as in Lemma 2.3.5. Let $N = T^{-1}(0)$, and consider the quotient space $\mathcal{M}_p(K)/N$ with quotient θ_p-topology, which is locally p-convex. Then, there is an induced injective mapping $\hat{T} : \mathcal{M}_p(K)/N \to X$. Restricted to $q(\delta(K))$ (where $q : \mathcal{M}_p(K) \to \mathcal{M}_p(K)/N$ is the quotient mapping), $\hat{T}$ is a homeomorphism onto K; thus, $\hat{T}^{-1}$ is the required embedding. This completes the proof. $\square$

Remark 2.3.7. In view of the proof for Theorem 2.3.6, we know that the required embedding mapping $\hat{T}^{-1}$ is indeed also continuous. Secondly, we could appeal to the results of Fuchssteiner [89, 90] to demonstrate the existence of p-extreme points and an analog of the Krein–Milman theorem. In fact, we may go further and establish a version of Choquet's theorem (improving Theorem 2 of Fuchssteiner [91]).

Theorem 2.3.8. *Let C be a compact p-convex subset $p \in (0, 1]$ of a topological vector space X, and let K be a closed subset of C such that C is the closure of $\Gamma_p(K)$. Then:*

(1) *$\partial_p C \subset K$;*
(2) *If $x \in C$, there is a sequence of distinct points $x_n \in \partial_p C$ and $a_n \ge 0$, with $\Sigma_{a_n}^p = 1$, such that $x = \Sigma_{n=1}^\infty a_n x_n$.*

Proof. Construct as in Lemma 2.3.5 the mapping $T : \mathcal{M}_p(K) \to X$. Then, $T(U_p^+)$ is a compact p-convex set containing K and is clearly the smallest such set, and hence $T(U_p^+) = C$. If $x \in \partial_p C$, then

$$x = \sum a_n c_n,$$

where $x_n \in K$ and $\sum a_n^p \le 1$. Since it is p-extreme, and using the fact that $0 \in C$, we see that this representation must be trivial, i.e., $x \in K$.

For (2), consider the mapping T as in (1) but with the case of $K = C$. For $x \in C$, the set $T^{-1}\{x\} \cap U_p^+$ is w^*-compact, and hence there exists $\nu \in T^{-1}\{x\} \cap U_p^+$ such that $\nu(C) \le \mu(C)$ whenever

$\mu \in T^{-1}\{x\} \cap U_p^+$. Let

$$\nu = \sum b_n \delta(y_n),$$

where y_n are distinct and each $b_n \neq 0$. Then, if some $y_k \notin \partial_p C$, we have $y_k = c_1 z_1 + c_2 z_2$, where $z_1, z_2 \in C$, $0 < c_1, c_2 < 1$ and $c_1^p + c_2^p = 1$. Consider

$$\nu' = \sum_{n \neq k} b_n \delta(y_n) + b_k c_1 \delta(y_1) + b_k c_2 \delta(y_2).$$

Then, $\nu' \in U_p^+ \cap T^{-1}\{x\}$, but

$$\nu'(C) = \sum_{n \neq k} b_n + b_k(c_1 + c_2) < \nu(C),$$

and we have a contradiction.

Now, we have $\sum b_n^p \leq 1$; if $\sum b_n^p = 1$, we get the conclusion. Suppose $0 \in \partial_p C$, then by the minimality of $\nu(C)$, $0 \notin \{y_n : n \in \mathbb{N}\}$. Hence, if $\sum b_n^p < 1$, the element x may be represented in the required form as follows:

$$x = \sum_{n=1}^{\infty} b_n y_n + \left(1 - \sum b_n^p\right)^{(1/p)} (0).$$

Next, suppose $0 \notin \partial_p C$, then $0 = c_1 z_1 + c_2 z_2$, where $z_1 \neq z_2 \in C$, $0 < c_1, c_2 < 1$, and $c_1^p + c_2^p = 1$. By the preceding argument, there exist non-zero measures $\nu_1, \nu_2 \in U_p^+$ such that $\nu_1(C \backslash \partial_p C) = \nu_2(C \backslash \partial_p C) = 0$ and $T(\nu_1) = z_1$, $T(\nu_2) = z_2$. Then, $T(c_1 \nu_1 + c_2 \nu_2) = 0$, and hence for any $\lambda \geq 0$,

$$T(\nu + \lambda(c_1 \nu_1 + c_2 \nu_2)) = x.$$

Then, $\lambda \to \|\nu + \lambda(c_1 \nu_1 + c_2 \nu_2)\|_p$ is continuous in λ and tends to infinity as $\lambda \to \infty$. Hence for suitable $\lambda > 0$,

$$\|\nu + \lambda(c_1 \nu_1 + c_2 \nu_2)\|_p = 1.$$

Letting

$$\nu + \lambda(c_1 \nu_1 + c_2 \nu_2) = \sum a_n \delta(x_n),$$

with x_n being distinct, we complete the proof. $\qquad \square$

Remark 2.3.9. To conclude this section, we would like to emphasize that the embedding theorem established above by Kalton [119]) states that each compact p-convex subset of a topological vector space can be linearly embedded into a locally p-convex topological vector space for $p \in (0, 1)$. This is a key tool for us to establish the existence of fixed point theorems for (single-valued) continuous mappings in general p-vector spaces and thus provides an affirmative answer for the conjecture raised by Schauder [237] in 1930. We discuss this in Chapter 4 by following the analysis method in functional analysis established by Ennassik and Taoudi [79] and Ennassik *et al.* [78]; see also Park [189–200] and Yuan [286–291] for related comprehensive references.

Chapter 3

Fixed Point Theorems in p-Normed Spaces

The goal of this chapter is to establish general fixed point theorems for continuous mappings and nonexpansive mappings in p-normed spaces for $p \in (0, 1]$ from finite-dimensional to infinite-dimensional spaces. These theorems include corresponding results in the Euclidean and general normed spaces as a special class. This chapter consists of four sections, which are explained as follows.

The first section discusses fixed point theorems in finite-dimensional p-normed spaces for single-valued continuous mappings and a new extension theorem for continuous mappings in p-normed spaces for $p \in (0, 1]$, which is a generalization of the classic Dugundji's extension theorem [73]. The second section aims to establish fixed point theorems for continuous mappings in p-normed spaces. The third section establishes general fixed point theorems and demiclosedness principles for nonexpansive mappings in p-normed spaces by introducing a new concept called the p-normal structure or p-Opial condition for $p \in (0, 1]$, and these results unify or improve the corresponding results in the existing literature. In particular, when $p = 1$, the corresponding results given in this section reduce to the classical Browder–Göhde–Kirk fixed theorem for nonexpansive mappings in Banach spaces with normal structures. The fourth section discusses a famous and very important tool in nonlinear functional analysis called the Caristi fixed point theorem, established by Caristi [39] in 1976 on complete metric spaces. The discussion in this section will demonstrate the power of the fixed point theorem developed in this book, as it shows that almost all key principles established in nonlinear analysis are equivalent to the Caristi fixed point theorem under the framework of metric spaces; in addition, the results established in this section will be used in the later chapters of this book to study fixed point theorems for non-self-contractive and nonexpansive (single- or set-valued) mappings.

3.1 Fixed Point Theorems in Finite-Dimensional p-Normed Spaces

The goal of this section is first to establish fixed point theorems in finite-dimensional p-normed spaces and then establish a new Dugundji-type extension theorem for continuous mappings in p-normed spaces for $p \in (0, 1]$, which is a generalization of the classic Dugundji's extension theorem [73], which will be used as a tool to derive fixed point theorems in p-normed spaces that are infinite-dimensional in general.

In order to establish the general fixed point theorems in p-normed spaces for $p \in (0, 1]$, our starting point in this chapter is the following classical Brouwer fixed point theorem. This theorem is presented without proof, as it is available in almost all books on (nonlinear) functional analysis (see, e.g., Granas and Dugundji [102], Rudin [232], Zeidler [297], and the references therein).

Brouwer Fixed Point Theorem in $\mathbb{R}^n$. Every continuous function from a closed ball of an n-dimensional Euclidean space $\mathbb{R}^n$ into itself has a fixed point, where $n \in \mathbb{N}$ is a positive integer. In addition, if the space $\mathbb{R}^n$ is replaced by a normed space X, the conclusion still holds when the closed ball is replaced by a nonempty compact convex of X.

We all know that the Brouwer fixed point theorem above has played a key role in the development of nonlinear analysis in functional analysis by following Shapiro [242]. It is the most easily stated and most insightful of the fixed point theorems widely used in the mathematical literature, including this book. The theorem was proved in 1912 by the Dutch mathematician Brouwer [29], and its initial setting was the closed unit ball B of an n-dimensional Euclidean space $\mathbb{R}^n$, where $n \in \mathbb{N}$.

We now first give some fundamental results in p-vector spaces, which will be used to support our study for the development of fixed point theory in p-vector spaces, where $p \in (0, 1]$. The first one is the following Lemma 3.1.1.

Lemma 3.1.1. *Let* $(X, \|\cdot\|_p)$ *be a p-normed space for* $p \in (0, 1]$ *and* $0 < s \le p \le 1$. *Then, we have the following:*

(a) *The ball* $B(\theta, r)$ *is s-convex, where* $r > 0$.
(b) *If* $C \subset X$ *is s-convex and* $\alpha \in K$, *then* αC *is s-convex.*
(c) *If* C_1 *and* $C_2 \subset X$ *are s-convex, then* $C_1 + C_2$ *is also s-convex.*
(d) *If* $C_i \subset X$, $i = 1, 2, \ldots$ *are all s-convex, then* $\cap_{i=1}^{\infty} C_i$ *is s-convex.*
(e) *If* $A \subset X$ *and* $\theta \in A$, *then* $co_s A \subset coA$, *where* coA *is the convex hull of* A *in* $(X, \|\cdot\|_p)$.
(f) *If* C *is a closed s-convex set and* $0 < k < s$, *then* C *is a closed k-convex set.*

Proof. They are true by the definition of s-convex subsets for $s \in (0, p]$, where $p \in (0, 1]$ (see also Lemma 1.4 of Xiao and Zhu [270]). Thus, we omit the details of this proof. The proof is complete. □

By Proposition 2.1.20, we have the following result, which will play a key role in discussions throughout the later chapters of this book.

Lemma 3.1.2. *Let* $(X, \|\cdot\|_p)$ *be a p-normed space for* $p \in (0, 1]$ *and* $0 < s \le p \le 1$, *and let* C *be an s-convex subset of* X *with the origin element* $\theta \in C$. *Let* q_c *be the Minkowski p-functional of* C. *Then, we have that:*

(a) $q_c(\theta) = 0$;
(b) q_c *is positively s-homogeneous, i.e.,* $q_c(tx) = t^s q_c(x)$, *for each* $x \in X$ *and* $t > 0$;
(c) q_c *is sub-additive, i.e.,* $q_c(x + y) \le q_c(x) + q_c(y)$, *for each* $x, y \in X$;
(d) *if* C *is bounded, then* $q_c(x) > 0$, *for each non-zero point* $x \in X$;
(e) *if* C *is closed, then* $q_c(x)$ *is lower semicontinuous and* $C := \{x \in X : q_c(x) \le 1\}$;
(f) *if* C *is absorbing, then* $q_c(x) < +\infty$, *for each* $x \in X$;
(g) *if* $\theta \in C^0$, *then* q_c *is continuous, and* $C^0 = \{x \in X : q_c(x) < 1\}$ *and* $\overline{C} = \{x \in X : q_c(x) \le 1\}$.

Proof.   Each of the statements listed above can be verified by using the Minkowski s-functional q_c for a given s-convex subset C in X, and thus we omit the details of the proof (see also Lemma 1.5 of Xiao and Zhu [270]). This completes the proof.   □

Though the following result is a special case of Lemma 2.2.10, we list it as Lemma 3.1.3 with proof in detail for the convenience of the readers, as it will be useful for the discussion in this chapter.

Lemma 3.1.3. *Let $(X, \| \cdot \|_p)$ be a p-normed space for $p \in (0, 1]$ and $0 < s \le p \le 1$, and let C be a bounded closed s-convex subset of X with $\theta \in C^0$. For each $x \in X$, we define an operator by*

$$r(x) = \frac{x}{max\{1, (q_c(x))^{\frac{1}{s}}\}},$$

where q_c is the Minkowski p-functional of C in X. Then, C is a retract of X and $r : X \to C$ is a continuous operator such that:

(a) *if $x \in C$, then $r(x) = x$;*
(b) *if $x \notin C$, then $r(x) \in \partial C$.*

Proof.   Since C is a closed and s-convex subset and $\theta \in C^0$, by Lemma 3.1.2, the Minkowski p-functional q_c of C is a positively p-homogeneous, sub-additive, and continuous functional, with $C = \{x \in X : q_c(x) \le 1\}$ and $C^0 = \{x \in X : q_c(x) < 1\}$. It is also clear that r is continuous. If $x \in C$, then $q_c(x) \le 1$, so we have $r(x) = x \in C$; if $x \notin C$, then $q_c(x) > 1$, and we also have

$$r(x) = \left(\frac{1}{q_c(x)}\right)^{\frac{1}{s}} x + \left(1 - \frac{1}{q_c(x)}\right)^{\frac{1}{s}} \theta \in C.$$

Hence, r is a continuous operator from X into C, where C is a retract of X, and (a) is valid.

If $x \notin C$, then, by $q_c(x) > 1$, we have

$$q_c(r(x)) = q_c\left(\frac{x}{(q_c(x))^{\frac{1}{s}}}\right) = \frac{q_c(x)}{q_c(x)} = 1,$$

which means that $r(x) \in \{x \in X : q_c(x) = 1\} = C \setminus C^0 = \partial C$. So, (b) is also true. This completes the proof.   □

In topology and functional analysis, the well-known Tietze extension theorem says that if a topological space X is normal and $f : A \to [0,1]$ is continuous, where $A \subset X$ is closed, then f has a continuous extension to all of X. The closed subset A in X maps into a finite-dimensional Euclidean space that is continuous if its component functions are each continuous, so the Tietzes theorem is adequate for the study under the framework of finite-dimensional spaces and related applications. However, in mathematics, we normally work and study with p-normed spaces $(X, \| \cdot \|_p)$ for $p \in (0,1]$, which are potentially infinite-dimensional. Thus, we need to establish the extension theorem for closed subsets in p-normed spaces as a useful tool and to include Dugundji's extension theorem [73] as a special class for the study of functional analysis and topology.

In this part, our goal is to establish the following extension theorem for closed subsets under the framework of p-normed spaces $(X, \| \cdot \|_p)$ for $p \in (0,1]$, which includes the classic Theorem 6.1 of Dugundgji [73] as a special case for $p = 1$. As mentioned above, we expect this new extension theorem in p-normed spaces would be a useful and powerful tool for the development of new results and theory in p-normed spaces which are either infinite-dimensional or finite-dimensional.

For the convenience of our discussion in the following, here we first introduce some mathematical notations or symbols uses in this section. For any $\epsilon > 0$ and a point $x \in (X, d)$, where, X is a metric space with metric d, we denote by $U_\epsilon(x)$ the open ball of radius ϵ at a point x in the metric space (X, d). For a subset $A \subset X$, we denote by $d(x, A)$ the distance of x and A, defined by $d(x, A) := \inf\{d(x, a) : a \in A\}$.

Lemma 3.1.4. *[Extension theorem in p-normed spaces] Let E be a metric space (E, d) with the metric d and $A \subset E$ be a nonempty closed subset, with $(Y, \| \cdot \|_p)$ being a p-normed space for $p \in (0,1]$. Then, every continuous mapping $f : A \to Y$ has a continuous extension $F : E \to Y$ such that $F(E) \subset co_s(f(A))$, and $F(x) = f(x)$ for each $x \in A$, where $co_s(f(A))$ is the s-convex hull of $f(A)$ in Y, and $0 < s \leq p \leq 1$.*

Proof. Let the family $\{U_{d(x,A)/2}(x)\}_{x \in X \setminus A}$ be an open cover of the space $X \setminus A$. As $X \setminus A$ is paracompact, it implies the existence of an open locally finite refinement $\{W_\alpha\}_{\alpha \in I}$ (see also Dugundji [73]). Then, we know that there is an existence of a partition of unity

$\{\psi_\alpha\}_{\alpha \in I}$ subordinate to $\{W_\alpha\}_{\alpha \in I}$ (e.g., see Theorem 4.2 in Chapter VIII of Dugundji [73] for a discussion on partitions). For each $\alpha \in I$, choose $a_\alpha \in A$ with $d(a_\alpha, W_\alpha) < 2d(A, W_\alpha)$, and define the extension mapping for f by $F : X \to co_s(f(A)) \subset Y$ as follows:

$$F(x) := \begin{cases} f(x), & \text{if } x \in A; \\ \sum_{\alpha \in I}(\psi_\alpha)^{\frac{1}{s}}(x) \cdot f(a_\alpha), & \text{if } x \in X \setminus A. \end{cases} \tag{3.1}$$

Clearly, F is continuous at all points of $X \setminus A$ and at every interior point of A. We only need to prove F is continuous in the boundary of A.

Here, we note that for any $s > 0$ with $0 < s \leq p \in (0, 1]$, the p-normed space $(Y, \| \cdot \|_p)$ is also a s-normed space with the s-norm $\| \cdot \|_s$, which can be defined by $\| \cdot \|_s = \| \cdot \|_p^{\frac{s}{p}}$.

Let a be a point in the boundary of the closed subset A in X (and we may assume $f(a) = 0$; otherwise, let $f_1(x) := f(x) - f(a)$ for each $x \in E$, then we use f_1 to replace f to prove its continuity at $f_1(a) = 0$), thus for any open ball $U_\epsilon(a)$, the set $U_\epsilon(a) \cap (X \setminus A) \neq \emptyset$. Now, let U be a p-convex neighborhood of the point $F(a)$ (actually, $F(a) = f(a)$, as a is a boundary point of a closed subset A) in the space $(Y, \| \cdot \|_p)$, as $f(a)$ is continuous in A, choose $\delta > 0$ small enough that $f(a') \in U$ whenever $a' \in U_\delta(a) \cap A$. Then, we will prove that for any $x \in U_{\delta/7}(a) \cap (X \setminus A)$, $F(x) \in U$. This would help us to claim that F is continuous at each boundary point of A, then we can complete the proof to show that F is continuous in X, and with the property in the statement of Lemma 3.1.4 above. Here, we note that the p-convex U is also s-convex since the p-normed space $(Y, \| \cdot \|_p)$ is also a s-normed space $(Y, \| \cdot \|_s)$, as mentioned above when $0 < s \leq p \leq 1$.

Indeed, for any $x \in U_{\delta/7}(x) \cap (X \setminus A)$, suppose $\alpha \in I$ such that $x \in W_\alpha$ and any $x' \in X$ such that $W_\alpha \subset U_{d(x,A)/2}(x')$, we first observe that

$$d(a_\alpha, W_\alpha) \geq d(a_\alpha, x') - d(x, A)/2 \geq d(a_\alpha, x') - d(x', a_\alpha)/2$$
$$= d(a_\alpha, x')/2,$$

and

$$d(x', x) \leq d(x', A)/2 \leq d(W_\alpha, A) \leq d(W_\alpha, a_\alpha),$$

then it implies that

$$d(a_\alpha, x) \le d(a_\alpha, x') + d(x, x') \le 3d(a_\alpha, W_\alpha) \le 6d(A, W_\alpha) \le 6d(a, x).$$

Therefore, we show that whenever $x \in W_\alpha$, the following is true:

$$d(a_\alpha, a) \le d(a_\alpha, x) + d(x, a) \le 7d(x, a) < \delta.$$

This means $a_\alpha \in U_\delta(a) \cap A$, thus $f(a_\alpha) \in U$ by the continuity of f at the point $a \in A$ with the associated neighborhood U under the condition and statement given above.

Now, by the fact that $\sum_{\alpha \in I}((\psi_\alpha)^{\frac{1}{s}})^s(x) = \sum_{\alpha \in I}(\psi_\alpha)(x) = 1$ for any $x \in X$ as it is a partition of unity subordinate to $\{W_\alpha\}_{\alpha \in I}$ given above, and U is s-convex, it follows that $F(x) = \sum_{\alpha \in I}(\psi_\alpha)^{\frac{1}{s}}(x) \cdot f(a_\alpha)$ is in the s-convex hull $co_s(f(a_\alpha)(x) : \alpha \in I)$ of elements $\{f(a_\alpha)(x) : \alpha \in I\}$, thus $F(x) \in U$. This means F is continuous at $a \in A$ (which is a boundary point of A). Therefore, we have proved that F is continuous in X. Moreover, by Definition (3.1) above for the mapping F, we do also have that $F(E) \subset co_s(f(A)) \subset Y$, where $co_s(f(A))$ is the s-convex hull of $f(A)$ in Y, for $0 < s \le p \le 1$. This completes the proof. $\qquad\square$

As an application of Lemma 3.1.4, we have the following retract result for an s-convex subset in p-normed spaces for $0 < s \le p \le 1$.

Lemma 3.1.5. *Let $(X, \|\cdot\|_p)$ be a p-normed space for $p \in (0, 1]$ and D be a closed subset of X for $0 < s \le p \le 1$, and $A := \overline{co_s}D$ (or $A := co_s D$). Then, D is a retract of A.*

Proof. For $0 < s \le p \le 1$, we denote by $A := \overline{co_s}D$ (or $A := co_s D$) in the space $(X, \|\cdot\|_p)$. Then, we define an identity mapping $Id_D : D \to D$ by $Id_D(x) := x$ for each $x \in D$. Then, the mapping Id_D is continuous. Now, by Lemma 3.1.4, the continuous mapping Id_D has one continuous extension mapping $f : A \to D$, which shows that D is a retract of A. The proof is complete. $\qquad\square$

Let (X, T_X) and (Y, T_Y) be two topological spaces with topology structures T_X and T_Y, respectively. We recall that a function $f : X \to Y$ is said to be bijective if and only if it is invertible; that is, a function $f : X \to Y$ is bijective if and only if there is a function $g : Y \to X$, the inverse of f, such that each of the two ways for composing the two functions produces an identity function: $g(f(x)) = x$ for each $x \in X$, and $f(g(y)) = y$ for each $y \in Y$.

For a positive integer $n \in \mathbb{N}$, the real field $\mathbb{R}$, and complex field $\mathbb{C}$, maybe the simplest finite-dimensional Banach spaces are $\mathbb{R}^n$ and $\mathbb{C}^n$, the standard n-dimensional vector spaces over $\mathbb{R}$ and $\mathbb{C}$, respectively, normed by the mean of the usual Euclidean metric as shown in the following.

First, we can have the norm $\| \cdot \|$ defined by $\|z\| = (|z_1|^2 + \cdots + |z_n|^2)^{\frac{1}{2}}$, for each $z_i \in \mathbb{C}$, $i = 1, 2, \ldots, n$, where $z = (z_1, \ldots, z_n)$ is a vector in $\mathbb{C}^n$.

Second, we can also define the norm $\| \cdot \|$, or $| \cdot |$, by $\|z\| = |z_1| + \cdots + |z_n|$, or $|z| = \max\{|z_i| : 1 \leq i \leq n\}$, for each $z = (z_1, \ldots, z_n) \in \mathbb{C}^n$.

These norms correspond, of course, to different metrics on $\mathbb{C}^n$ (when $n > 1$), but one can see very easily that they all induce the same topology on $\mathbb{C}^n$. Indeed, a stronger result holds, as shown by the following result by Theorem 3.1.6. But first, we recall the following definition for homeomorphism between two spaces.

We recall that if X is a topological vector space over $\mathbb{C}$ (the complex field) and $dim X = n$, then every basis of X induces an isomorphism of X onto $\mathbb{C}$. The following result shows that this isomorphism must be a homeomorphism. In other words, this result states that the topology of $\mathbb{C}^n$ is the only vector topology that an n-dimensional complex ($\mathbb{C}$) topological vector space can have. In addition, we shall also see that finite-dimensional subspaces are always closed. Moreover, all the following results discussed remain true when real scalars ($\mathbb{R}$) are used in place of complex ones ($\mathbb{C}$).

As we will see in Chapter 5 and thereafter, in order to study the existence of fixed point theorems for upper semi-continuous (USC) set-valued mappings in locally p-convex spaces by taking s-convex values (instead of p-convex values) for $s, p \in (0, 1]$, we need to discuss the equivalence of s-norm and p-norm in $l^s(n)$ and $l^p(n)$ spaces with respect to the standard n-dimensional Euclidean space $\mathbb{R}^n$ with its standard norm $\| \cdot \|$ for $s, p \in (0, 1]$ (see their notions/definitions in the following). Indeed, we recall that for a finite n-dimensional topological vector space (TVS) V_n, where $n \in \mathbb{N}^+$, Tychonoff [260] in 1935 proved that each n-dimensional TVS V is isomorphic to $\mathbb{R}^n$ with its usual topology. More precisely, if $\{e_1, \ldots, e_n\}$ is a basis of V_n, then the mapping $\mathbb{R}^n \to V_n$, given by $(a_1, \ldots, a_n) \to a_1 e_1 + \cdots, a_n e_n$, is an isomorphism (see also Theorem 1.21 of Rudin [232, p. 16] or Theorem 3.5.6 of Jarchow [116, p. 66]), which means that all s- and p-norms on a finite-dimensional vector space over $\mathbb{R}$ (or $\mathbb{C}$) are equivalent for $s, p \in (0, 1]$. Here, we recall that the "equivalence" between the p-norm (of $l^p(n)$) and the standard Euclidean norm $\| \cdot \|$ of $\mathbb{R}^n$ means both p-norm and $\| \cdot \|$ generate the same topological structures (i.e., they generate all the same open sets) for $p \in (0, 1]$. The discussion and observation here lead us to the following proposition.

Theorem 3.1.6. *For a given $n \in \mathbb{N}^+$, Y is an n-dimensional subspace of a topological vector space X with real $\mathbb{R}$ (or complex $\mathbb{C}$) scalars, then we have that:*

(a) *every isomorphism of $\mathbb{R}^n$ (or $\mathbb{C}^n$) onto Y is a homeomorphism;*
(b) *Y is closed; and*
(c) *the s-normed space $(Y, \| \cdot \|_s)$ (including $l^s(n)$, $\mathbb{R}^n$ and any n-dimensional vector spaces) is equivalent to any p-normed space $(Y, \| \cdot \|_p)$ for $s, p \in (0, 1]$, and all finite n-dimensional spaces are locally convex.*

Proof. Parts (a) and (b) are given by Theorem 1.21 of Rudin [232, p. 16] (see also Theorem 3.5.6 of Jarchow [116, p. 66]); part (c) follows by Tychonoff's isomorphism theorem [260], which states that any n-dimensional TVS V_n is an isomorphism to $\mathbb{R}^n$, and thus all norms of V_n are equivalent to the norm of $\mathbb{R}^n$ in the usual topology, and the s-norm $\| \cdot \|_s$ of the s-normed space $(Y, \| \cdot \|_s)$ is equivalent to the p-norm $\| \cdot \|_p$ of the p-normed space $(Y, \| \cdot \|_p)$ for any $s, p \in (0, 1]$. This also implies that all n-dimensional topological vector spaces are locally convex. The proof is complete. $\square$

Theorem 3.1.6 states that each finite n-dimensional $(X, \| \cdot \|_p)$ p-normed space ($p \in (0, 1]$) is (linear) homeomorphic to the n-dimensional Euclidean space $\mathbb{R}^n$, where the integer $n > 0$ and $p \in (0, 1]$. In particular, the result (c) of Theorem 3.1.6 tells us that for any given finite n-dimensional spaces, by using the notions of p-norm and s-norm for such $l^p(n)$ and $l^s(n)$ spaces with $s, p \in (0, 1]$, it says that the family of s-convex subsets generated by s-norm in $l^s(n)$ is equivalent to the family of p-convex subsets generated by p-norm in $l^p(n)$ spaces (or with any other n-dimensional vector spaces such as $\mathbb{R}^n$).

It is known that every Hausdorff topological vector spaces of finite dimension n over $\mathbb{K}$ (either a real field, or a complete field) is linear homeomorphic with the Euclidean space $\mathbb{K}^n$ by Theorem 3.1.6 above, and so is $(X, \|\cdot\|_p)$. By following the argument used by Rudin [232], or Xiao and Wang [269], we have the following result.

Lemma 3.1.7. *Let $(X, \|\cdot\|_p)$ be an n-dimensional p-normed space, where $p \in (0,1]$. Then, there is a linear homeomorphism L of X into $l^1(n)$ with positive constants m and M such that*

$$m\|x\|_p \le \|Lx\|^* \le M\|x\|_p,$$

for all $x \in X$, where $\|\cdot\|^*$ denotes the norm in terms of the Euclidean metric discussed above. In addition, let L be a linear homeomorphism of $(X, \|\cdot\|_p)$ $(0 < p \le 1)$ into $l^1(n)$. It is easy to see that if C is a closed s-convex set, then so is $L(C)$, and vice versa.

Proof. By following the argument used by Rudin [232], or see the proof given by Xiao and Wang [269, pp. 149–150], the conclusion is true. The proof is complete. □

Lemma 3.1.8. *Let $(X, \|\cdot\|_p)$ be a p-normed space for $p \in (0,1]$ and C be a compact s-convex subset of X, where $0 < s \le p$. Let $\theta \ne x_0 \in C$ and $D = C - \overline{co}_s\{x_0\}$. Then, D is a compact s-convex set, $\theta \in D$, and $C \subset D$. Let $x \in D$ and $C_x = \{y \in C : x = y - tx_0\}$ for some $t \in [0,1]\}$ and*

$$\alpha(x) = \min_{y \in C_x}[\|x - y\|_p / \|x_0\|_p].$$

Then, $\alpha(x) \in [0,1]$ is continuous on D, and each $x \in D$ has a unique decomposition:

$$x = y(x) - \alpha x_0, \text{ for some } y(x) \in C.$$

For a given operator $T : C \to C$, the operator $T_0 : D \to D$ is defined by

$$T_0(x) = T[x + \alpha(x)x_0] - \alpha(x)x_0,$$

for each $x \in D$. Then, the operator T_0 has the following properties:

(a) *T is continuous if and only if T_0 is continuous.*
(b) *T has a fixed point if and only if T_0 has a fixed point.*

Proof. Since C and $\overline{co}_s\{x_0\}$ are all compact and s-convex, D is compact and S-convex by Lemma 3.1.1. Clearly, we have $\theta = x_0 - x_0 \in D$. Since $\theta \in \overline{co}_s\{x_0\}$, it follows that $C = C - \theta \subset C - \overline{co}_s\{x_0\} = D$.

Now, let $x \in D$. It is clear that C_x is a closed subset of C. If $x = y - tx_0$, then $t = \|x - y\|_p/\|x_0\|_p$. Since C_x is also compact, we see that $\alpha(x) \in [0,1])$ is well defined. Let $y(x) = x + \alpha(x)x_0$. Then, x has the unique decomposition $x = y(x) + \alpha(x)x_0$.

Assume that $\{x_n\} \subset D$ is a sequence with $x_n \to x$. For each fixed $y \in C_x$, $f_y(x) = \|x - y\|_p/\|x_0\|_p$ is a continuous functional. Since the infimum of a collection of continuous functionals is upper semicontinuous, we have $\limsup_{n\to\infty} \alpha(x_n) \leq \alpha(x)$. Let $\liminf_{n\to\infty} \alpha(x_n) = \beta(x)$. By the compactness of C, there is a subsequence $\{x_{n_j}\}$ of $\{x_n\}$ such that $\lim_{j\to\infty} \alpha(x_{n_j}) = \beta(x)$ and $\lim_{j\to\infty} y(x_{n_j}) = y_0 \in C$. It follows from $x_{n_j} = y(x_{n_j}) - \alpha(x_{n_j})x_0$ that $x = y_0 - \beta(x)x_0$. This shows that $y_0 \in C_x$. Since $x = y(x) - \alpha(x)x_0$, by the definition of $\alpha(x)$, we have $\alpha(x) \leq \beta(x)$. Hence, we have $\lim_{n\to\infty} \alpha(x_n) = \alpha(x) = \beta(x)$ and $y(x) = y_0$. Thus, $\alpha(x)$ is continuous on D.

(a) If $T : C \to C$ is continuous, then the continuity of T_0 follows from the continuity of T and $\alpha(\cdot)$ by the fact that if $\lim_{n\to\infty} = x$, then we have that
$$\lim_{n\to\infty} T_0(x_n) = \lim_{n\to\infty} T(x_0 + \alpha(x_n)x_0] - \alpha(x_n)x_0$$
$$= T[x + \alpha(x)x_0] - \alpha(x)x_0 = T_0(x).$$

Conversely, if T_0 is continuous, then for $x \in C \subset D$, we have $x = x - \theta \cdot x_0$, so that $\alpha(x) = 0$. Thus, $T(x) = T(x + \alpha(x)x_0) - \alpha(x)x_0 = T_0(x)$, which means that T is continuous.

(b) If T has a fixed point $z \in C$, then $z = T(z) = T(z + \theta) - \theta = T_0(z)$, where z is a fixed point of T_0. Conversely, if T_0 has a fixed point $z \in D$, i.e., $z = T_0(z)$, then by the definition of T_0, we have $z = T[z + \alpha(z)x_0] - \alpha(z)x_0$. This shows that $z + \alpha(z)x_0$ is a fixed point of T. The proof is complete. $\qquad\square$

Lemma 3.1.9. *Let B_p and B_1 be unit closed balls with center θ in $l^p(n)$, for $(0 < p \leq 1)$, and $l^1(n)$, respectively. Then, there is a homeomorphism H of $l^p(n)$ into $l^1(n)$ with $H(B_p) = B_1$.*

Proof. For $\alpha \in K$, we denote by $sgn\alpha$ for $(sgn\alpha)|\alpha| = \alpha$. Define $H : l^p(n) \to l^1(n)$ by $H(\alpha_1, \ldots, \alpha_n) := (\beta_1, \ldots, \beta_n)$, for each

$(\alpha_1, \ldots, \alpha_n) \in l^p(n)$, where $\beta_i = (sgn\alpha_i)|\alpha_i|^p$, for $i = 1, \ldots, n$. It is easy to verify that H is a bijective mapping. As $\Sigma_{i=1}^n |\alpha_i|^p \leq 1$ if and only if $\Sigma_{i=1}^n |\beta_i| \leq 1$, it follows that $H(B_p) = B_1$. Since the functionals $f_i(\alpha_i) = (sgn\alpha_i)|\alpha_i|^p$ and $g_i(\beta_i) = (sgn\beta_i)|\beta_i|^{\frac{1}{p}}$ are continuous in K for $i = 1, \ldots, n$, we conclude that H and H^{-1} are continuous, so H is a homeomorphism. This completes the proof. $\square$

Lemma 3.1.10. *Let B_p be unit closed balls with center θ in $l^p(n)$ for $p \in (0, 1]$ and $n \in \mathbb{N}$. Assume that $T : B_p \to B_p$ is a continuous operator. Then, there exists $u \in B_p$ such that $T(u) = u$.*

Proof. Let H be a homeomorphism which is as defined in the proof of Lemma 3.1.9 and B_1 be a closed unit ball with center θ in $l^1(n)$. Then, $HTH^{-1} : B_1 \to B_1$ is continuous. By the classical Brouwer fixed point theorem in a finite n-dimensional space $\mathbb{R}^n$, where $n \in \mathbb{N}$, there exists one $x \in B_1$ such that $HTH^{-1}(x) = x$. Now, let $H^{-1}(x) = u$. Then, $u \in B_p$, which is a fixed point of T, i.e., $T(u) = u$. The proof is complete. $\square$

Theorem 3.1.11 (Brouwer fixed point in finite-dimensional p-normed spaces). *Let $(X, \|\cdot\|_p)$ be a finite-dimensional p-normed space and C be a bounded closed s-convex subset of X, where $0 < s \leq p \leq 1$. If $T : C \to C$ is continuous, then T has a fixed point in C, i.e., there exists $z \in C$ such that $T(z) = z$.*

Proof. Without loss of generality, we assume that $span(C) = X$ and $dim X = n$. Since X is linear homeomorphic with $l^p(n)$ by Lemma 3.1.7, we also assume that $X = l^p(n)$. As $(X, \|\cdot\|_p)$ is finite-dimensional, without loss of generality, we may assume $C^0 \neq \emptyset$, and $\theta \in C^0$ by Lemma 3.1.8 and Remark 2.1.17. Then, the Minkowski p-functional q_c of C is a positively p-homogeneous, sub-additive, and continuous functional with $C = \{x \in l^p(n) : q_c(x) \leq 1\}$. Since C is bounded, for $\theta \neq x \in l^p(n)$, we have $p_c > 0$, i.e., q_c is positive definite. Now, define a mapping $S : B_p \to C$ for each $x \in B_p$ by

$$y = S(x) = \begin{cases} (\frac{\|x\|_p}{q_c(x)})^{\frac{1}{s}} x & \text{if } x \neq \theta, \\ \theta & \text{if } x = \theta, \end{cases} \tag{3.2}$$

where B_p is a unit closed ball with center θ in $l^p(n)$. Since $q_c(y) = (\|x\|_p/q_c(x))q_c(x) = \|x\|_p$, we have

$$\left(\frac{q_c(y)}{\|y\|_p}\right)^{\frac{1}{p}} = \left(\frac{\|x\|_p}{(\frac{\|x\|_p}{q_c(x)})^{\frac{p}{s}}\|x\|_p}\right)^{1/p} = \left(\frac{q_c(x)}{\|x\|_p}\right)^{\frac{1}{s}}, \quad \text{for all } \theta \neq y \in C.$$

Hence, there exists the inverse $S^{-1} : C \to B_p$ given by, for each $y \in C$,

$$x = S^{-1}(y) = \begin{cases} (\frac{q_c(y)}{\|y\|_p})^{\frac{1}{p}} y & \text{if } y \neq \theta, \\ \theta & \text{if } y = \theta. \end{cases} \tag{3.3}$$

Clearly, by the continuity of q_c and $\|\cdot\|_p$, from (3.2) and (3.3) above, we see that S is continuous at $x \neq \theta$ and S^{-1} is continuous at $y \neq \theta$. Suppose that $x_n \to \theta$. By (3.2), we have $q_c(S(x_n)) = \|x_n\|_p \to \theta$. Since C is bounded, there is $M > 0$ such that $\|x\|_p \leq M$, for all $x \in C$. For each $\epsilon > 0$, there is a positive integer N such that $q_c(S(x_n)) < \epsilon$, for all $n \geq N$. By the definition of q_c, there is t_n such that $0 < t_n < \epsilon$ and $S(x_n) \in t_n^{\frac{1}{s}} C$. So, we have $\|S(x_n)\|_p \leq t_n^{\frac{p}{s}} M < \epsilon^{\frac{p}{s}} M$, for all $n \geq N$. This shows that $S(x_n) \to \theta$. Suppose that $y_n \to \theta$. By (3.3) and the continuity of q_c, we have $\|S^{-1}(y_n)\|_p = q_c(y_n) \to \theta$. Therefore, S and S^{-1} are all continuous, and so $S : B_P \to C$ is a homeomorphism. Since $S^{-1}TS : B_p \to B_p$ is continuous, by Lemma 3.1.10, there exists $u \in B_p$ such that $S^{-1}TS(u) = u$. Let $z := S(u)$. Then, we have $z \in C$ and $T(z) = z$, which is the fixed point of T. The proof is complete. $\square$

Remark 3.1.12. In this section, by establishing the extension theorem for continuous mappings in p-norm spaces for $p \in (0,1]$, which is a generalization of the famous and classic Theorem 6.1 of Dugundji [73] as a special case, we then establish the Schauder fixed points in p-norm spaces which are either infinite-dimensional or finite-dimensional. Moreover, we do expect our extension result Lemma 3.1.4 to play a key role in the development of new results and related theory in the study of nonlinear analysis for p-normed spaces in the years to come.

3.2   Fixed Point Theorems for Continuous Mappings in p-Normed Spaces

The goal of this section is to establish fixed point theorems for s-convex subsets in p-normed spaces by applying the existence of homeomorphisms for s-convex subsets in p-normed spaces, where $s, p \in (0, 1]$. Here, we would like to point out that the approach used in this section is different from those used in the existing literature.

For a complete linear metric space (which is also called a F-space), the results established below in this section show that any F-space having l^p as a continuous linear image for some $p \in (0, 1)$ must contain proper closed subspaces which are dense in the weak topology. In particular, when $p \in (0, 1]$, such subspaces are present in l^p, and the same is shown to hold in certain F-spaces of analytic functions which nevertheless have enough continuous linear functionals to separate points. In other words, every non-locally convex F-space must contain a proper, closed, weakly dense (PCWD) subspace.

The following result, which is actually Proposition 2 of Shapiro [241], states that every complete, separable p-normed space is a continuous linear image of l^p, for each $p \in (0, 1]$. This property is often called the universal property of F-spaces in the mathematical literature.

Lemma 3.2.1. *Every complete, separable p-normed space E, where $p \in (0, 1]$, is a continuous linear image of l^p.*

Proof.   Let $\mathcal{S} := \{e_n : n \geq 0\}$ be a countable dense subset of the set $U := \{f \in E : \|f\| \leq 1\}$. If $x = (\xi_j)$ belongs to l^p and $n, k > 0$, then we have

$$\|\Sigma_{=n}^{n+k}\xi_j e_j\| \leq \Sigma_{j=n}^{n+k}|\xi_j|^p.$$

The sequence $(\Sigma_{j=0}^n \xi_j e_j : n \geq 0\}$ is therefore Cauchy in E, so it converges to an element $\Sigma_{j=0}^\infty \xi_j e_j$ of E. Now, define the mapping $T : l^p \to E$ by

$$T(x) = \Sigma_{j=0}^\infty \xi_j e_j$$

for each $x = (\xi_j)$ in l^p. Then, we have that if x belongs to l^p, $\|Tx\| \leq \|x\|$, so T is a continuous linear mapping. We next show that $T(l^p) = E$. Suppose f is in E with $\|f\| \leq 1$. Since $\mathcal{S}$ is dense in U, there exists an index $n(0)$ such that $\|f - e_{n(0)}\| \leq 2^{-p}$. Thus, $f - e_{n(0)}$ belongs to $\frac{1}{2}U$, which contains $\frac{1}{2}S$ as a dense subset. Consequently, there exists an index $n(1)$ such that $\|f - e_{n(0)} - \frac{1}{2}e_{n(1)}\| \leq 4^{-p}$. By continuing in this manner, we obtain a sequence $\{n(k) : k = 0, 1, \ldots, \}$ of indices such that

$$\|f - \Sigma_{j=0}^{k} 2^{-j} e_{n(j)}\| \leq 2^{-(k+1)p},$$

for $k = 0, 1, \ldots,$. Now, let $x = (\xi_n^0)$ be the sequence such that

$$\xi_n^0 = \begin{cases} 2^{-k} & \text{if } n = n(k), \\ 0 & \text{if } n \neq n(k). \end{cases} \tag{3.4}$$

Then, $x = (\xi_n^0)$ belongs to l^p, and we have $Tx = f$. This completes the proof. $\square$

By Lemma 3.2.1 and the space decomposition approach, we have the following result.

Lemma 3.2.2. *Let $(X, \|\cdot\|_p)$ be a complete and separable p-normed space, where $p \in (0, 1]$. Then, there exists a closed subspace l_X^p of l^p and a linear operator $T : l^p \to X$ such that $T : l_X^p \to X$ is a homeomorphism.*

Proof. By Lemma 3.2.1, there exists a continuous linear operator $T : l^p \to X$ such that $T(l^p) = X$. We denote the kernel (null) space of T by $kerT := \{x \in l^p : Tx = 0\}$. Then, $kerT$ is a closed subspace of l^p, as T is linear and continuous. Now, by following the same approach as that of Bonsall [24, Chapter, p. 35], the quotient space l_X^p, defined by $l_X^p := l^p/KerT$ with the p-norm defined by $\|\bar{x}\|_p := \|T(x)\|_p$ for each $x \in l^p$ with $\bar{x} = x + KerT \in l^p/KerT$, is a p-norm on $l_X^p (= l^p/KerT)$, and the quotient space l_X^p is also closed since $Ker\bar{T}$ is closed. By the fact that the mapping $\pi : l^p \to l_X^p$ by $\pi(x) := \bar{x}$ for each $x \in l^p$, with $\bar{x} = x + KerT \in l_X^p$, is a continuous homomorphism from l^p to the quotient space l_X^p, endowed with the topology induced by the p-norm $\|\cdot\|_p$, as $\|\pi(x)\|_p = \|\bar{x}\|_p = \|T(x)\|_p$ for each $x \in l^p$, l_X^p, is closed in l^p, and $T(l^p) = X$. Thus, the mapping $T : l_X^p \to X$ is one to one and onto; therefore, T is a homeomorphism from the closed subspace l_X^p of l^p to X. This completes the proof. $\square$

Now, we have the following existence result for the homeomorphism of a s-convex compact subset in l^p spaces for $0 < s \leq p \in (0,1]$.

Theorem 3.2.3. *Let* $(X, \| \cdot \|_p)$ *be a complete p-normed space and* D *be a s-convex compact subset of* X, *where* $p \in (0,1]$ *and* $s \in (0,p]$. *Then, there exists a linear (continuous) mapping* $F : D \to l^p$ *such that* $F(D) \subset C_s$ *and* $F : D \to F(D)$ *is a homeomorphism, where*

$$C_s := \{x = (\alpha_1, \ldots, \alpha_n, \ldots) \in l^p : |\alpha_n|^s \leq 1/n^2\}.$$

Proof. Let $Y = \overline{\mathrm{span}D}$. Since D is compact, Y is separable. As Y is a closed subspace of X, Y is also complete. By Lemma 3.2.2, there exists a closed subspace l_Y^p of l^p and a linear operator $T : l^p \to Y$ such that $T : l_Y^p \to Y$ is a homeomorphism. Let $F = T^{-1}$. Then, $F : D \to l^p$ is a linear operator such that $F : D \to F(D)$ is a homeomorphism. In addition, it is easy to verify that $F(D) \subset C_s$ (e.g., see the argument given by Xiao and Zhu [270, p. 1744]). This completes the proof. □

Here, we remark that for $0 < s < p \leq 1$, there is no linear continuous mappings from l^p to l^s (resp., $L^p[0,1]$ to $L^s[0,1]$), thus the so-called identity mapping $Id : l^1 \to l^s$, defined by $Id(x) := x$ for each $x \in l^1$ to l^s, is not continuous (e.g., see Lemma 2.7 of Kalton *et al.* [122, p. 24] or Corollary 2 of Wang [263, p. 79]), but we do have the existence result on the homeomorphisms of a s-convex compact subset in l^p spaces for $0 < s \leq p \in (0,1]$, which is due to the universal property of l^p space itself.

We also need following extension result in p-normed spaces for $p \in (0,1]$.

Lemma 3.2.4. *Let* $(X, \|\cdot\|_p)$ *be a complete separable p-normed space for* $0 < s \leq p \in (0,1]$ *and* D *be a (bounded) closed subset of* X. *Let* $T : D \to X$ *be a continuous mapping. Then,* T *has a continuous extension* $S : X \to X$ *such that* $S(X) \subset co_s(T(D))(\subset \overline{co}_s(T(D)))$.

Proof. For $0 < s \leq p \in (0,1]$, the conclusion follows by Lemma 3.1.4 above. The proof is complete. □

Now, by applying Theorem 3.1.11, we have the following fixed point theorem in l^p space for $p \in (0, 1]$.

Lemma 3.2.5. *Let C_s be the closed cuboid in l^p defined by*

$$C_s := \{x = (\alpha_1, \ldots, \alpha_n, \ldots) \in l^p : |\alpha_n|^s \leq 1/n^2\}$$

and $T : C_s \to C_s$ be a continuous operator, where $0 < p \leq 1$, $0 < s \leq p$. Then, C_s is s-convex and compact, and there exists $z \in C_s$ such that $T(z) = z$.

Proof. By its definition, we can verify that the set C_s is a compact subset of l^p and is also s-convex (see also [270]). For each n, we define an operator P_n on l^p by

$$P_n(\alpha_1, \ldots, \alpha_n, \ldots) = (\alpha_1, \ldots, \alpha_n, 0, \ldots),$$

for each $(\alpha_1, \ldots, \alpha_n, \ldots) \in l^p$. Then, P_n is linear and continuous, and $P_n(C_s) \subset C_s$. It follows from the s-convexity and compactness of C_s that $P_n(C_s)$ is an s-convex and compact subset of $l^p(n)$. Since P_nT is continuous and $P_nT(P_n(C_s)) \subset P_nT(C_s) \subset P_n(C_s)$, by Theorem 3.1.11 (which is the Brouwer fixed point theorem in a finite-dimensional space $l^p(n)$), there is a point $x_n \in P_n(C_s)$ such that $P_nT(x_n) = x_n$. Let $T(x_n) = \{\alpha_i^{(n)}\}_{i=1}^{\infty}$. Then, by the definition of P_n, we have

$$\|x_n - T(x_n)\|_p = \|P_nT(x_n) - T(x_n)\|_p$$

$$= \Sigma_{i=n+1}^{\infty}|\alpha_i^{(n)}|^p \leq \Sigma_{i=n+1}^{\infty}|\alpha_i^{(n)}|^s \leq \Sigma_{i=n+1}^{\infty}\frac{1}{i^2}.$$

Now, $\{T(x_n)\}$ is a sequence in the compact set C_s, so there is a point $z \in C_s$ and a subsequence $\{T(x_{n_j})\}$ such that $T(x_{n_j}) \to z$. By the above inequality, it follows that $x_{n_j} \to z$. Since T is continuous, we have $T(z) = \lim_{j \to \infty} T(x_{n_j}) = z$, which is the desired conclusion. The proof is complete. □

Now, as an application of Theorem 3.2.3, we have the following fixed point theorem in complete p-normed spaces for $s, p \in (0, 1]$.

Theorem 3.2.6. *Let $(X, \|\cdot\|_p)$ be a complete p-normed space and C be a compact s-convex subset of X, where $s, p \in (0, 1]$. If $T : C \to C$ is continuous, then there exists $z \in C$ such that $Tz = z$.*

Proof. We prove this result in two steps as follows.

Step 1: We prove the conclusion for $p \in (0, 1]$ and $0 < s \leq p \leq 1$. By Theorem 3.2.3, there exists a linear homeomorphism $F : C \to F(C)$ such that $F(C) \subset C_s$, where C_s is an s-convex compact subset of l^p defined by Lemma 3.2.5. Then, $F(C)$ is an s-convex compact. Since T is continuous, the mapping $FTF^{-1} : F(C) \to F(C)$ is also continuous. By Lemma 3.2.4, FTF^{-1} has a continuous extension $S : C_s \to C_s$ such that $S(C_s) \subset \overline{co}_s F(C)$. By Lemma 3.2.5, which is a fixed point theorem for a compact s-convex subset C_s in l^p spaces (where, $0 < s \leq p \leq 1$), there is $u \in C_s$ such that $u = S(u) \in F(C)$, and so $u = S(u) = FTF^{-1}(u)$. Let $z = F^{-1}(u)$. It follows that $z \in C$ and $T(z) = z$, which is a fixed point of the mapping T.

Step 2: We prove the case of $0 < p < s < 1$ and $0 < p < s = 1$. For the case $0 < p < s < 1$, by Lemma 3.1.1(f), C is also p-convex. By applying the conclusion given in the first step above, it follows that there exists $x \in C$ such that $T(x) = x$. Now, the only case left to prove is for $0 < p < s = 1$, which means C is convex. We choose an arbitrary $x_0 \in C$, and let $C_0 := \{x - x_0 : x \in C\}$. Then, it is clear that C_0 is a compact convex subset of X, which contains the zero element. By Lemma 3.1.1(e) again, we conclude that C_0 is s-convex, for any $s \in (0, 1)$. Now, define the mapping $S : C_0 \to C_0$ by $S(x - x_0) = T(x) - x_0$, for each $x - x_0 \in C_0$. Clearly, S is continuous, and applying the result of the first step to S, we conclude that there exists $x \in C$ such that $S(x - x_0) = x - x_0$. It implies that $x = T(x)$, which is a fixed point of T. This completes the proof. $\square$

Remark 3.2.7. In the proof of Theorem 3.2.6, we may or may not assume $\theta \in C$. Indeed, if $\theta \notin C$, take $x_0 \in C$ and let $D := C - \overline{co}_s \{x_0\}$. Then, we have $\theta \in D$, and D is compact and s-convex. In addition, by Lemma 3.1.9(a), the translation operator $T_0 : D \to D$ (of T) is also continuous. This implies that T_0 has a fixed point, and thus T has a fixed point, too, by Lemma 3.1.9.

In order to prove that Theorem 3.2.6 also holds for a p-normed space which may not be complete for $p \in (0, 1]$, we need the following result.

Lemma 3.2.8. *Let* $(X, \| \cdot \|_p)$ *be a p-normed space, where* $p \in (0, 1]$. *Then, there is a complete p-normed space* $\hat{X}$ *and a linear isometry mapping i from* X *onto the subspace* $W := i(X)$, *which is the dense in* $\hat{X}$.

Proof. It is Theorem 2.2 of Ennassik and Taoudi [79]. Its proof completely follows the way for the proof of Theorem 2.3.2 by Kreyszig [146]. This completes the proof. $\square$

As applications of Theorem 3.2.6 and Lemma 3.2.8, we have the following general fixed point theorem in p-normed spaces, which may not be complete.

Theorem 3.2.9. *Let* $(X, \| \cdot \|_p)$ *be a p-normed space and* C *be a compact s-convex subset of* X, *where* $s, p \in (0, 1]$. *If* $T : C \to C$ *is continuous, then there exists* $z \in C$ *such that* $Tz = z$.

Proof. Let $\hat{X}$ be the completion of X. By Lemma 3.2.8, there exists a linear isometric embedding $i : X \to \hat{X}$ with $i(X)$ dense in $\hat{X}$. We define $\hat{T} : i(C) \to i(C)$ by $\hat{T}(i(x)) = i(T(x))$, for each $x \in C$. Then, this mapping is easily checked to be well defined, and it is continuous since i is a linear isometry and T is continuous on C. Furthermore, the set $i(C)$ is compact, being the image of a compact set under a continuous mapping. It is also s-convex, as it is the image of an s-convex set under a linear mapping.

Now, by Theorem 3.2.6, there exists $x \in C$ such that $\hat{T}(i(x)) = i(x)$. Thus, $i(T(x)) = i(x)$, so $T(x) = x$, which means T has a fixed point in C. This completes the proof. $\square$

Remark 3.2.10. By applying a new extension theorem in p-norm spaces for $p \in (0, 1]$, which is a generalization of the theorem given by Dugundji [73] and Theorem 3.2.3 above, we establish general fixed point theorems for s-convex subsets in p-normed spaces, which is Theorem 3.2.9, where $s, p \in (0, 1]$. Here, we would like to share with readers that the proof for the existence of homeomorphisms for s-convex subsets in p-normed spaces is new (and uses a different idea in comparison to those given by Xiao and Zhu [270]). By the fact that each p-normed space includes normed spaces as a special class (with $p = 1$), Theorem 3.2.9 indeed unifies the corresponding results in the existing literature. For more in details, see Agarwal *et al.* [1–3], Ben-El-Mechaiekh and Mechaiekh [17], Ben-El-Mechaiekh and Saidi [18], Browder [30–36], Cauty [42, 43], Chang *et al.* [48], Ennassik *et al.* [78], Fan [81], Goebel and Kirk [96], Granas and Dugundji [102], Kirk [130–134], Mauldin [167], Park [184–190], Takahashi [255], Xiao and Zhu [270], Zeidler [297], and the references therein. In addition, for fixed point theorems in p-vector and locally p-convex spaces, we also refer to Ennassik *et al* . [78], Ennassik and Taoudi [79], Yuan [286–291], Yuan and Xiao [294, 295] for more recently results in details.

3.3 Fixed Point Theorems for Nonexpansive Mappings in p-Normed Spaces with p-Normal Structure

The goal of this section is to establish general fixed point theorems and demiclosedness principles for nonexpansive mappings in p-normed spaces by introducing a new concept called p-normal structure or p-Opial condition for $p \in (0, 1]$. These results unify or improve the corresponding results in the existing literature. In particular, when $p = 1$, the corresponding results given in this section reduce to the classical Browder–Göhde–Kirk fixed point theorem for nonexpansive mappings in Banach spaces with the normal structures. The results proved in this section are fundamental tools for the study of nonlinear analysis and related topics on the geometry of p-normed spaces and related fixed point theory.

It is known that the class of p-seminorm spaces for $p \in (0, 1]$ is an important generalization of usual normed spaces with rich topological and geometrical structures, and its related studies have received considerable attention. However, to the best of our knowledge, the corresponding basic tools and associated results in the category of nonlinear functional analysis for nonexpansive mappings have not been well developed in p-normed spaces, where $p \in (0, 1]$.

Our fixed point theorems given in this section unify or improve the corresponding results in the literature, and when $p = 1$, the corresponding results reduce to the classical Browder–Göhde–Kirk fixed theorem for nonexpansive mappings in normed (Banach) space with a normal structure. The results established in this section are fundamental tools for the study of nonlinear analysis and related topics on the geometry of general p-normed spaces and fixed point theory; e.g., see Goebel and Kirk [96], Kirk [134], Xiao and Zhu [270],

Yuan [286–291], and related references in them for more detailed discussions.

Now, we first recall some notions and concepts used by Goebel and Kirk [96], Kirk [133], Penot [202], and the references therein for the study of metric fixed point theory and applications.

Let (M, d) be a metric space. For a subset D of M and $u \in D$, we define

$$\delta(D) := \sup\{d(x, y) : x, y \in D\}, r_u(D) := \sup\{d(u, y) : y \in D\},$$

and

$$r(D) := \inf\{r_u(D) : u \in D\}, \text{ and } D_c := \{u \in D, r_u(D) = r(D)\},$$

where $\delta(D)$ denotes the diameter of a subset D (also sometimes denoted by $diam(D)$ unless specified otherwise in this book), $r_u(D)$ is the Chebyshev radius of D with respect to u, $r(D)$ is the Chebyshev radius of D, and D_c is the Chebyshev center of D in M. In addition, we define

$$h(D) := \begin{cases} r(D)/\delta(D) & \text{if } \delta(D) > 0, \\ 1 & \text{if } \delta(D) = 0. \end{cases}$$

According to Kirk [133] (see also Kirk [134], Penot [202], and the references therein), convexity is not essential for the existence of fixed points for nonexpansive mappings in Banach spaces (or metric spaces in general) with a normal structure. As introduced by Brodskii and Milman [28], we thus first present the following concept called p-normal structure for p-normed spaces, where $p \in (0, 1)$.

Definition 3.3.1 (p-normal structure). A class $\mathcal{S}$ of subsets of a p-normed spaces $(X, \| \cdot \|_p)$ with $p \in (0, 1)$ is said to be normal if for each $D \in \mathcal{S}$, with $\delta(D) > 0$, we have $h(D) \in (0, 1)$. Secondly, the class $\mathcal{S}$ is said to be (countably) compact if each (countable) subfamily of $\mathcal{S}$ possessing the finite intersection property has a nonempty intersection.

Remark 3.3.2. Please note that by comparing with the classical definition for a normal structure of normed spaces (i.e., the case for p-normed spaces with $p = 1$) given by Brodskii and Milman [28] (see also Kirk [130] and Goebel and Kirk [96]), Definition 3.1 above for the p-normal structure of p-normed space for $p \in (0, 1)$ drops the

"convexity condition". This is partially due to the fact that each closed p-convex subset contains at least the zero element in p-normed spaces for $p \in (0,1)$, as given by Lemma 2.2 in Section 3.2 above, which seems larger than the usual convex subset defined in normed spaces.

Let D be a nonempty subset of a metric space (M, d). We also recall that a mapping $F : D \to M$ is said to be nonexpansive if $d(F(u), F(v)) \le d(u, v)$, for each $u, v \in M$. In addition, we use $B(u; r)$ to denote a closed ball centered at $u \in M$ with radius $r > 0$, i.e., $B(u; r) = \{y \in M : d(u, y) \le r\}$. In the case where $(M, \| \cdot \|_p)$ is a p-normed space for $p \in (0, 1]$, its distance d is defined by $d(x, y) := \|x - y\|_p$ for each $x, y \in M$ throughout this book unless specified otherwise.

We also recall that by following Brodskii and Milman [28] in a traditional way, for a given subset D with more than one point in a metric space (X, d), a point $x \in D$ is called diametral if $r_x(D)$ $(= \sup\{d(x, y) : y \in D\}) = \delta(D)$, where $\delta(D)$ denotes the diameter of D defined above, and sometimes the diameter of D is also denoted by $diam(D)$ in the book whenever there is no risk of confusion.

We now have the following fixed point theorem for contractive mappings in complete p-normed spaces for $p \in (0, 1]$ by following the traditional iteration technique used for the proof of the classic Banach contractive principle in complete metric spaces. (We omit its detailed proof here, as it is available in the literature; see Goebel and Kirk [96] and Nadler [173]).

Lemma 3.3.3. *Let $(X, \| \cdot \|_p)$ be a complete p-normed space for $p \in (0, 1]$. Assume that $F : X \to X$ is a contractive mapping (i.e., there exists a constant $k \in (0, 1)$ such that $\|F(x) - F(y)\|_p \le k\|x - y\|_p$, for all $x, y \in X$). Then, F as a unique fixed point in K. In particular, let K be a nonempty closed s-convex subset of $(X, \|\cdot\|_p)$. If $F : K \to K$ is contractive, then F has a unique fixed point in K, where $s, p \in (0, 1]$.*

For a given p-normed space $(X, \| \cdot \|_p)$ for $p \in (0, 1]$, we use the symbol $\hat{X}$ to denote the completion of the space $(X, \| \cdot \|_p)$. Lemma 3.2.8 states that every p-normed space $(X, \| \cdot \|_p)$ for $p \in (0, 1]$ can be complete (by following the proof of Theorem 2.3.2 given by Kreyszig [146], and also see Theorem 2.2 of Ennassik and Taoudi [79]).

Theorem 3.3.4. *Let K be a nonempty closed s-convex subset of a p-normed space $(X, \| \cdot \|_p)$, where $s, p \in (0, 1]$. Then, the contractive mapping $F : K \to K$ has a unique fixed point.*

Proof. Let $\hat{X}$ be the completion of X. By Lemma 3.2.8, there exists a linear isometric embedding $i : X \to \hat{X}$ with $i(X)$ dense in $\hat{X}$. Define a mapping $\hat{F} : i(K) \to i(K)$ by $\hat{F}(i(x)) = i(F(x))$, for each $x \in K$. Then, it is easy to check that the mapping $\hat{F}$ is well defined, and it is contractive on $i(K)$ since i is a linear isometry and F is contractive on K. In addition, the set $i(K)$ is closed and s-convex, being the image under a continuous and linear mapping.

By Lemma 3.3.3, there exists a unique $x \in K$ such that $\hat{F}(i(x)) = i(x)$. Thus, we have $i(F(x)) = i(x)$, implying that $F(x) = x$, which is a unique fixed point of F. This completes the proof. $\qquad\square$

For the convenience of our subsequent discussion, we assume that all p-normed spaces are complete for $p \in (0, 1]$ throughout this book. Now, we recall some important facts and results.

Lemma 3.3.5. *If F is a nonempty of bounded closed weakly compact subset of a p-nomed space $(X, \| \cdot \|_p)$ for $p \in (0, 1]$, then F's Chebyshev center F_c is nonempty and closed.*

Proof. Let $F(x, n) := \{y \in F : d(x, y) \le r(F) + \frac{1}{n}\}$. It is easily seen that the family of subsets $C_n = \cap_{x \in F} F(x, n)$ for $n = 1, 2, \ldots$, forms a decreasing sequence of nonempty closed subsets, and hence $F_c = \cap_{n=1}^{\infty} C_n$ is closed. Thus, we have $F_c \ne \emptyset$ by the nonempty property for the finite nonempty intersection of the family of nonempty closed weakly compact subsets under the weak topology (see Kelley [126] or Goebel and Kirk [96]). This completes the proof. $\qquad\square$

The following result is indeed Lemma 2 of Kirk [130]; we include its short proof here.

Lemma 3.3.6. *Let F be a nonempty bounded closed subset of a metric space (X, d), which contains more than one point. Assume that F has a normal structure and its Chebyshev center F_c is not empty. Then, $\delta(F_c) < \delta(F)$, where F_c is the Chebyshev center of F.*

Proof. By the normal structure of F, we may assume that F contains at least one nondiametral point x. Hence, $r_x(F) < \delta(F)$. If z and w are any two points of F_c, then we have $d(z, w) \le r_z(F) = r(F)$. Thus, we have $\delta(F_c) = \sup\{d(z, w) : w, z \in F_c\} \le r(F) \le$

$r_x(F) < \delta(F)$, as x is a nondiametral point in F. This completes the proof. $\square$

Now, by the definition of the p-normal structure of (the family of) subsets in p-normed spaces for $p \in (0, 1)$ above and applying a similar argument used by Kirk [130] (see also Goebel and Kirk [96] and Kirk [133] and Penot [202]), we obtain the following Browder–Göhde–Kirk-type fixed point theorem for nonexpansive mappings in p-normed spaces with a p-normal structure for $p \in (0, 1]$ (where, when $p = 1$, we apply the traditional definition for "normal structure", introduced by Brodskii and Milman [28] and used widely in the literature; e.g., see Goebel and Kirk [96] and the references therein).

Theorem 3.3.7. *Let K be a nonempty, bounded, closed weakly compact subset (and may not be s-convex for $s \in (0,1]$) of a (complete) p-normed space $(X, \|\cdot\|_p)$, and suppose K has the p-normal structure, where $p \in (0, 1]$. If $T : K \to K$ is a nonexpansive mapping, then T has a fixed point in K.*

Proof. The conclusion is true for $p = 1$ by Theorem of Kirk [130], with the definition of normal structure given by Brodskii and Milman [28] (see also Goebel and Kirk [96]). Thus, without loss of generality, we may only consider the case of $p \in (0, 1)$ and apply Definition 3.3.1 for the definition of the p-normal structure for p-normed spaces.

Let $\mathcal{F}$ denote the collection of all nonempty closed subsets of K, with the property that each of which is mapped into itself by the mapping T. Now, order this family by set inclusion: For $K_1, K_2 \in \mathcal{F}$, we define the order " $\leq$ " by "$K_1 \leq K_2$" provided "$K_2 \subset K_1$". Then, by weak compactness (actually by the Cantor intersection theorem under the weak topology), any chain (linearly ordered family) of sets in $\mathcal{F}$ has a nonempty intersection, and hence an upper bound relative to the order " $\leq$ " is given by $\cap_{K \in \mathcal{F}} K \neq \emptyset$. Now, by Zorn's lemma, $\mathcal{F}$ has a minimal element, which we denote by F, with the property $T(F) \subset F$.

We now complete the proof by showing that F consists of a single point $\{x_0\}$ which is indeed the fixed point of T, i.e., $T(x_0) = x_0$. By Lemma 3.3.5 (and see also Lemma 3.3.6), F_c is nonempty. Now, for a given $x \in F_c$, we have $\|T(x) - T(y)\|_p \leq \|x - y\|_p \leq r_x(F) = r(F)$, for all $y \in F$, and hence $T(F)$ is contained in the spherical ball $\overline{U}$ centered at $T(x)$ with radius $r(F)$, i.e., $\overline{U} = \{w \in F : \|T(x) - w\| \leq r(F)\}$. By the fact that for each $w \in \overline{U}$, we have $\|T(x) - T(w)\|_p \leq$

$\|x - w\|_p \leq r_x(F) = r(F)$ by noting that $x \in F_c$, it shows that $T(w) \in \overline{U}$, which implies that $T(\overline{U}) \subset \overline{U}$. By the fact that we know $T(F) \subset F$, it implies that $T(F \cap \overline{U}) \subset F \cap \overline{U}$. Now, by the minimality of F in $\mathcal{F}$, we should have $F \subset \overline{U}$.

Now, we prove that $T(F_c) \subset F_c$. Indeed, by the proof above, for a given $x \in F_c$, we have $F \subset \overline{U}$, which means $\|T(x) - w\| \leq r(F)$, for all $w \in F$. This implies that $r_{T(x)}(F) = \sup_{w \in F} \|T(x) - w\| \leq r(F) \leq r_{T(x)}(F)$, and thus we show that $r_{T(x)}(F) = r(F)$, which means $T(x) \in F_c$. This proves that $T(F_c) \subset F_c$, which means $F_c \in \mathcal{F}$.

Now, if F is not a singleton, i.e., $\delta(F) > 0$, as $F_c \subset F$, then by Lemma 3.3.6, F_c is a proper subset contained in F. This contradicts the minimality of F, and thus we must have $\delta(F) = 0$, which means F consists of a single point x_0, i.e., $F = \{x_0\}$, which is a fixed point of T such that $T(x_0) = x_0$ as $T(F) \subset F$. This completes the proof. □

By letting $p = 1$ in Theorem 3.3.7, we have the following famous Browder–Göhde–Kirk fixed point theorem for nonexpansive mappings in reflexive space $(X, \|\cdot\|)$ with a normal structure.

Corollary 3.3.8 (Browder–Göhde–Kirk theorem in normed spaces). *Let K be a nonempty bounded closed (may not be convex) weakly compact subset of a Banach space $(X, \|\cdot\|)$ with a normal structure. Assume that $T : K \to K$ is a nonexpansive mapping. Then, T has a fixed point in K.*

By the fact that the "Opial condition" (the case under a complete p-normed space with $p = 1$) in Banach spaces introduced by Opial [179] in 1967, which is independent of uniform convexity (see Goebel and Kirk [96, p. 107]) and also as remarked by Karlovitz [123, 124], it is possible for the existence of fixed points for nonexpansive mappings in Banach spaces which may not have a normal structure. Thus, we also introduce the following new concept called "p-Opial condition" in p-normed spaces, where $p \in (0, 1]$.

Definition 3.3.9 (p-Opial condition for p $\in$ (0, 1]). *Let $(X, \|\cdot\|_p)$ be a p-normed space, where $p \in (0, 1]$. Then, X is said to satisfy p-Opial condition if whenever a sequence $\{x_n\}$ in X converges weakly to x_0, then for any $x \neq x_0$, we have*

$$\liminf_{n \to \infty} \|x_n - x_0\|_p < \liminf_{n \to \infty} \|x_n - x\|_p.$$

Now, we can establish the following new fixed point theorem for nonexpansive mappings defined in p-normed spaces which satisfy the p-Opial condition introduced above for $p \in (0, 1]$.

Theorem 3.3.10. *Let K be a nonempty, bounded, closed weakly compact subset (and may not be s-convex for $s \in (0, 1]$) of a p-normed space $(X, \| \cdot \|_p)$ satisfying the p-Opial condition, where $p \in (0, 1]$. Then, for a given nonexpansive mapping $T : K \to K$, T has a fixed point in K.*

Proof. As $K \subset X$ is weakly compact, by a standard argument based on Banach's contraction principle in p-normed spaces (see Lemma 3.3.3 above), there exists in K a sequence $\{x_n\}$ of approximate fixed points, that is, $\|T(x_n) - x_n\|_p \to 0$. Passing to a subsequence if necessary, we may assume that x_n weakly converges to x_0. Then, we claim that $T(x_0) = x_0$. Indeed as T is nonexpansive, it follows that

$$\liminf_{n \to \infty} \|T(x_0) - x_n\|_p = \liminf_{n \to \infty} \|T(x_0) - T(x_n)\|_p \leq \liminf_{n \to \infty} \|x_0 - x\|_p,$$

so $T(x_0) \neq x_0$ would contradict the p-Opial condition, thus we have $T(x_0) = x_0$. The proof is complete. $\qquad\square$

Remark 3.3.11. By letting $p = 1$ in Theorem 3.3.10, the corresponding results are given in Banach spaces with the Opial condition by Opial [179], Karlovitz [123, 124], and van Dulst [261], and also see Goebel and Kirk [96] for more details. In addition, similar to the (traditional) Opial condition in Banach spaces, it has another nice property called the **demiclosedness principle**, which was first given by Browder [32] and heavily related to fixed point theory for nonexpansive mappings.

Here, we recall the definition of demiclosedness for mappings under the framework of p-normed spaces, where $p \in (0, 1]$: A mapping f defined on a subset D of a p-normed space $(X, \| \cdot \|_p)$ to X is said to be **demiclosed** if for any sequence $\{u_j\}$ in D, the following implication holds:

$$\text{w-}\lim_{j \to \infty} u_j = u \text{ and } \lim_{j \to \infty} \|f(u_j) - w\|_p = 0,$$

implying that $u \in D$ and $f(u) = w$, where the symbol "w-lim" means weakly convergence (also denoted by " $\overset{*}{\rightharpoonup}$ " if there is no confusion).

Now, we have the following general demiclosedness principle for nonexpansive mappings, for $p \in (0,1]$ in p-normed spaces which satisfy the p-Opial condition.

Theorem 3.3.12 (demiclosedness principle). *Let K be a nonempty closed (and may not be convex) subset of a p-normed space $(X, \| \cdot \|_p)$ which satisfies the p-Opial condition, where $p \in (0,1]$, and suppose $T : K \to X$ is nonexpansive. Then, the mapping $f = I - T$ is demiclosed on K.*

Proof. Suppose $\{u_n\}$ in K satisfies that u_n weakly converges to u, i.e., $w - \lim_{u \to \infty} u_n = u$ and $\|u_n - T(u_n) - w\|_p = 0$.

For the convenience of our discussion, we define a new mapping $T_w : K \to X$ by $T_w(x) := T(x) + w$ for each $x \in K$. Then, it is clear that $\lim_{n \to \infty} \|u_n - T_w(u_n)\|_p = 0$ and T_w is also nonexpansive. We now have that

$$\|T_w(u) - u_n\|_p \le \|T_w(u) - T_w(u_n)\|_p + \|T_w(u_n) - u_n\|_p,$$

hence

$$\liminf_{n \to \infty} \|T_w u - u_n\|_p \le \liminf_{n \to \infty} \|u - u_n\|_p.$$

As X satisfies the p-Opial condition, it follows that $T_w(u) = u$ (otherwise, if $T_w(u) \ne u$, if will contradicts the p-Opial condition), i.e., $u - T(u) = w$. This completes the proof. $\square$

If $p = 1$ in Theorem 3.3.12, it reduces to Theorem 10.3 of Goebel and Kirk [79], which plays a very important role in the study of fixed point theory. Actually, when Banach spaces are uniformly convex, the demiclosedness principle still holds for nonexpansive mappings, but we leave its discussion to Chapter 7 associated with fixed point theorems for nonexpansive mappings there.

Here, we also recall that when $p = 1$, each p-normed space $(X, \| \cdot \|_p)$ reduces to the usually normed vector space $(X, \| \cdot \|)$. By following Goebel and Reich [98] and the references therein, we recall the following definition, which will be used in the remaining part of this book.

Definition 3.3.13. A normed vector space $(X, \| \cdot \|)$ is said to be uniformly convex if for every $\epsilon \in (0,2]$, there is some $\delta > 0$ such that for any two vectors $x, y \in X$ with $\|x\| = \|y\| = 1$, the condition $\|x - y\| \ge \epsilon$ implies that $\|\frac{x+y}{2}\| \le 1 - \delta$.

As the concept of uniformly convexity is one of the key notions in the study of geometry for Banach spaces, we also recall the following related concept called the "modulus of convexity" of Banach spaces.

Definition 3.3.14. The modulus of convexity of a Banach space X is the function $\delta_X : [0,2] \to [0,1]$ defined by

$$\delta_X(\epsilon) := \inf \left\{ 1 - \left\| \frac{x+y}{2} \right\| : \|x\| \leq 1, \|y\| \leq 1, \|x - y\| \geq \epsilon \right\}.$$

Note that for any $\epsilon > 0$, the value of $\delta_X(\epsilon)$ is the largest number for which the following implication always holds:

For any $x, y \in X$, if $\|x\| \leq 1$, $\|y\| \leq 1$ and $\|x - y\| \geq \epsilon$, we have $\left\| \frac{x+y}{2} \right\| \leq 1 - \delta_X(\epsilon)$.

Obviously, a space X is uniformly convex if and only if its modulus of convexity satisfies $\delta_X(\epsilon) > 0$ for $\epsilon \in (0,2]$. Secondly, it is clear that if $\{x_n\}$ and $\{y_n\}$ are sequences in a uniformly convex space, then we have the following fact:

If $\lim_{n\to\infty} \|x_n\| = \lim_{n\to\infty} \|y_n\| = \lim_{n\to\infty} \frac{1}{2}\|x_n + y_n\| = 1$, then we have $\lim_{n\to\infty} \|x_n - y_n\| = 0$.

For the convenience of later reference and discussion, we list its following equivalent formulation for the concept of uniform convexity:

For any $x, y, p \in X$ and $r \in [0, 2R]$, if $\|x - p\| \leq R$, $\|y - p\| \leq R$, $\|x - y\| \geq \epsilon$, we have $\|p - \frac{1}{2}(x + y)\| \leq (1 - \delta_X(\frac{r}{R}))R$, where R is a given positive number.

Indeed, by following Clarkson [60], we recall that a Banach space $(X, \| \cdot \|)$ is said to be strictly convex if whenever x and y are not collinear vectors of X, then $\|x + y\| < \|x\| + \|y\|$. Thus, all uniformly convex Banach spaces are strictly convex, but not all Banach spaces are strictly convex, as shown by the following example.

Example 3.3.15. $(\mathbb{R}^n, \| \cdot \|_\infty)$ is not strictly convex, while the spaces $(\mathbb{R}^n, \| \cdot \|_p)$ for $p \in (1, \infty)$ are strictly convex.

Let C be a convex set in a Banach space X. A point $z \in C$ is said to be an extreme point for C if whenever $z = tx + (1 - t)y$ for some $t \in (0, 1)$ and some x and y in C, then $x = y$. Then, we have the following characterizations for strictly convex spaces (see their detailed proof in Theorem 1.3 of Ayerbe Toledano *et al.* [12]).

Theorem 3.3.16. *Let $(X, \| \cdot \|)$ be a Banach space. The following assertions are equivalent:*

(1) X *is strictly convex.*
(2) *If* $\|x\| = \|y\| = 1$ *and* $x \neq y$, *then* $\|\frac{x+y}{2}\| < 1$.
(3) *For every* $p \in (1, +\infty)$ *and for all* $x \neq y$, *we have* $\|\frac{x+y}{2}\|^p < \frac{1}{2}(\|x\|^p + \|y\|^p)$.
(4) *For every* $x \in X$, *with* $\|x\| = 1$, *there is an extreme point of the closed unit ball of* X.
(5) *For each non-zero* $f \in X^*$, *there is at most one point* $x \in X$ *in the closed unit ball at which* f *attains its norm, that is,* $f(x) = \|f\|$, *where* X^* *is the dual space of the space* X.

We also list the following facts for Banach spaces which are either uniformly convex or strictly convex.

Theorem 3.3.17. *Let* $(E, \|\cdot\|)$ *be a Banach space. Then, we have the following:*

(1) *If* E *is uniformly convex, then it is reflexive.*
(2) *Suppose* E *is strictly convex, let* D *be a nonempty closed convex subset of* E, *with* $T : D \to E$ *being a nonexpansive mapping, and assume that the set of fixed points (denoted by)* $F(T)$ *for* T *is nonempty. Then, the set* $T(F)$ *is closed and convex.*

Proof. Here, we only give the outline of the proof's related references. When E is uniformly convex, it is reflexive (see, by Theorem 2 of Diestel [64, p. 37]). This is the so-called Milman–Pettis theorem, which was proved independently by Milman [170] in 1938 and Pettis [209] in 1939). Secondly, if E is strictly convex and if the set of fixed points denoted by $F(T)$ for T is nonempty, then $F(T)$ is closed by the continuity of T. In addition, the convexity of the set $F(T)$ follows by the definition of strict convexity; e.g., see Lemma 3.4 of Goebel and Kirk [96] for a simple and smart argument. This completes the proof. $\square$

3.4 Caristi Fixed Point Theorem and Related Principles in Nonlinear Analysis

The goal of this section is to discuss the famous and a very important tool in nonlinear functional analysis called Caristi fixed point theorem established by Caristi [39] in 1976 for complete metric spaces. The discussion in this section demonstrates the power of fixed point theorem developed in this book, as it shows that almost all key principles established in nonlinear analysis are equivalent to the Caristi fixed point theorem under the framework of metric spaces. In addition, the results established in this section will be used in the later chapters of this book to study point theorems for non-self-contractive and nonexpansive (single or set-valued) mappings.

The discussion in this section, in particular, shows that Caristi's fixed point theorem [39] is equivalent to Ekeland's variational principle [75], Takahashi's nonconvex minimization theorem [255], Daneš' drop theorem [62] (see also Rolewicz [229]), the flower petal theorem discussed by Penot [203], and Oettli–Théra theorem [177]; see also Almezel *et al.* [6], Kirk [134], Park [200], and the references therein for more details.

Following the notions and concepts used by Goebel and Kirk [96], let (M, d) be a complete metric space. Then, a mapping $T : M \to M$ is said to be a contraction mapping with the Lipschitz constant $k \in (0, 1)$ if

$$d(T(x), T(y)) \leq kd(x, y)$$

for each $x, y \in M$. If we define a mapping $\phi : M \to [0, +\infty)$ by

$$\phi(x) := (1 - k)^{-1} d(x, Tx)$$

for each $x \in M$, as T is k contractive, where $k \in (0, 1)$, then we have

$$d(x, Tx) - kd(x, Tx) \leq d(x, Tx) - d(Tx, T^2 x).$$

Then, it follows that

$$d(x, Tx) \leq \phi(x) - \phi(Tx),$$

for each $x \in M$, which implies that T has a unique fixed point in M, as T is a contraction mapping with the Lipschitz constant $k \in (0, 1)$.

Actually, if the mapping ϕ above is assumed to be only lower semicontinuous and bounded below, for any mapping $T : M \to M$ which may not be a contraction mapping, but satisfies the above inequality with ϕ, it then has at least one fixed point, as shown by the Caristi fixed point theorem. This theorem was established by Caristi [39] in 1976, which we list as Theorem 3.4.1 as follows.

Theorem 3.4.1 (Caristi fixed point theorem for single-valued mappings). *Let (X, d) be a complete metric space. Assume that $T :*

$X \to X$ is a mapping and $\phi : X \to \mathbb{R}$ is a lower semicontinuous mapping which is bounded below such that for any $x \in X$,

$$d(x, T(x)) \leq \phi(x) - \phi(T(x)).$$

Then, T has a fixed point in X.

Proof. It is indeed Theorem 2.1' of Caristi [39]. The key idea for the proof of this remarkable result has been discussed in the theorem of Siegel [246], Brézis and Browder [27], and Goebel and Kik [96]. Later, a refinement argument was given by Wong [267], and a direct proof was provided recently by Du [70, 71] without using Zorn's lemma, transfinite induction, or any other well-known principle (see also a direct proof given by Kozlowski [145] and related discussion there).

Here, we provide a direct proof which is based on an elegant argument given by Du [70, 71] mentioned above.

We define a set-valued mapping $F : X \to 2^X$ by

$$F(x) := \{y \in X : d(x, y) \leq \phi(x) - \phi(y)\}$$

for each $x \in X$. Clearly, we have that x and $Tx \in F(x)$, and hence $F(x) \neq \emptyset$ for each $x \in X$.

We now claim that for each $y \in F(x)$, $\phi(y) \leq \phi(x)$ and $F(y) \subset F(x)$. Let $y \in F(x)$ be any given point. Then, $y \neq x$ and $d(x, y) \leq \phi(x) - \phi(y)$, thus $\phi(y) \leq \phi(x)$. As $F(y) \neq \emptyset$, let $z \in F(y)$. Then, $d(y, z) \leq \phi(y) - \phi(x)$. It follows that $\phi(z) \leq \phi(y) \leq \phi(x)$, and hence

$$d(x, z) \leq d(x, y) + d(y, z) \leq \phi(x) - \phi(z).$$

Thus, $z \in F(x)$; therefore, we prove that $F(y) \subset F(x)$ for each $y \in F(x)$.

We now construct a sequence $\{x_n\}$ in X by induction, starting with any point $x_1 \in X$ as follows: Suppose $x_n \in X$ is known for $n \in \mathbb{N}$, then choose $x_{n+1} \in F(x_n)$ such that

$$\phi(x_{n+1}) \leq \inf_{z \in F(x_n)} \phi(z) + \frac{1}{n}.$$

For each $n \in \mathbb{N}$, since $x_{n+1} \in F(x_n)$, we obtain

$$d(x_n, x_{n+1}) \leq \phi(x_n) - \phi(x_{n+1}).$$

Thus, $\phi(x_{n+1}) \leq \phi(x_n)$ for each $n \in \mathbb{N}$.

Since ϕ is bounded below, the number $\lambda := \lim_{n \in \mathbb{N}} \phi(x_n)$ exists. Now, for $m > n$, with $m, n \in \mathbb{N}$, by the inequalities above, we obtain

$$d(x_n, d_m) \leq \Sigma_{i=n}^{m-1} d(x_i, x_{i+1}) \leq \phi(x_n) - \lambda.$$

Since $\lim_{n \to \infty} \phi(x_n) = \lambda$, we have

$$\sup_{n \to \infty} \{d(x_x, x_m) : m > n\} = 0.$$

Hence, $\{x_n\}$ is a Cauchy sequence in X. By the completeness of X, there exists some $\nu \in X$ such that $\lim_{n \to \infty} x_n = \nu$. As ϕ is lower semicontinuous, it follows that for all $i \in \mathbb{N}$,

$$\phi(\nu) \leq \liminf_{n \to \infty} \phi(x_n) = \inf_{n \in \mathbb{N}} \phi(x_n) \leq \phi(x_i).$$

Now, we claim that $\cap_{n=1}^{\infty} F(x_n) = \{\nu\}$. Indeed, for $m > n$ with $m, n \in \mathbb{N}$, by the inequalities above, we have

$$d(x_n, x_m) \leq \Sigma_{i=n}^{m-1} d(x_i, x_{i+1}) \leq \phi(x_n) - \phi(\nu).$$

Since $\lim_{m \to \infty} x_m = \nu$, the inequality above implies for each $n \in \mathbb{N}$,

$$d(x_n, \nu) \leq \phi(x_n) - \phi(\nu).$$

It implies that $\nu \in \cap_{n=1}^{\infty} F(x_n)$, and thus $\cap_{n=1}^{\infty} F(x_n) \neq \emptyset$. Moreover, we can see that

$$F(\nu) \subset \cap_{n=1}^{\infty} F(x_n).$$

Now, for any $c \in \cap_{n=1}^{\infty} F(x_n)$, by the inequality in the beginning above, we have

$$d(x_n, c) \leq \phi(x_n) - \phi(c) \leq \phi(x_n) - \inf_{z \in F(x_n)} \phi(z)$$

$$\leq \phi(x_n) - \phi(x_{n+1}) + \frac{1}{n},$$

for all $n, m \in \mathbb{N}$. Therefore, we have $\lim_{n \to \infty} d(x_n, c) = 0$, or, equivalently, $\lim_{n \to \infty} x_n = \nu$. By the uniqueness of the limit of a convergent sequence, we have $c = \nu$. This means $\cap_{n=1}^{\infty} F(x_n) = \{\nu\}$.

On the other hand, since $F(\nu) \neq \emptyset$, and $F(\nu) \subset \cap_{n=1}^{\infty} F(x_n) = \{\nu\}$. Thus, we have $F(\nu) = \{\nu\}$. For the element $\nu \in X$, as $T\nu \in F(\nu)$, it implies that $\nu = T\nu$, which is a fixed point of T. Therefore, T must have a fixed point in X, and this completes the proof. $\square$

Here, we would like to remark that another proof without using the transfinite induction technique has been given by Brézis and Browder [27]. Further, Park [195–200] gave a comprehensive review of the study of the Caristi fixed point theorem and related metric fixed pint theory under the general framework of quasi-metric spaces; see also Feng and Liu [88], Lau and Yao [152], and the references therein for some new results in a different direction, which are not covered by the current version of the book.

Now, as a consequence of Theorem 3.4.1, we have the following conclusion, which is Proposition of Wong [267].

Theorem 3.4.2. *Let (X, d) be a nonempty complete metric space. Let $T : X \to X$ be a single-valued mapping with nonempty values. Suppose there exists a lower semicontinuous function $\phi : X \to [0, +\infty)$ such that for any $x \in X$, there exists $y \in X \setminus \{x\}$ satisfying*

$$d(x, y) \leq \phi(x) - \phi(y).$$

Then, T has a fixed point.

Proof. Suppose T does not have a fixed point. Then, for each $x \in X$, $T(x) \neq x$, but there exists $y \in X$ with $y \neq x$ such that $d(x, y) \leq \phi(x) - \phi(y)$. We now define a new mapping $T^* : X \to X$ with $T^*(x) := y$ for each $x \in X$, where y is fixed in $X \setminus \{x\}$ satisfying $d(x, y) \leq \phi(x) - \phi(y)$, as mentioned by the assumption. It is then clear that T^* is well defined, and we have $d(x, T^*(x)) \leq \phi(x) - \phi(T^*(x))$, for each $x \in X$. By Theorem 3.4.1, T^* has a fixed point $x^* \in X$ such that $T^*(x^*) = x^*$. However, by the definition of T^*, $x^* \neq T^*(x^*)$, this is a contradiction. Thus, T must have a fixed point in X. This completes the proof. $\square$

Now, we have the following Caristi fixed point theorem for set-valued mappings in complete metric spaces.

Theorem 3.4.3 (Caristi fixed point theorem for set-valued mappings). *Let (X, d) be a nonempty complete metric space. Let $T : X \to 2^X$ be a set-valued mapping with nonempty values. Suppose there exists a lower semicontinuous function $\phi : X \to [0, +\infty)$ such that for any $x \in X$, there exists $y \in T(x)$ such that*

$$d(x, y) \leq \phi(x) - \phi(y).$$

Then, T has a fixed point.

Proof. Following Ansari [7] (for more details, see Almezel *et al.* [6]), for each $x \in X$, we define a single-valued mapping $f : X \to X$ by $f(x) := y$, where we choose one $y \in X$ such that $y \in T(x)$ and $d(x, y) + \phi((y) \leq \phi(x)$.

Then, f is well defined and $f(x) \in T(x)$ for each $x \in X$, and we also have

$$d(x, f(x)) + \phi(f(x)) \leq \phi(x).$$

Now, by Theorem 3.4.1 for the single-valued mapping $f : X \to X$, it follows that there exists $x \in X$ such that $x = f(x)$. By the definition of f, we have $x = f(x) \in T(x)$, and thus x is a fixed point of the set-valued mapping T. This completes the proof. $\square$

As an application of the Caristi fixed point theorem, we can prove the following Nadler's fixed point theorem for contraction set-valued mappings. We note that as an application of Takahashi's minimization theorem (presented later in this section), Nadler's fixed point theorem can also be proved without using the axiom of choice (but we do not include its proof here in this book for space considerations).

Theorem 3.4.4 (Nadler's fixed point theorem for contraction set-valued mappings). *Let (X, d) be a complete metric space. Then, each contraction set-valued mapping $T : X \to CB(X)$ has a fixed point, where $CB(X)$ denotes the family of all nonempty closed bounded subsets of X.*

Proof. Let $T : X \to CB(X)$ be a Nadler contraction set-valued mapping with contraction constant h. Choose a real number $\alpha \in \mathbb{R}$ such that $h < \alpha < 1$. Let $x \in X$. Then, the set $\{y \in T(x) : \alpha d(x, y) \leq d(x, T(x))\} \neq \emptyset$.

By the axiom of choice, there is a singe-valued mapping $f : X \to X$ such that $f(x) \in T(x)$ for each $x \in X$. and $\alpha d(x, f(x)) \leq d(x, T(x))$. Thus, we have

$$d(f(x), T(f(x))) \leq d_H(T(x), T(f(X))) \leq h \cdot d(x, f(x)).$$

Note that

$$d(x, f(x)) = \frac{1}{\alpha - h}(\alpha d(x, f(x)) - h \cdot d(x, f(x)))$$

$$\leq \frac{1}{\alpha - h}(d(x, T(x)) - d(f(x), T(f(x)))).$$

Now, set $\phi(x) := \frac{1}{\alpha-h}d(x, T(x))$ for each $x \in X$. Then, we know that ϕ is continuous from X to $[0, \infty)$ and

$$d(x, f(x)) \leq \phi(x) - \phi(f(x))$$

for each $x \in X$. Now, by the Caristi fixed point theorem, there exists $x \in X$ such that $x = f(x)$, which is a fixed point of f, so $x = f(x) \in T(x)$ is a fixed point of T. This completes the proof. □

Now, we prove Ekeland's variational principle by using Theorem 3.4.3, which is also called the Caristi and Kirk theorem in the literature [40].

Theorem 3.4.5 (Strong form of Ekeland's variational principle). *Let (X, d) be a complete metric space and $f : X \to \mathbb{R} \cup \{+\infty\}$ be a proper, bounded below, and lower semicontinuous functional. Let $\epsilon > 0$ and $\hat{x} \in X$ be given such that $f(\hat{x}) \leq \inf_{x \in X} f(x) + \epsilon$. Then, for a given $\lambda > 0$, there exists $\bar{x} \in X$ such that:*

(1) $f(\bar{x}) \leq f(\hat{x})$;
(2) $d(\hat{x}, \bar{x}) \leq \lambda$; *and*
(3) $f(\bar{x}) < f(x) + \frac{\epsilon}{\lambda}d(x, \bar{x})$, *for all $x \in X \setminus \{\bar{x}\}$.*

Proof. Without loss of generality, we may assume that $\lambda = 1$. Let $\epsilon > 0$ be given and choose $\hat{x} \in X$ such that $f(\hat{x}) \leq \inf_{x \in X} f(x) + \epsilon$.

Now, let $X' := \{x \in X : f(x) \leq f(\hat{x}) - \epsilon d(\hat{x}, x)\}$. Then, X' is nonempty. By the lower semicontinuity of f, it follows that X' is a closed subset of a complete metric space and, hence, a complete metric space. For each $x \in X'$,

$$S(x) := \{y \in X : y \neq x, f(y) \leq f(x) - \epsilon d(x, y)\},$$

for each $x \in X$. Now, we define a set-valued mapping $T : X \to 2^X$ by

$$T(x) = \begin{cases} x & \text{if } S(x) = \emptyset, \\ S(x) & \text{if } S(x) \neq \emptyset. \end{cases} \tag{3.5}$$

Then, T is a set-valued mapping from X' to itself with nonempty values. Indeed, $T(x) = x \in X'$ if $S(x) = \emptyset$. If not, since $T(x) = S(x)$, we have for all $y \in T(x)$,

$$\epsilon d(\hat{x}, y) \le \epsilon d(\hat{x}, x) + \epsilon(x, y) \le f(\hat{x}) - f(x) + f(x) - f(y)$$
$$\le f(\hat{x}) - f(y),$$

and hence, $y \in X'$. Moreover, for all $x \in X$ and $y \in T(x)$, we have

$$\frac{1}{\epsilon} f(y) + d(x, y) \le \frac{1}{\epsilon} f(x).$$

Now, by Theorem 3.4.3, T has a fixed point $\bar{x} \in X'$. Consequently, $S(\bar{x}) = \emptyset$, that is, we have

$$f(x) > f(\bar{x}) - \epsilon d(\bar{x}, x),$$

for all $x \in X$ with $x \ne \bar{x}$. Since $\bar{x} \in X'$, we have

$$f(\bar{x}) \le f(\hat{x}) - \epsilon(\hat{x}, \bar{x}) \le f(\hat{x}).$$

Further, we have

$$\epsilon d(\hat{x}, \bar{x}) \le f(\hat{x}) - f(\bar{x}) \le f(\hat{x}) - \inf_{x \in X} f(x) \le \epsilon.$$

and hence, we have $d(\hat{x}, \bar{x}) \le 1$. This completes the proof. $\qquad \square$

Now, as an application of Theorem 3.4.5, we have the following Takahashi's minimization theorems given by Takahashi in [254, 255].

Theorem 3.4.6 (Takahashi's minimization theorem). *Let (X, d) be a complete metric space and $f : X \to \mathbb{R} \cup \{\infty\}$ be a proper, bounded below, and lower semicontinuous functional. Suppose that, for each $\hat{x} \in X$ with $\inf_{x \in X} f(x) < f(\hat{x})$, there exists $z \in X$ such that $z \ne \hat{x}$ and $f(z) + d(\hat{x}, z) \le f(\hat{x})$. Then, there exists $\bar{x} \in X$ such that $f(\bar{x}) = \inf_{x \in X} f(x)$.*

Proof. We prove the conclusion by using the strong form of Ekeland's variational principle as a tool. By Theorem 3.4.5 above, for any given $\epsilon > 0$, there exists $\bar{x} \in X$ such that

$$f(\bar{x}) < f(x) + \epsilon d(x, \bar{x}),$$

for all $x \in X$ with $x \ne \bar{x}$. We now claim that $f(\bar{x}) = \inf_{x \in X} f(x)$.

Assume to the contrary that there exists $w \in X$ such that $f(w) > \inf_{x \in X} f(x)$. Now, by this hypothesis, there exists $z \in X$ such that $z \neq w$ and

$$f(z) + d(w, z) \leq f(w),$$

which contradicts the inequality above. Hence, we have some $\bar{x} \in X$ such that $f(\bar{x}) = \inf_{x \in X} f(x)$. This completes the proof. $\square$

In what follows, we also give the following theorem that characterizes the completeness of the underlying metric spaces which is as an application of Takahashi's minimization theorem [254].

Theorem 3.4.7. *A metric space (X, d) is complete if for every uniformly continuous function $f : X \to \mathbb{R} \cup \{+\infty\}$ and every $\hat{x} \in X$ with $\inf_{x \in X} f < f(\hat{x})$, there exists $z \in X$ such that $z \neq \hat{x}$ and*

$$f(z) + d(\hat{x}, z) \leq f(\hat{x}),$$

then there exists $\bar{x} \in X$ such that $f(\bar{x}) = \inf_{x \in X} f(x)$.

Proof. Let $\{x_n\}$ be a Cauchy sequence in X. Consider the function $f : X \to \mathbb{R} \cup \{+\infty\}$ defined by $f(x) := \lim_{n \to \infty} d(x_n, x)$ for each $x \in X$. f is uniformly continuous and $\inf_{x \in X} f(x) = 0$. Let $f(\hat{x}) = 0$. Then, there exists an $x_m \in X$ such that

$$x_m \neq \hat{x}, f(x_m) < \frac{1}{3} f(\hat{x}) \text{ and } d(x_m, \hat{x}) - f(\hat{x}) < f(\hat{x}).$$

Thus, we have

$$3f(x_m) + d(x_m, \hat{x}) < f(\hat{x}) + 2f(\hat{x}) = 3f(\hat{x}).$$

Therefore, there exists an $\bar{x} \in X$ such that $f(\bar{x}) = \inf_{x \in X} f(x) = 0$, and so, $0 = f(\bar{x}) = \lim_{n \to \infty} d(x_n, \bar{x})$. Thus, $\{x_n\}$ converges to $\bar{x}$, and hence, X is complete. This completes the proof. $\square$

Here, we also refer interested readers to Cobzaş [61], Jachymski *et al.* [117], Park [195], Park and Rhoades [201], and related references therein for a comprehensive review of the various circumstances in which fixed point results imply completeness for metric spaces by involving Ekeland's variational principle and its equivalent, the Caristi fixed point theorem. The results also include other fixed point results having this property in metric spaces, including quasi-metric and partial metric spaces; in addition, the aforementioned contains discussion on topology and order and on fixed points in ordered structures and their completeness properties.

Now, by using the Cantor intersection theorem in complete metric spaces as a tool, we can establish the following so-called extended version of Ekeland's variational principle, which is equivalent to the Caristi–Kirk-type fixed point theorem, as shown in the following.

Theorem 3.4.8 (Extended Ekeland's variational principle).
Let K be a nonempty closed subset of a complete metric space (X, d) and $F : K \times K \to \mathbb{R}$ be a bifunction mapping. Assume that $\epsilon > 0$ and the following assumptions are satisfied:

(i) *For all $x \in K$, the set $L := \{y \in K : F(x, y) + \epsilon d(x, y) \leq 0\}$ is closed;*
(ii) *$F(x, x) = 0$ for all $x \in K$;*
(iii) *$F(x, y) \leq F(x, z) + F(z, y)$ for all $x, y, z \in K$.*

Now, if $\inf_{y \in K} F(x_0, y) > -\infty$ for some $x_0 \in K$, then there exists $\bar{x} \in K$ such that:

(a) *$F(x_0, \bar{x}) + \epsilon d(x_0, \bar{x}) \leq 0$; and*
(b) *$F(\bar{x}.x) + \epsilon d(\bar{x}, x) > 0$ for all $x \in K$ with $x \neq \bar{x}$.*

Proof. For the sake of convenience, we set $d_\epsilon(u, v) = \epsilon d(u, v)$. Then, d_ϵ is equivalent to d, and (X, d_ϵ) is complete. For each $x \in X$, we define a set

$$S(x) := \{y \in K : F(x, y) + d_\epsilon(x, y) \leq 0\}.$$

By condition (i), $S(x)$ is closed for each $x \in K$. From condition (ii), $x \in S(x)$ for all $x \in K$, and thus $S(x)$ is nonempty for each $x \in X$. So, there exists $y \in S(x)$, that is,

$$F(x, y) + d_\epsilon(x, y) \leq 0,$$

and also there exists $z \in S(y)$ such that

$$F(y, z) + d_\epsilon(z, y) \leq 0.$$

By adding the above inequalities together and using condition (iii), we obtain

$$0 \geq F(x,y) + d_\epsilon(x,y) + F(y,z) + d_\epsilon(z,y) \geq F(x,z) + d_\epsilon(x,z).$$

Therefore, $z \in S(x)$, which implies that $S(y) \subset S(x)$. Now, we show that there is a sequence $\{x_n\}$ of K such that for each $n \in \mathbb{N}$,

$$x_{n+1} \in S(x_n), F(x_n, x_{n+1}) < \inf_{z \in S(x_n)} F(x_n, z) + \frac{1}{n+1}.$$

Let $\nu(x_0) := \inf_{z \in S(x_0)} F(x_0, z) > -\infty$, and construct a sequence in the following manner:

For x_1, let $x_1 \in S(x_0)$ such that $F(x_0, x_1) < \nu(x_0) + \frac{1}{1}$.

Since $x_1 \in S(x_0)$, we have $S(x_1) \subset S(x_0)$. By condition (iii), we have

$$\nu(x_1) = \inf_{z \in S(x_1)} F(x_1, z) \geq \inf_{z \in S(x_1)} F(x_0, z) - F(x_0, x_1)$$

$$\geq \nu(x_0) - F(x_0, x_1) > -\infty.$$

Then, there exists $x_2 \in S(x_1)$ such that

$$F(x_1, x_2) \leq \nu(x_1) + \frac{1}{2}.$$

Continuing in this way, we obtain a sequence $\{x_n\}$ such that for each $n \in \mathbb{N}$,

$$x_{n+1} \in S(x_n), F(x_n, x_{n+1}) < \nu(x_n) + \frac{1}{n+1},$$

and

$$\nu(x_{n+1}) \geq \nu(x_n) - F(x_n, x_{n+1}),$$

which implies that

$$-\nu(x_n) \leq -F(x_n, x_{n+1}) + \frac{1}{n+1} \leq \nu(x_{n+1}) - \nu(x_n) + \frac{1}{n+1}.$$

Thus, we have that for $n \in \mathbb{N}$,

$$\nu(x_{n+1}) + \frac{1}{n+1} \geq 0. \tag{3.6}$$

Now, if $z_1, z_2 \in S(x_n)$, then

$$d(z_1, z_2) \leq d(x_n, z_1) + d(x_n, z_2) \leq -F(x_n, z_1) - F(x_n, z_2) - 2\nu(x_n).$$

Thus, it implies that for the diameter of $S(x_n)$, denoted by $diam(S(x_n))$, we have $diam(S(x_n)) \leq -2\nu(x_n)$. Thus, by inequality (3.8), we have $\lim_{n \to \infty} diam(S(x_n)) = 0$.

By the fact that $\{S(x_n)\}$ is a family of closed sets such that $S(x_{n+1}) \subset S(x_n)$, for all $n \in \mathbb{N}$, and $\lim_{n \to \infty} diam(S(x_n)) = 0$, by Cantor's intersection theorem in complete metric spaces, there exists exactly one point $\bar{x} \in X$ such that $\cap_{n=1}^{\infty} S(x_n) = \{\bar{x}\}$.

This implies that $\bar{x} \in S(x_0)$, that is, we have

$$F(x_0, \bar{x}) + d_\epsilon(x_0, \bar{x}) \leq 0,$$

which shows that conclusion (a) holds.

Secondly, $\bar{x} \in \cap_{n=1}^{\infty} S(x_n)$ and since $S(\bar{x}) \subset S(x_n)$ for all $n \in \mathbb{N}$, we thus have $S(\bar{x}) = \{\bar{x}\}$. It then follows that any $x \notin S(\bar{x})$ whenever $x \neq \bar{x}$, implying that

$$F(\bar{x}, x) + d_\epsilon(\bar{x}, x) > 0,$$

that is, conclusion (b) holds. This completes the proof. $\qquad\square$

As mentioned above, the following result shows the equivalences among extended Ekelend's variational principle, extended Takahashi's minimization theorem, Caristi–Kirk fixed point theorem for set-valued mappings, and Oettli–Théra theorem in complete metric spaces.

Theorem 3.4.9. *Let K be a nonempty closed subset of a complete metric space (X, d) and $F : K \times K \to \mathbb{R}$ be a (bifunction) mapping such that it is lower semicontinuous in the second argument and the following conditions hold:*

(i) *$F(x, x) = 0$ for all $x \in X$; and*
(ii) *$F(x, y) \leq F(x, z) + F(z, y)$ for all $x, y, z \in X$.*

Assume that there exists $\hat{x} \in X$ such that $\inf_{x \in X} F(\hat{x}, x) > -\infty$. Let

$$\hat{S} := \{x \in X : F(\hat{x}, x) + d(\hat{x}, x) \leq 0. \qquad (3.7)$$

Then (as, by (i), it follows that $\hat{x} \in \hat{S} \neq \emptyset$), the following statements are equivalent:

(a) (*Extended Ekeland's variational principle*): *There exists* $\bar{x} \in \hat{S}$ *such that*

$$F(\bar{x}, x) + d(\bar{x}, x) > 0, \ \text{for all } x \neq \bar{x}. \tag{3.8}$$

(b) (*Extended Takahashi's minimization theorem*): *Assume that*

$$\begin{cases} \text{for every } \hat{x} \in \hat{S} \text{ with } \inf_{x \in X} F(\hat{x}, x) < 0 \text{ there exist} \\ x \in X \text{ such that } F(\hat{x}, x) + d(\hat{x}, x) \leq 0 \text{ for all } x \neq \hat{x}. \end{cases} \tag{3.9}$$

Then, there exists $\bar{x} \in \hat{S}$ *such that* $F(\bar{x}, x) \geq 0$ *for all* $x \in X$.

(c) (*Caristi–Kirk fixed point theorem*): *Let* $T : X \to 2^X$ *be a set-valued mapping such that*

$$\begin{cases} \text{for every } \hat{x} \in S \text{ there exists} \\ x \in T(\hat{x}) \text{ satisfying } F(\hat{x}, x) + d(\hat{x}, x) \leq 0. \end{cases} \tag{3.10}$$

Then, there exists $\bar{x} \in S$ *such that* $\bar{x} \in T(\bar{x})$.

(d) (*Oettli–Théra theorem*): *Let* $D \subset X$ *have the property that*

$$\begin{cases} \text{for every } \hat{x} \in \hat{S} \setminus D \text{ there exists} \\ x \in X \text{ such that } F(\hat{x}, x) + d(\hat{x}, x) \leq 0, \ \text{for all } x \neq \hat{x}. \end{cases} \tag{3.11}$$

Then, there exists $\bar{x} \in \hat{S} \cap D$.

Proof. We first prove $(a) \Rightarrow (d)$: Let (a) and the hypothesis of (d) hold. Then, (a) giving any $\bar{x} \in \hat{S}$ such that $F(\bar{x}, x) + d(\bar{x}, x) > 0$, for all $x \neq \bar{x}$. From (3.13), we have $\bar{x} \in D$. Hence, $\bar{x} \in \hat{S} \cap D$, and (d) holds.

$(d) \Rightarrow (a)$: Let (d) hold. For all $\hat{x} \in X$, define

$$\Gamma(\hat{x}) = \{x \in X : F(\hat{x}, x) + d(\hat{x}, x) \leq 0, x \neq \hat{x}\}.$$

Choose $D := \{\hat{x} \in X : \Gamma(\hat{x}) = \emptyset\}$. If $\hat{x} \notin D$, then from the definition of D, there exists $x \in \Gamma(\hat{x})$. Hence, (3.13) is satisfied, and by (d), there exists $\bar{x} \in \hat{S} \cap D$. Then, $\Gamma(\bar{x}) = \emptyset$, that is, $F(\hat{x}, x) + d(\hat{x}, x) > 0$ for all $x \neq \bar{x}$. Hence, (a) holds.

$(b) \Rightarrow (d)$: Suppose that both (b) and the hypothesis of (d) hold. Assume to the contrary that $\hat{x} \notin D$ for all $\hat{x} \in \hat{S}$. Then, by (3.13) for all $\hat{x} \in \hat{S}$,

$$\text{there exists } x \neq \hat{x} \text{ with } F(\hat{x}, x) + d(\hat{x}, x) \leq 0. \qquad (3.12)$$

Hence, equation (3.12) is satisfied. By (b), there exists $\bar{x} \in \hat{S}$ such that $F(\bar{x}, x) \geq 0$, for all $x \in X$. This implies that $F(\bar{x}, x) + d(\bar{x}, x) > 0$, for all $x \in X$, with $x \neq \bar{x}$, a contradiction with (3.14). Hence, $\hat{x} \in D$ for some $\hat{x} \in \hat{S}$, and (d) holds.

$(d) \Rightarrow (b)$: Suppose that both (d) and the hypothesis of (b) hold. Choose $D := \{\hat{x} \in X : \inf_{x \in X} F(\hat{x}, x) \geq 0\}$. Then, (3.13) follows from (3.11), and (d) furnishes some $\bar{x} \in \hat{S} \cap D$. It follows from the definition of D that $\inf_{x \in X} F(\bar{x}, x) \geq 0$. Hence, (b) holds.

$(c) \Rightarrow (d)$: Let (c) and the hypothesis of (d) hold. Define a set-valued mapping $T : X \to 2^X$ by $T(\hat{x}) := \{x \in X : x \neq \hat{x}\}$ for each $x \in X$. Assume to the contrary that $\hat{x} \notin D$ for all $\hat{x} \in \hat{S}$. Then, (3.12) follows from (3.13), and by (c), there exists $\bar{x} \in T(\bar{x})$. But this is clearly impossible from the definition of T. Hence, $\hat{x} \in D$ for some $\hat{x} \in \hat{S}$, and (d) holds.

$(d) \Rightarrow (c)$: Suppose that both (d) and the hypothesis of (c) hold. Choose $D := \{\hat{x} \in X : \hat{x} \in T(\hat{x})\}$. Then, (3.13) follows from (3.12), and (d) furnishes some $\bar{x} \in \hat{S} \cap D$, which, from the definition of D, necessarily belongs to $T(\bar{x})$. Hence, (c) holds.

This completes the entire proof. $\qquad\qquad\qquad\qquad\qquad \Box$

Actually, we have that Ekeland's variational principle holds in locally complete convex spaces, as discussed in the following, but we first need to recall a few notions and concepts for locally complete convex spaces.

Let X be a real locally convex Hausdorff space (LCS) and X^* be its topological dual. By following Pérez Carreras and Bonet [204, Chapter 3] and also Horváth [109], we call a bounded absolutely convex set B in X a disk, denote by "span[B]" the linear subspace spanned by B, and denote by p_B the Minkowski functional of B. Then, $E_B := (\text{span[B]}, p_B)$ is a normed space. If E_B is a Banach space, then B is called a Banach disk. A sequence (x_n) in X is said to be locally convergent to an element x if there exists a disk B in X such that (x_n) is convergent to x in E_B and (x_n) is said to be a locally Cauchy sequence if there exists a disk B in X such that (x_n) is a Cauchy sequence in E_B.

By following Pérez Carreras and Bonet [204, Chapter 5] and Horváth [109], we now have the following definitions.

Definition 3.4.10. A locally convex space X is said to be sequentially complete if every Cauchy sequence in X is convergent. A locally convex space X is said to be a locally complete convex space if every locally Cauchy sequence is locally convergent. This is equivalent to saying that each bounded subset of X is contained in a certain Banach disk. Let $A \subset X$ be nonempty. Then, A is said to be locally closed if for any locally convergent sequence in A, its local limit point belongs to A.

We know that every (sequentially) complete locally convex space is locally complete and every (sequentially) closed set is locally closed. But none of the converses are true; for further details, please refer to Horváth [109], Pérez Carreras and Bonet [204], and Qiu [215–220]. By using the notion of locally closed sets, we introduce the following class of functions, which is wider than the class of (sequentially) lower semicontinuous functions.

Definition 3.4.11. Let X be a locally convex space and $f : X \to (-\infty, +\infty]$ be a proper function (i.e., dom $f := \{x \in X : f(x) \neq +\infty\} \neq \emptyset$). Then, f is said to be locally lower semicontinuous if for each $r \in \mathbb{R}$, the set $\{x \in X : f(x) \leq r\}$ is locally closed in X.

We now recall the following result, which is Corollary 3.1 of Qiu [218].

Theorem 3.4.12. *Let X be a locally complete convex space, $\{p_\lambda\}_{\lambda \in I}$ be a family of seminorms defining the topology on X, and $\{\alpha_\lambda\}_{\lambda \in I}$ be a family of positive real numbers. Let $f : X \to (-\infty, +\infty]$ be a locally lower semicontinuous, bounded from below, proper function, and let $\{\alpha_\lambda\}_{\lambda \in I}$ be a family of positive real numbers. Let $x_0 \in dom f$. Then, there exists $z \in X$ such that for all $\lambda \in I$:*

(i) *$f(z) + \alpha_\lambda p_\lambda(z - x_0) \leq f(x_0)$; and*
(ii) *for any $x \neq z$, there exists $\mu \in I$ such that $f(z) < f(x) + \alpha_\mu p_\mu(x - z)$.*

Proof. The proof is given by Theorem 1.1 of Qiu [219], which is indeed Corollary 3.1 of Qiu [218]. Thus, we omit it here. The proof is complete. □

Now, as an application of Theorem 3.4.12, we have the following Caristi fixed point theorem in locally complete convex spaces, which will be used later in Chapter 8.

Lemma 3.4.13. *Let X be a locally complete convex space, $\{p_\lambda\}_{\lambda \in I}$ be a family of seminorms defining the topology on X, and $\{\alpha_\lambda\}_{\lambda \in I}$ be a family of positive real numbers. Let $f : X \to (-\infty, +\infty]$ be a locally lower semicontinuous, bounded from below, proper function, and let $\{\alpha_\lambda\}_{\lambda \in I}$ be a family of positive real numbers. Suppose that $T : X \to X$ is a single-valued mapping with nonempty values. Assume for each $x \in X$ such that for all $\lambda \in I$, we have*

$$\alpha_\lambda p_\lambda(x - Tx) + f(Tx) \leq f(x).$$

Then, there exists $z \in dom f \subset X$ such that $T(z) = z$, i.e., T has a fixed point $z \in X$.

Proof. Take any $x_0 \in dom f$. Then, by Theorem 3.4.12, we know that there exists $z \in X$ such that for all $\lambda \in I$, we have

$$f(z) + \alpha_\lambda p_\lambda(z - x_0) \leq f(x_0),$$

and for any $x \in X$ with $x \neq z$,

$$f(z) < f(x) + \sup_{\lambda \in I} \alpha_\lambda p_\lambda(x - z).$$

Thus, $z \in dom f$. We now prove z is a fixed point of T. If not, $T(z) \neq z$, and then by the inequality above, we have

$$f(z) < f(T(z)) + \sup_{\lambda \in I} \alpha_{\lambda \in I} p_\lambda(T(z) - z).$$

Thus, there exists $\mu \in I$ such that $f(z) < f(T(z)) + \alpha_\mu p_\mu(T(z) - z)$. Now, by the assumption for the mapping T, we have

$$f(z) < f(T(z)) + \alpha_\mu p_\mu(T(z) - z) \leq f(z),$$

which is a contradiction; therefore, we must have $T(z) = z$. This completes the proof. □

Now, we have the following general fixed point theorem, which shows that the conclusion of Lemma 3.4.13 is true when mapping T is a set-valued mapping in locally complete convex spaces.

Theorem 3.4.14. *Let X be a locally complete convex space, $\{p_\lambda\}_{\lambda \in I}$ be a family of seminorms defining the topology on X, and $\{\alpha_\lambda\}_{\lambda \in I}$ be a family of positive real numbers. Let $f : X \to (-\infty, +\infty]$ be a locally lower semicontinuous, bounded from below, proper function, and let $\{\alpha_\lambda\}_{\lambda \in I}$ be a family of positive real numbers. Suppose that $T : X \to 2^X$ is a set-valued mapping with nonempty values. Assume for each $x \in X$, there exists $y \in T(x)$ such that for all $\lambda \in I$, we have*

$$\alpha_\lambda p_\lambda(x - y) + f(y) \leq f(x).$$

Then, there exists $z \in \operatorname{dom} f \subset X$ such that $z \in T(z)$, i.e., T has a fixed point $z \in X$.

Proof. For each $x \in X$, we define a single-valued mapping $T_1 : X \to X$ by $T_1(x) := \{y\}$, where y is an element in $T(x)$ such that for all $\lambda \in I$, the following inequality holds:

$$\alpha_\lambda p_\lambda(x - y) + f(y) \leq f(x).$$

By the assumption for the mapping T, we have $T_1(x) \neq \emptyset$, well-defined, and $T_1(x) \subset T(x)$ for all $x \in X$. In addition, for each $x \in X$, such that for all $\lambda \in I$,

$$\alpha_\lambda p_\lambda(x - T_1 x) + f(T_1 x) \leq f(x).$$

Now, applying Lemma 3.4.13 to the mapping T_1, it implies that T_1 has one fixed point $z \in X$ such that $z = T_1(z) \subset T(z)$, which means z is a fixed point of T. This completes the proof. □

Remark 3.4.15. We note that as each complete p-normed space $(E, \| \cdot \|_p)$ for $p \in (0, 1]$ is a metric space, with the metric d defined by $d(x, y) := \|x - y\|_p$ for each $x, y \in E$, the conclusion of Theorem 3.4.1 also holds for complete p-normed spaces. For further discussion of the Caristi fixed point theorem [39] (1976) and some key related principles in nonlinear analysis, such as the variational principle by Ekeland [75] (1972), Takahashi's nonconvex minimization theorem [254, 255] (1991) and others topics, we refer to Almezel *et al.* [6], Ansari [7], Browder [34], Caristi and Kirk [40], Ekeland [76], Kirk [131, 132], Feng and Liu [88], Kirk and Caristi [135], Lau and Yao [152], Park [200], and the references therein. In addition, for a discussion on the equivalence of Ekeland's principle with Caristi fixed point theorem, we refer to Oettli and Théra [177] and related references therein. Moreover, for a comprehensive review and survey on the contraction mappings related to the Caristic fixed point theorems on metric fixed point theory and applications, we refer to Cho [58] and Ruzhansky *et al.* [234], Park [189–200], Zhang [299], and the references therein. Also, we suggest the references given by Kirk and Shahzad [137] for a comprehensive review and discussion on the four classical fixed point theorems with metric extensions: (1) the Banach contraction mapping principle, (2) Nadler's well-known set-valued extension of that theorem, (3) the extension of Banach's theorem to nonexpansive mappings, and (4) Caristi's theorem. Those comparisons form a significant component of the survey on the development of fixed point theory.

Before concluding this chapter, we would like to point out that when X is a Banach space in Theorem 3.3.12, its conclusion also holds in all uniformly convex spaces (in spite of the fact that not all such spaces have weakly sequentially continuous duality mappings, as pointed out by Browder [33]). Thus, the study of general demiclosedness principle under the framework of p-normed spaces and its related geometry and fixed point theory would be a very important topic; see Goebel and Kirk [96] and related references for a comprehensive discussion on some recent developments of metric fixed point theory and the associated geometry of Banach spaces.

Secondly, we would like to share with readers that a new concept called "p-normal structure" for p-normed spaces has been introduced, where $p \in (0, 1)$, which does not require the convexity condition by comparing with traditional definition of normal structure for normed spaces (i.e., for the case of a p-normed space with $p = 1$), which is used to establish the Browder–Göhde–Kirk theorem in p-normed spaces for $p \in (0, 1]$. We do believe more research should be conducted and developed for the geometry of p-normed spaces for $p \in (0, 1]$, and this should be a very exciting topic which would play an important role in the study of fixed point theory and associated nonlinear analysis in general.

Finally, we also like to summarize that in this chapter, some new fixed point theorems for nonexpansive mappings are established in p-normed spaces with the p-normal structure or by assuming that the p-Opial condition is satisfied for $p \in (0, 1]$. In addition, Xiao and Zhu [271] recently established a fixed point theorem for nonexpansive mappings for p-normed spaces under the traditional definition of normal structure with some additional conditions in terms of p-convexity by including the zero element. For more on the study of fixed point theorems for general continuous mappings instead of nonexpansive mappings in p-normed spaces, see also Agarwal *et al.* [1–3], Ben-El-Mechaiekh and Mechaiekh [17], Ben-El-Mechaiekh and Saidi [18], Browder [30–36], Cauty [42, 43], Chang *et al.* [48], Ennassik *et al.* [78], Fan [81], Goebel and Kirk [96], Granas and Dugundji [102], Kirk [130–134], Mauldin [167], Park [184–190], Takahashi [255], Xiao and Zhu [270], Zeidler [297], and the references therein. In addition, for fixed point theorems in p-vector and locally p-convex spaces, we also refer to Ennassik *et al.* [78], Ennassik and Taoudi [79], Yuan [286–291], and Yuan and Xiao [294, 295] for details on more recent results.

In addition, in the fields of functional analysis and topology in mathematics, we normally work and study with p-normed spaces $(X, \| \cdot \|_p)$ for $p \in (0, 1]$, which are infinite-dimensional rather than finite-dimensional. So, a new Dugundji-type extension theorem for closed subsets in p-normed spaces is established, including Dugundji's extension theorem [73] as a special case, which can serve as a useful tool for the development of new results and related theory for nonlinear functional analysis and topology in the years to come.

Chapter 4

Fixed Point Theorems for Single-Valued Mappings in p-Vector Spaces

The goal of this chapter is to develop a general fixed point theory for single-valued continuous mappings in p-vector spaces by using the method of functional analysis combined with Kalton's (1977) remarkable embedding theorem [119] for compact p-convex sets in topological vector spaces. This approach allows us to provide a positive answer to the Schauder conjecture [237], raised in 1930, for $p \in (0, 1]$.

This chapter consists of two sections. The first section aims to develop fixed point theorems for single-valued continuous mappings defined on s-convex compact subsets of p-vector and locally p-convex spaces. The second section studies fixed point theorems for compact single-valued mappings, where $s, p \in (0, 1]$. Actually, based on the results established in Chapter 3, we are able to prove the Schauder fixed point theorem for continuous single-valued mappings in p-vector spaces for $p \in (0, 1]$. Thus, the fixed point theorems established in this chapter resolve the Schauder conjecture with the support of Kalton's embedding theorem given in Chapter 2.

4.1 Fixed Point Theorems for Single-Valued Mappings

In this section, we aim to develop fixed point theorems for single-valued continuous mappings defined on s-convex compact subsets of

p-vector and locally p-convex spaces and for compact single-valued mappings, where $s, p \in (0, 1]$. We use the method of functional analysis developed by Ennassik and Taoudi [79] and Ennassik *et al.* [78] to establish these fixed point theorems, thereby providing a positive answer to the Schauder conjecture; see also the works of Park [189–200] and Yuan [286–291] for a comprehensive list of related references.

For the convenience of our discussion in this section, we first list some necessary definitions and results given by Ennassik and Taoudi [79] (see also those given by Yuan [286–291]).

Definition 4.1.1. Let X and Y be two topological vector spaces and M be a subset of X. A mapping $T : M \to Y$ is called uniformly continuous if, for every neighborhood V of the origin element θ in Y, there exists a θ-neighborhood U in X such that $T(x) - T(y) \in V$ holds for all $(x, y) \in M \times M$, with $x - y \in U$.

By the definition above and the compactness of a space, it is clear that the following result is true (see also Jarchow [116, p. 57]).

Proposition 4.1.2. *Let X and Y be two topological vector spaces and K be a compact subset of X. Then, every continuous mapping $T : K \to Y$ is uniformly continuous.*

For this chapter, we also need the following result to prove the fixed point theorems in p-vector spaces.

Lemma 4.1.3. *Let q be an r-seminorm on a vector space X, and let $F := \{x \in X : q(x) = 0\}$. Then, we have that:*

(1) *F is a linear subspace of X, and the functional $\|\cdot\|_r$ defined on the quotient space X/F by $\|\bar{x}\|_r = q(x)$ for each $\bar{x} = x + F \in X/F$ is an r-norm on X/F.*
(2) *The mapping $\pi : X \to \bar{x}$ is a continuous homomorphism from X equipped with the q-topology to X/F endowed with the topology induced by the r-norm $\| \cdot \|_r$.*

Proof. The proof of (1) follows a similar approach to that of statement (k) given by Bonsall [24, p. 35 of Chapter 3]. Indeed, for each $x, y \in F$, $0 \leq q(x + y) \leq q(x) + q(y) = 0$. Thus, $x + y \in F$. By the fact that $q(x) = 0$ implies $q(\lambda x) = 0$ for any $\lambda \in \mathbb{R}$, F is a linear subspace of X. The definition of $q(x)$ is, in fact, free from ambiguity, which means that if $x, x' \in F$, then $|q(x) - q(x')| \leq q(x - x') = 0$, implying $q(x) = q(x')$. Finally, it is entirely straightforward that the mapping $\|\bar{x}\|_r := q(x)$ for each $\bar{x} = x + F \in X/F$ satisfies the axioms of an r-norm.

For (2), since π is linear, then π is continuous if and only if (in short, iff) it is continuous at the origin 0. The continuity of π at 0 follows immediately from the equality $\|\pi\|_r = q(x)$ for all $x \in X$. $\qquad\qquad\square$

Referring to Bonsall [24], we also recall the following definitions and notions.

Definition 4.1.4. Let K be a compact subset of a topological vector space X. Let $T : K \to K$ be a continuous mapping, and let $\mathcal{S}$ and $\mathcal{S}'$ be two sets of continuous r-seminorms on X. We say that $\mathcal{S}'$ dominates $\mathcal{S}$ with respect to T if the following are satisfied:

(1) For any $q \in \mathcal{S}'$ and any $x \in K$, we have $q(x) \leq 1$.
(2) For any $p \in \mathcal{S}$ and any $\epsilon > 0$, there exist $q \in \mathcal{S}'$ and $\alpha > 0$ such that, for all $x, y \in K$, we have $p(T(y) - T(x)) < \epsilon$ when $q(y - x) < \alpha$.

If $\mathcal{S}' = \mathcal{S}$, we say that $\mathcal{S}$ is self-dominating, and in the case when $\mathcal{S}$ has a single element $\{p\}$, i.e., $\mathcal{S} = \{p\}$, we say that $\mathcal{S}'$ dominates p.

We also need for later the following technical lemma which is Lemma 3.2 of Ennassik and Taoudi [79].

Lemma 4.1.5. *Let K be a compact subset of a locally r-convex space X, with $r \in (0, 1]$. Let $T : K \to K$ be a continuous mapping and p_0 be a continuous r-seminorm on X. Then, there exists an r-seminorm q on the vector subspace $\mathrm{lin}(K)$ (i.e., the linear hull of K) of X satisfying the following:*

(1) *$p_0(x) \leq q(x)$ for each $x \in \mathrm{lin}(K)$.*
(2) *q is continuous on $K - K$.*
(3) *K is compact with respect to the q-topology.*
(4) *T is uniformly continuous in K with respect to the topology generated by q, i.e., for any $\epsilon > 0$, there exists $\delta > 0$ such that, for all $x, y \in K$, we have $q(T(x) - T(y)) < \epsilon$ when $q(y - x) < \delta$.*

Proof. The proof is derived by following the idea given in the proof of Theorem 3.1 by Bonsall [24, Chapter 3]. Indeed, as K is compact, p_0 is continuous, and thus p_0 is bounded on K. Then, without loss of generality, we may assume that $p_0(x) \leq 1$ for all $x \in K$. We wish to construct a countable self-dominating set of continuous r-seminorms containing p_0. To achieve this, take any continuous r-seminorm p on X and any integer $n \geq 1$. It is easy to see that the set $V_n = \{x \in X : p(x) < \frac{1}{n}\}$ is a neighborhood of the origin 0 in X. By Proposition 4.1.2, there exists U_n, an open, absolutely r-convex neighborhood of the origin 0 in X, such that $p(T(y - T(x))) < \frac{1}{n}$ holds for all $(x, y) \in K \times K$, with $y - x \in U_n$. In addition, it is clear that for the open, absolutely r-convex subset U_n containing the origin 0, its Minkowski r-functional P_{U_n}, which is an r-seminorm and also continuous, and we have $U_n = \{x \in X : p_{U_n}(x) < 1\}$.

Now, multiplying p_{U_n} by an appropriate constant $\alpha_n > 0$, we obtain a continuous r-seminorm q_n satisfying: (1) $q_n(x) \leq 1$; and (2) $p(T(y) - T(x)) < \frac{1}{n}$ for $(x, y) \in K \times K$, with $q_n(x - y) < \alpha_n$.

Thus, we have that the family $\{q_n : n \in \mathbb{N}\}$ is a countable set of continuous r-seminorms which dominates p. It follows that, for any given countable set $\mathcal{S}$ of continuous r-seminorms, there exists a countable set $\mathcal{S}'$ of continuous r-seminorms which dominates $\mathcal{S}$. Then, the set $\{p_0\}$ is dominated by a countable set $\mathcal{S}_1$, $\mathcal{S}_1$ is dominated by a countable set $\mathcal{S}_2$, and so on. Accordingly, $\mathcal{S}_\infty = \{p_0\} \cup_{n=1}^{\infty} \mathcal{S}_n$, which is a countable self-dominating set of continuous r-seminorms satisfying $p(x) \leq 1$ for all $p \in \mathcal{S}_\infty$ and all $x \in K$.

Now, let $\{p_n : n \in \mathbb{N}\}$ be an enumeration of $\mathcal{S}_\infty$, and we define q as the following:

$$q(x) := \Sigma_{n=0}^{\infty} 2^{-n} p_n(x)$$

for each $x \in \mathrm{lin}(K)$.

Keeping in mind that $p_n(x) \leq 1$ for all $x \in K$ and all $n \in \mathbb{N}$, it is easy to see that the q defined above converges on $\mathrm{lin}(K)$ (the linear hull of K), and indeed q is an r-seminorm on $\mathrm{lin}(K)$. Moreover, we have the following facts:

(i) It follows immediately from the definition of q above that $p_0(x) \leq q(x)$ for all $x \in \mathrm{lin}(K)$.

(ii) Since $p_n(x) \leq 2$ for all $x \in K - K$ and all $n \in \mathbb{N}$, as q converges uniformly on $K - K$ and so q is continuous on $K - K$.

(iii) Since q is continuous on $K - K$, then for any $x \in K$ and for any $\rho > 0$, the set $B(x, \rho) := \{y \in K : q(y - x) < \rho\}$ is an open subset of K in the topology τ on $\mathrm{lin}(K)$ induced from the initial topology on X. Hence, each open subset of K in the q-topology (the r-seminorm topology induced by q on $\mathrm{lin}(K)$) is also open in the topology induced by τ. Thus, K is compact with respect to the q-topology.

(iv) Let $\epsilon > 0$, and choose an integer $N \in \mathbb{N}$ such that $2^{-N} < \frac{\epsilon}{4}$. Then, for $x, y \in K$, we have

$$\Sigma_{n=N+1}^{\infty} 2^{-n} p_n(y - x) \leq \Sigma_{n=N+1}^{\infty} 2^{-n+1} = 2 \cdot 2^{-N} < \frac{\epsilon}{2},$$

so that

$$q(y - x) = \Sigma_{n=0}^{N} 2^{-n} p_n(y - x) + \Sigma_{n=N+1}^{\infty} 2^{-n} p_n(y - x)$$

$$< \Sigma_{n=0}^{N} 2^{-n} p_n(y - x) + \frac{\epsilon}{2}.$$

In addition, we note that for $T(x), T(y) \in K$, it follows that

$$q(T(x) - T(y)) < \Sigma_{n=0}^{N} 2^{-n} p_n(T(y) - T(x)) + \frac{\epsilon}{2}.$$

By the fact that $\mathcal{S}_\infty$ is self-dominating, for each integer $n \in \mathbb{N}$, there exists an integer k_n and a real $\alpha_n > 0$ such that, for each $x, y \in K$, we have $p_n(T(x) - T(y)) < \frac{\epsilon}{4}$ when $p_{k_n}(y - x) < \alpha_n$.

Now, put $N' = \max\{k_0, \ldots, k_N\}$ and $\alpha = 2^{-N'} \min\{\alpha_0, \ldots, \alpha_N\}$. Since $p_{k_n} \leq 2^{N'} q$ for all $n \in \{0, \ldots, N\}$, then for all $x, y \in K$, we have $p_{k_n}(y - x) < \alpha_n$ when $q(y - x) < \alpha$ for $n = 0, 1, \ldots, N$. By the definition of q above, we also have $p_n(T(y) - T(x)) < \frac{\epsilon}{4}$ when $q(y - x) < \alpha$ for $n = 0, 1, \ldots, N$.

Bringing all the above together, we then have $q(T(y) - T(x)) < \epsilon$ when $q(y - x) < \alpha$ for all $x, y \in K$. This completes the proof. $\square$

As an application of the above, we can develop a general fixed point theory for single-valued continuous mappings in p-vector spaces, where $p \in (0, 1]$. These results provide a positive answer for the Schauder conjecture with the support of Kalton's embedding theorem, which is Theorem 2.3.6 established in Chapter 2 (see also Theorem 1 by Kalton [119]).

Theorem 4.1.6. *Let K be a nonempty compact s-convex subset of a Hausdorff locally r-convex space X (where $s, r \in (0, 1]$), and let $T : K \to K$ be a continuous mapping. Then, T has a fixed point in K.*

Proof. The proof given here is an adaptation of the proof of Theorem 3.2 given by Bonsall [24, Chapter 3]. To keep things simple, we assume, without loss of generality, that $\lin(K) = X$. Suppose that T has no fixed point in K. Then, for each $x \in K$, $T(x) - x \neq 0$. It follows that for each point x of K, there exists a continuous r-seminorm p_x such that $p_x(T(x) - x) > 0$.

In view of the continuity of T and p_x, there exists a neighborhood U_x of x such that, for all $y \in U_x$, we have

$$p_x(T(y) - y) > 0.$$

Since K is compact, there is a finite covering of K by such neighborhoods, say $U_{x_1}, \ldots, U_{x_m}$. Let $p := p_{x_1} + \cdots + p_{x_m}$. Then, p is a continuous r-seminorm, and we have

$$p(T(x) - x) > 0$$

for all $x \in K$.

Now, let q be the r-seminorm constructed in Lemma 4.1.5 above, with $p_0 = p$. Then, q is defined on $\lin(K)(= X)$, we have $q \geq p$, and K is compact in the q-topology. Given $r > 0$, there exists $\delta > 0$ such that $q(T(x) - T(y)) < \epsilon$ when $q(x - y) < \delta$ for $x, y \in K$.

By following the idea used in Lemma 4.1.5 above, let $F := \{x \in X : q(x) = 0\}$. It is known that F is a closed subspace of X and the quotient space X/F is an r-normed space when it is endowed with the r-norm given by $\|\bar{x}\|_r = q(x)$ for each $\bar{x} = x + F \in X/F$.

Let $\overline{K} := \{\bar{x} : x \in K\}$. Then, $\overline{K}$ is an r-normed space. As the mapping $x \to \bar{x}$ is a continuous homomorphism from X equipped with the q-topology to X/F endowed with the r-norm topology, $\overline{K}$ is a compact s-convex set in X/F, we also have that $q(T(x) - T(y)) = 0$, when $q(x - y) =$ for all $x, y \in K$.

Note that for each $\bar{x} \in \overline{K}$, there exists a point $x \in \bar{x} \cap K$ where we can define a mapping $\overline{T} : \overline{K} \to \overline{K}$ by

$$\overline{T}(\bar{x}) := \overline{T(x)}$$

for each $\bar{x} = x + F$. It is easy to see that this mapping $\overline{T}$ is well defined, and we know that $\|\bar{x} - \bar{y}\|_r = q(x - y)$ for each $\bar{x} = x + F$ and $\bar{y} = y + F$, where $x, y \in K$.

As the mapping $\overline{T} : \overline{K} \to \overline{K}$ is also continuous, then by Theorem 3.2.9, we deduce that there is an $\bar{x} \in \overline{K}$ such that $\overline{T}(\bar{x}) = \bar{x}$.

Therefore, we have $T(x) - x \in F$ and, consequently, $q(T(x) - x) = 0$. By the fact that $T(x) - x \in K - K$ and $p \leq q$ on $\mathrm{lin}(K)$, we deduce that $p(T(x) - x) = 0$, which contradicts our assumption at the beginning of the proof above. Thus, T must have a fixed point in K. This completes the proof. $\square$

As an application of Kalton's embedding theorem for compact p-convex sets [119], we have the following fixed point theorem in Hausdorff topological vector spaces for continuous mappings, which affirms the Schauder conjecture.

Theorem 4.1.7. *If K is a nonempty compact r-convex subset of a Hausdorff topological vector space X, with $r \in (0, 1]$, then any continuous mapping $T : K \to K$ has a fixed point.*

Proof. We give the proof in two steps, as follows:

Step 1: Assume that K is r-convex, with $r \in (0, 1)$. By Theorem 2.3.6, the set K can be linearly embedded in a Hausdorff locally r-convex space E, which means that there exists a linear mapping $L : \mathrm{lin}(K) \to K$ whose restriction to K is a homeomorphism. Now, we define a mapping $S : L(K) \to L(K)$ by $S(Lx) = L(T(x))$ for each $x \in K$. This mapping is easily checked to be well defined, and it is continuous since L is a homeomorphism and T is continuous on K. In addition, the set $L(K)$ is compact, as it is the image of a compact set under a continuous mapping, and it is also r-convex since it is the image of an r-convex set under a linear mapping. Now, by Theorem 4.1.6, there exists $x \in K$ such that $S(L(x)) = L(x)$, which means that $L(T(x)) = L(x)$. As L is a homeomorphism, it follows that $T(x) = x$, which means T has a fixed point in K.

Step 2: Assume that K is convex. Choose an arbitrary $x_0 \in K$ and define $K_0 := K - x_0$. Then, K_0 is a compact convex subset of X, which contains the zero element. By Lemma 2.2.4, we also conclude that K_0 is r-convex for any $r \in (0, 1)$. Now, define a mapping $R : K_0 \to K_0$ by $R(x - x_0) := T(x) - x_0$ for each $x - x_0 \in K_0$. Clearly, R is continuous. Now, applying the result of the first step to the mapping R, we conclude that there exists $x \in K$ such that $R(x - x_0) = x - x_0$, and thus x is a fixed point of T, i.e., $T(x) = x$. This completes the proof. □

An immediate consequence of Theorem 4.1.7 is the following result, which gives an affirmative answer to the Schauder conjecture.

Corollary 4.1.8. *If K is a nonempty compact convex subset of a Hausdorff topological vector space X, then any continuous mapping $T : K \to K$ has a fixed point.*

Proof.   Apply Theorem 4.1.7 with $r = 1$. The proof is complete. □

Remark 4.1.9. Theorem 4.1.7 provides a constructive method to solve the Schauder conjecture regarding the existence of fixed points for continuous mappings in topological vector spaces. The analysis method used above was essentially established by Ennassik and Taoudi [79] as an easily accessible approach for the majority of the mathematical community. In addition, for more information on the path to resolving the Schauder conjecture, we refer the reader to Remark 4.2.9, which mentions a number of remarkable results of the past 90 years!

In addition, noting that each p-normed space includes normed spaces as a special class (with $p = 1$), Theorem 4.1.7 and Corollary 4.1.8 unify the corresponding results in the existing literature. For more details, see Agarwal *et al.* [1–3], Ben-El-Mechaiekh and Mechaiekh [17], Ben-El-Mechaiekh and Saidi [18], Browder [30–36], Cauty [42, 43], Ennassik *et al.* [78], Fan [81–86], Granas and Dugundji [102], Li [157, 158], Mauldin [167], Park [184–200], Zeidler [297], and the references therein. In addition, for a more in-depth coverage of the fixed point theorem in p-vector and locally p-convex spaces, see the recent works of Ennassik and Taoudi [79] and Yuan [286–291].

4.2   Fixed Point Theorems for Compact Single-Valued Mappings

The goal of this section is to establish general fixed point theorems for compact single-valued continuous mapping in Hausdorff

p-vector spaces. These results also provide a solution to the Schauder conjecture in the affirmative under the general settings of p-vector and locally p-convex spaces for continuous single-valued mappings defined on s-convex subsets, which is fundamental for nonlinear functional analysis in mathematics, where $s, p \in (0, 1]$.

For the convenience of our discussion, throughout this section, we assume that all topological vector spaces, p-vector spaces, and locally p-convex spaces are Hausdorff, unless specified otherwise, for $p \in (0, 1]$. Here, we first gather together the definitions, notations and known facts needed for this section.

Definition 4.2.1. Let X and Y be two topological spaces. A set-valued mapping (also called a multifunction) $T : X \longrightarrow 2^Y$ is a point to a (nonempty) set function such that, for each $x \in X$, $T(x)$ is a nonempty subset of Y. The mapping T is said to be upper semicontinuous (in short, USC) if the subset $T^{-1}(B) := \{x \in X : T(x) \cap B \neq \emptyset\}$ (equivalently, the set $\{x \in X : T(x) \subset B\}$) is closed (equivalently, open) for any closed (resp., open) subset B in Y. The function $T : X \to 2^Y$ is said to be lower semicontinuous (in short, LSC) if the set $T^{-1}(A)$ is open for any open subset A in Y.

Definition 4.2.2. We recall that, for two given topological spaces X and Y, a set-valued mapping $T : X \to 2^Y$ is said to be compact if there is a compact subset C in Y such that $T(X)(:= \{y \in T(x), x \in X\})$ is contained in C, i.e., $F(X) \subset C$.

In order to develop fixed point theorems for compact single-valued continuous mappings in p-vector spaces, we first recall the following result, which is Proposition 4.2.3 (equivalent to Proposition 6.7.2 given by Jarchow [116]) presented without its proof.

Proposition 4.2.3. *Let K be compact in a topological vector X and $p \in (0, 1]$. Then, the closure of the p-convex hull, denoted by $\overline{C}_p(K)$, and the closure of the absolutely p-convex hull of K, denoted by $\overline{AC}_p(K)$, are compact if and only if $\overline{C}_p(K)$ and $\overline{AC}_p(K)$ are complete, respectively.*

We now have the following fixed point theorems for a compact single-valued continuous mapping defined in Hausdorff topological vector spaces and locally p-convex spaces for $p \in (0, 1]$.

Similar to the notation used in Definition 2.3.1, for a given set A in vector space X, we denote by "lin(A)" the "linear hull" of A in X. Then, we also recall the following definition.

Definition 4.2.4. Let A be a subset of a topological vector space X and Y be another topological vector space. We say that A can be linearly embedded in Y if there is a linear mapping $L : \text{lin}(A) \to Y$ (not necessarily continuous) whose restriction to A is a homeomorphism.

For the convenience of our discussion, we first mention that Theorem 2.3.6 in Chapter 2 above is a significant embedding result for compact convex subsets in topological vector spaces due to Theorem 1 by Kalton [119], which states that, in general, although not every compact convex set in topological vector spaces can be linearly embedded in a locally convex space (e.g., see Roberts [224] and Kalton *et al.* [122]), for p-convex sets when $p \in (0, 1)$, every compact p-convex set in topological vector spaces can be considered a subset of a locally p-convex vector space, and hence every such set has sufficiently many p-extreme points.

Secondly, by property (ii) in Lemma 2.2.4 above, we know that each convex subset containing a origin 0 in a topological vector space is always p-convex for $p \in (0, 1]$. Thus, it is possible for us to transfer the problem involving p-convex subsets from topological vector spaces into the locally p-convex vector spaces. This indeed allows us to establish the existence of fixed points for compact single-valued mappings defined on non-compact p-convex subsets in locally p-convex spaces and p-vector spaces to cover the case when the underlying is just a topological vector space. Then, this helps us derive the solution for the Schauder conjecture in the affirmative for single-valued continuous mapping under the setting of topological p-vector spaces by following the analysis method established by Ennassik and Taoudi [79] for $p \in (0, 1]$. For the convenience of our discussion, here we restate Kalton's embedding result [119] (which is Theorem 2.3.6 above) as the following lemma.

Lemma 4.2.5. *Let K be a compact p-convex subset for $p \in (0, 1)$ of a topological vector space X. Then, K can be linearly embedded in a locally p-convex space.*

Proof. Refer to Theorem 2.3.6 in Chapter 2 (which is indeed Theorem 1 given by Kalton [119] in 1977). This completes the proof. □

Remark 4.2.6. At this point, it is important to note that Lemma 4.2.5 does not hold for $p = 1$. By Theorem 9.6 given by Kalton *et al.* [122], it was shown that the spaces $L_p = L_p(0, 1)$, where $p \in (0, 1)$, contain compact convex sets with no extreme points, which thus cannot be linearly embedded in a locally convex space; see also Roberts [224].

We now have the following result, which is a generalization of Theorems 3.1 and 3.3 due to Ennassik and Taoudi [79] defined on non-compact s-convex subsets of locally p-convex spaces or topological vector spaces for $s, p \in (0, 1]$.

Theorem 4.2.7. *If K is a nonempty closed s-convex subset of either a Hausdorff locally p-convex space for $s, p \in (0, 1]$ or a Hausdorff topological vector space X, then each compact single-valued continuous mapping $T : K \to K$ has at least one fixed point.*

Proof. As T is compact, there exists a compact subset A in K such that $T(K) \subset A$. Let $K_0 := \overline{C}_s(A)$ be the closure of the s-convex hull of the subset A in K. Then, K_0 is compact s-convex by Proposition 4.2.3, and the restriction of the mapping $T : K_0 \to K_0$ is also continuous.

First, if K is a nonempty closed s-convex subset of a locally p-convex space, where $s, p \in (0, 1]$, the conclusion is obtained by considering the self-mapping T on K_0 as an application of Theorems 4.1.6 and 4.1.7.

Second, if K is a nonempty closed s-convex subset of a Hausdorff topological vector space X, we prove the conclusion by applying Lemma 4.2.5 with the following two cases.

Case 1: For $0 < s < 1$, K_0 is a nonempty compact s-convex subset of a topological vector space X for $s \in (0, 1)$. By Lemma 4.2.5, it follows that K_0 can be linearly embedded in a locally s-convex space X, which means that there exists a linear mapping $L : \text{lin}(K_0) \to X$ whose restriction to K_0 is a homeomorphism. Define the mapping $S : L(K_0) \to L(K_0)$ by $S(Lx) := L(Tx)$ for each $x \in K_0$. Then, this mapping is easily checked to be well defined. The mapping S is continuous since L is a (continuous) homeomorphism and T is continuous on K_0. Furthermore, the set $L(K_0)$ is compact, as it is the image of a compact set under a continuous mapping L, and $L(K_0)$

is also s-convex since it is the image of a s-convex set under a linear mapping. Then, by the conclusion given in the first part above, T has a fixed point, $x \in K_0$. Thus, there exists $x \in K_0$ such that $Lx = S(Lx) = L(Tx)$; therefore, it implies that $x = T(x)$ since L is a homeomorphism, which is the fixed point of T.

Case 2: For $s = 1$, consider any point $x_0 \in K_0$, and let $K_0' := K_0 - \{x_0\}$. Now, define a new mapping $T_0 : K_0' \to K_0'$ by $T_0(x - x_0) := T(x) - x_0$ for each $x - x_0 \in K_0'$. By the fact that now K_0' is compact and s'-convex by Lemma 2.2.4(ii) for some $s' \in (0, 1)$ and that T_0 is also continuous, T_0 has a fixed point in K_0' according to the proof in Case 1 above. So, T has a fixed point in K_0. The proof is complete. □

Remark 4.2.8. The reader should note that when K in Theorem 4.2.7 is compact, we do not need to assume that the space X is complete. Indeed, Lemma 3.2.8 states that each p-normed space can be completed; therefore, the completion of p-normed spaces may not be needed for the existence of fixed points for continuous mappings, where $p \in (0, 1]$. In addition, for the fixed point theorem in p-vector and locally p-convex spaces, we refer to the proof of Theorem 2.3.2 by Kreyszig [146] and Ennassik and Taoudi [79], as well as some recent work by Yuan [286–291] for more details.

Secondly, Theorem 4.2.7 states that each compact single-valued mapping defined on a closed p-convex subset of topological vector spaces for $p \in (0, 1]$ has the fixed point property (FPP), which include or improve most available results for fixed point theorems in the existing literature as special cases (to mention a few, Ben-El-Mechaiekh and Mechaiekh [17], Ben-El-Mechaiekh and Saidi [18], Ennassik and Taoudi [79], Mauldin [167], Granas and Dugundji [102], O'Regan and Precup [182], Reich [222], Park [189–200], Yuan [286–291], and the comprehensive references therein).

Once again, as mentioned above, as an application of the embedding theorem for compact p-convex sets in p-vector spaces established by Kalton [119] in 1977 for $p \in (0, 1)$, the positive answer to the conjecture proposed by Schauder [237] in 1930 for a single-valued continuous mapping was obtained by Ennaassik and Taoudi [79] defined on nonempty compact p-convex subsets in topological vector spaces, using the method of functional analysis, for $p \in (0, 1]$. However, the actual method that is used to resolve the Schauder conjecture emerged through several developments since 1930, creating a rich narrative, which we briefly recount in the following remark.

Remark 4.2.9 (the path to resolving the Schauder conjecture). Here, we briefly recount the path that led to the resolution of the Schauder conjecture and related developments of the study of the FPP over nearly a century, beginning in the early 20th century. In 1930, Schauder [237] conjectured that every compact convex set in an arbitrary Hausdorff linear topological space has the FPP, which is well known as Problem 54 in [167]. In 1935, Tychonoff [260] proved that the Schauder conjecture holds in a Hausdorff locally convex space. During 1961–1964, Fan [82,83] proved that the conjecture holds in a Hausdorff linear topological space admitting sufficiently many continuous linear functionals. In 2001, Cauty [41] claimed that he completely solved the conjecture for linear topological metric spaces. However, it was later found out that Cauty's proof given in Ref. [41] as well as its elaboration with supplementary revised proof by Dobrowolski [67] contained a gap; see also comments given by Mauldin [167, p. 131] and Isac [118, p. 91]. Thereafter, by filling the gaps through his subsequent work [42–44], he claims affirmatively resolving the Schauder conjecture under the framework of linear metric spaces based on his new concept involving the class of algebraic absolute neighborhood retracts (ANRs). But unfortunately, Professor Robert Cauty unexpectedly passed away on July 8, 2013. Since then, it seems that not much work has followed up in this line by applying his approach or methodology.

However, in 2021, by applying the functional analysis method, Ennassik and Taoudi [79] provided an elegant and easily understandable alternative proof for the Schauder conjecture for general linear topological spaces and p-vector spaces under the support of the embedding theorem of compact p-convex subsets in locally p-convex spaces, established by Kalton [119] in 1977, and the fixed point theorem for continuous mappings defined in p-normed spaces which are infinite-dimensional, established by Xiao and Zhu [270] in 2011 for $p \in (0, 1]$ (see also Ennalssik et al. [78]). In addition, as we will see in Section 5.6 of Chapter 5, the general fixed point theorems of upper semi-continuous set-valued mappings defined on s-convex subsets in Hausdorff p-topological vector spaces by using the functional analysis method without requiring the metric on the underlying p-vector spaces for $s, p \in (0, 1]$ are systemically established for the first time. These results include or unify the corresponding results in the existing literature and also provide a positive answer to Schauder's open question through a unified approach. We also emphasize that the method used in this and the following chapter is mainly based on tools from functional analysis and topology, unlike Cauty's approach [42]– [44] (and related references therein), by using the new concept of algebraic ANRs, which comes under the category of algebraic topology in mathematics.

We would like to mention that, in this book, the fixed point theorems for continuous mappings defined in p-normed spaces are proved by using our new Dugundji-type extension theorem for locally p-convex spaces, established for the first time. We hope that the presentation used in this book, following the functional analysis and topology approach, gives readers a clear understanding of how we answer the Schauder conjecture on the existence of fixed point theorems in topological vector spaces without a metric. For more details on the recent developments in nonlinear analysis and related topics, we refer interested readers to Park [189]–[200], Yuan [286]–[291], and the related comprehensive references therein.

On the other hand, in the course of solving the Schauder conjecture, the problem of whether the FPP of a convex set in a Hausdorff linear topological space also implies its compactness was first studied by Klee [139], where an affirmative answer was given for the case of a metrizable locally convex space and eventually solved by Dobrowolski and Marciszewski [68]. In this work, the authors proved that the FPP of a convex set in a metrizable linear topological space implies its compactness, and they also demonstrated that the FPP of a convex set in a Hausdorff locally convex space does not always imply its compactness. Therefore, the results by Cauty [42] and Dobrowolski and Marciszewski [68] can be combined to yield the result that a convex set in a metrizable linear topological space has the FPP if and only if the convex set is compact. For more detailed discussions, and related topics see Guo *et al.* [104] and the references therein.

Chapter 5

Fixed Point Theorems for Set-Valued Mappings in Locally p-Convex and p-Vector Spaces

The goal of this chapter is to develop a new general fixed point theory for set-valued mappings in locally p-convex spaces. This chapter consists of five sections, introduced as follows.

The first section establishes some basic results by applying the Knaster–Kuratowski–Mazurkiewicz (KKM) principle in abstract convex spaces. Section 5.2 establishes fixed point theorems for upper semicontinuous (in short, USC) set-valued mappings in locally p-convex spaces, where $p \in (0, 1]$. Section 5.3 presents graph approximation results for quasi USC set-valued mappings in p-vector spaces. Section 5.4 establishes fixed point theorems for quasi USC set-valued mappings in locally p-convex spaces. Section 5.5 derives fixed point theorems for the classes of 1-set contractive and condensing mappings using the concept of the noncompactness measure in locally p-vector spaces. Section 5.6 establishes fixed point theorems of USC set-valued mappings defined on s-convex subsets in Hausdorff p-topological vector spaces by using the functional analysis method without requiring the metric on the underlying p-vector spaces for $s, p \in (0, 1]$. These results include or unify the corresponding results in the existing literature, and also provide a positive answer to Schauder's open question. We emphasize that the method established in this section is mainly based on the tools from functional analysis and topology, unlike Cauty's approach [42]–[44] (and related references therein) by introducing a new concept called "algebraic absolute neighborhood retracts" (ANRs) under the category of algebraic topology in mathematics.

In particular, we would like to point out that the selection theorem established in Section 5.5 and the proof method used in Section 5.6 for the proof of fixed point theorems for USC set-valued mappings under the framework of general p-vector spaces would be useful tools/approaches for the development of new theories in nonlinear analysis and topology.

Other updated developments for the study of the general topological fixed point theory can also be found from those references provided by Park [189]–[200], Yuan [286]–[291], and the references therein.

5.1 The KKM Principle in Abstract Convex Spaces

The goal of this section is to establish some basic results by applying the KKM principle in abstract convex spaces, which will be used in this chapter later.

It is well known that abstract convex spaces play a very important role in the development of the KKM principle and related applications. Once again, for the corresponding comprehensive discussion on the KKM theory and its various applications to nonlinear analysis and related topics, we refer to Mauldin [167], Granas and Dugundji [102], Park [189–191], Yuan [284–291], and a considerable number of references provided in these works.

As mentioned above, Knaster, Kuratowski, and Mazurkiewicz (in short, KKM) [142] in 1929 obtained the so-called KKM principle to give a new proof for the Brouwer fixed point theorem in finite-dimensional spaces. Later, in 1961, Fan [82] (see also Fan [85, 86], Granas and Dugundji [102], Shih and Tan [243, 244], Tarafdar and Yuan [259], Yuan [283, 284], and the related references therein) extended the KKM principle to any topological vector spaces and applied it to various results, including the Schauder fixed point theorem. Then, there appears to be a large number of works devoted to applications of the KKM principle. In 1992, the associated research field was called the KKM theory for the first time by Park [184]. Subsequently, the KKM theory has been extended to general abstract convex spaces by Park [187] (see also Park [188,189]), which actually include locally p-convex spaces for $p \in (0, 1]$ as a special class.

Now, for the convenience of readers and for the book to be self-contained, we recall some notions and definitions on the abstract convex spaces, which play an important role for the development of the KKM principle and its related applications. Once again, for the corresponding comprehensive discussion on the KKM theory and its various applications to nonlinear analysis and related topics, we refer to Mauldin [167], Granas and Dugundji [102], Park [189–191], Yuan [284–291], and related comprehensive reference therein.

The KKM principle is a powerful tool in many areas of mathematics. We first recall the original KKM principle with the following version.

Let $n \in \mathbb{N}$ be an integer number. We recall that an n-dimensional simplex is the convex hull of $n + 1$ points in general position in $\mathbb{R}^n$, called the vertices of the simplex. For $k = 0, \ldots, n$, we define the k-skeleton of the simplex to be the union of all convex hulls of $k + 1$ vertices. In this book, Δ_n will be the n-dimensional simplex, generated by $(0, 0, \ldots, 0), (1, 0, \ldots, 0), \ldots, (1, 1, \ldots, 1)$. Note that $(x_1, x_2, \ldots, x_n)$ is in Δ_n if and only if $0 \leq x_1 \leq x_2 \leq \cdots \leq x_n \leq 1$.

The KKM principle (given by KKM [142] in 1929). For every vertex g of Δ_n, let Γ_g be a closed subset in Δ_n. If for each set W of vertices of Δ_n, the convex hull of W is contained in $\cup_{g \in W} \Gamma_g$, then $\cap_{g \in W} \Gamma_g \neq \emptyset$.

Here, we note that the KKM principle is also true if we use open rather than closed subsets for each Γ_g for $g \in W$ (the set of vertices of Δ_n) (as given by Shih and Tan [243]; see also further references therein).

Let $\langle D \rangle$ denote the set of all nonempty finite subsets of a given nonempty set D, and 2^D denotes the family of all subsets of D. We have the following definition for abstract convex spaces, essentially given by Park [187].

Definition 5.1.1. An abstract convex space $(E, D; \Gamma)$ consists of a topological space E, a nonempty set D, and a set-valued mapping $\Gamma : \langle D \rangle \to 2^E$ with nonempty values $\Gamma_A := \Gamma(A)$ for each $A \in \langle D \rangle$, such that the Γ-convex hull of any $D' \subset D$ is denoted and defined by $\mathrm{co}_\Gamma D' := \cup \{ \Gamma_A | A \in \langle D' \rangle \} \subset E$.

A subset X of E is said to be a Γ-convex subset of $(E, D; \Gamma)$ relative to D' if for any $N \in \langle D' \rangle$, we have $\Gamma_N \subseteq X$, that is, $\text{co}_\Gamma D' \subset X$. For the convenience of our discussion, in the case of $E = D$, the space $(E, E; \Gamma)$ is simply denoted by $(E; \Gamma)$ unless specified otherwise.

Definition 5.1.2. Let $(E, D; \Gamma)$ be an abstract convex space and Z be a topological space. For a set-valued mapping (or, say, set-valued mapping) $F : E \to 2^Z$ with nonempty values, if a set-value mapping $G : D \to 2^Z$ satisfies $F(\Gamma_A) \subset G(A) := \bigcup_{y \in A} G(y)$ for all $A \in \langle D \rangle$, then G is called a KKM mapping with respect to F. A KKM mapping $G : D \to 2^E$ is a KKM mapping with respect to the identity mapping 1_E.

Definition 5.1.3. The partial KKM principle for an abstract convex space $(E, D; \Gamma)$ is that, for any closed-valued KKM mapping $G : D \to 2^E$, the family $\{G(y)\}_{y \in D}$ has the finite intersection property. According to the KKM principle, the same property also holds for any open-valued KKM mapping.

An abstract convex space is called a (partial) KKM space if it satisfies the (partial) KKM principle. We now gave some known examples of (partial) KKM spaces (see Park [187, 188]) as follows.

Definition 5.1.4. A ϕ_A-space $(X, D; \{\phi_A\}_{A \in \langle D \rangle})$ consists of a topological space X, a nonempty set D, and a family of continuous functions $\phi_A : \Delta_n \to 2^X$ (that is, Δ_n is a singular n-simplex) for $A \in \{D\}$ with $|A| = n + 1$. By putting $\Gamma_A := \phi_A(\Delta_n)$ for each $A \in \langle D \rangle$, the triple $(X, D; \Gamma)$ becomes an abstract convex space.

Remark 5.1.5. For a ϕ_A-space $(X, D; \{\phi_A\})$, we see easily that any set-valued mapping $G : D \to 2^X$ satisfying $\phi_A(\Delta_J) \subset G(J)$ for each $A \in \langle D \rangle$ and $J \in \langle A \rangle$ is a KKM mapping.

By the definition, it is clear that every ϕ_A-space is a KKM space, and thus we have the following fact (see Lemma 1 of Park [188]).

Lemma 5.1.6. *Let $(X, D; \Gamma)$ be a ϕ_A-space and $G : D \to 2^X$ be a set-valued (multimap) with nonempty closed (resp., open) values. Suppose that G is a KKM mapping, then $\{G(a)\}_{a \in D}$ has the finite intersection property.*

Proof. Let $A = \{a_0, a_1, \ldots, a_n\} \in \langle D \rangle$. Then, there exists a continuous function $\phi_A : \Delta_n \to \Gamma_A$ such that, for any $0 \le i_0 < i_1 <$

$\cdots < i_k \leq n$, we have

$$\phi_A(\mathrm{co}\{e_{i0}, e_{i_1}, \ldots, e_{i_k}\}) \subset \Gamma(\{a_{i_0}, a_{i_1}, \ldots, a_{i_k}\}) \cap \phi_A(\triangle_n).$$

Since G is a KKM mapping, it follows that

$$\mathrm{co}\{e_{i_0}, e_{i_1}, \ldots, e_{i_k}\} \subset \phi_A^{-1}(\Gamma(\{a_{i_0}, a_{i_1}, \ldots, a_{i_k}\}) \cap \phi_A(\triangle_n))$$

$$\subset \cup_{j=0}^k \phi_A^{-1}(G(a_{i_j}) \cap \phi_A(\triangle_n)).$$

Since $G(a_{i_j}) \cap \phi_A(\triangle_n)$ is closed (resp., open) in the compact subset $\phi_A(\triangle_n)$ of Γ_A, $\phi_A^{-1}(G(a_{i_j}) \cap \phi_A(\triangle_n))$ is closed (resp., open) in $\triangle_n$. Note that $e_i \to \phi_A^{-1}(G(a_i) \cap \phi_A(\triangle_n))$ is a KKM mapping on $\{e_0, e_1, \ldots, e_n\}$. Hence, by the original KKM principle above, we have

$$\cap_{i=0}^n \phi_A^{-1}(G(a_i) \cap \phi_A(\triangle_n)) \neq \emptyset,$$

which readily implies that $\cap_{i=0}^n G(a_i) \neq \emptyset$. This completes the proof.

□

By following Definition 2.1.10, we recall that a topological vector space is said to be locally p-convex if the origin has a fundamental set of absolutely p-convex θ-neighborhoods. Now, we have a new KKM space as follows, inducted by the concept of p-convexity (see Lemma 2 of Park [188]).

Lemma 5.1.7. *Suppose that X is a subset of topological vector space E and $p \in (0, 1]$, and D is a nonempty subset of X such that $C_p(D) \subset X$. Let $\Gamma_N := C_p(N)$ for each $N \in \langle D \rangle$. Then, $(X, D; \Gamma)$ is a ϕ_A-space.*

Proof. Since $C_p(D) \subset X$, Γ_N is well defined. For each $N = \{x_0, x_1, \ldots, x_n\} \subset D$, we define $\phi_N : \triangle_n \to \Gamma_N$ by $\sum_{i=0}^n t_i e_i \mapsto \sum_{i=0}^n (t_i)^{\frac{1}{p}} x_i$. Then, clearly $(X, D; \Gamma)$ is a ϕ_A-space. This completes the proof. □

5.2 Fixed Point Theorems for USC Set-Valued Mappings

In this section, we establish fixed point theorems for USC and condensing mappings for p-convex subsets under the general framework of p-vector spaces, which will be a tool used in Chapters 6 and 7 to establish the best approximation, fixed points, the principle of nonlinear alternatives, Birkhoff–Kellogg problems, and the Leray–Schauder alternative. All of these will be useful tools in nonlinear analysis for the study of nonlinear problems arising from theory as well as practice. Here, we first gather together the necessary definitions, notations, and known facts needed in this section.

Definition 5.2.1. Let X and Y be two topological spaces and for a set-valued mapping (also called a multifunction) $T : X \longrightarrow 2^Y$ being a point to set function such that for each $x \in X$, $T(x)$ is a subset of Y. We recall that the mapping T is said to be USC if the subset $T^{-1}(B) := \{x \in X : T(x) \cap B \neq \emptyset\}$ (resp., the set $\{x \in X : T(x) \subset B\}$) is closed (resp., open) for any closed (resp., open) subset B in Y. The function $T : X \to 2^Y$ is said to be lower semicontinuous (in short, LSC) if the set $T^{-1}(A)$ is open for any open subset A in Y.

As an application of the KKM principle for general abstract convex spaces, using the embedding lemma for Hausdorff compact p-convex subsets from topological vector spaces into locally p-convex vector spaces, we have the following general existence result for the "approximation" of fixed points for USC and LSC set-valued mappings in p-convex vector spaces for $p \in (0, 1]$ (see also the corresponding results given by Theorem 2.7 of Gholizadeh *et al.* [94], Theorem 5 of Park [188], and related discussion therein).

The following result was originally given by Yuan [286]. Here, we provide a sketch of its proof to ensure that the text is self-contained for the reader.

Theorem 5.2.2. *Let A be a p-convex compact subset of a locally p-convex vector space X, where $p \in (0, 1]$. Suppose that $T : A \to 2^A$ is lower (resp., upper) semicontinuous with nonempty p-convex values. Then, for any given U which is a p-convex neighborhood of zero in X, there exists $x_U \in A$ such that $T(x_U) \cap (x_U + U) \neq \emptyset$.*

Proof. Suppose U is any given element of $\mathfrak{U}$, there is a symmetric open neighborhood V of zero for which $\overline{V} + \overline{V} \subset U$ in the locally p-convex neighborhood of zero, we prove the results for the two cases of T, namely (1) LSC and (2) USC.

In Case (1), we assume T to be LSC: As X is a locally p-convex vector space, suppose that $\mathfrak{U}$ is the family of neighborhoods of 0 in X. For any element U of $\mathfrak{U}$, there is a symmetric open neighborhood

V of zero for which $\overline{V} + \overline{V} \subset U$. Since A is compact, there exist $x_0, x_1, \ldots, x_n$ in A such that $A \subset \cup_{i=0}^{n} (x_i + V)$. By using the fact that A is p-convex, we find $D := \{b_0, b_2, \ldots, b_n\} \subset A$, for which $b_i - x_i \in V$ for all $i \in \{0, 1, \ldots, n\}$, and we define C by $C := C_p(D) \subset A$. By the fact that T is LSC, it follows that the subset $F(b_i) := \{c \in C : T(c) \cap (x_i + V) = \emptyset\}$ is closed in C (as the set $x_i + V$ is open) for each $i \in \{0, 1, \ldots, n\}$. For any $c \in C$, we have $\emptyset \neq T(c) \cap A \subset T(c) \cap \cup_{i=0}^{n} (x_i + V)$; it follows that $\cap_{i=0}^{n} F(b_i) = \emptyset$. Now, applying Lemmas 5.1.6 and 5.1.7, it implies that that there is $N := \{b_{i_0}, b_{i_1}, \ldots, b_{i_k}\} \in \langle D \rangle$ and $x_U \in C_p(N) \subset A$ for which $x_U \notin F(N)$, and so $T(x_u) \cap (x_{i_j} + V) \neq \emptyset$ for all $j \in \{0, 1, \ldots, k\}$. As $b_i - x_i \in V$ and $\overline{V} + \overline{V} \subset U$, which implies that $x_{i_j} + \overline{V} \subset b_{i_j} + U$, meaning that $T(x_U) \cap ((b_{i_j} + U) \neq \emptyset$. It follows that $N \subset \{c \in C : T(x_U) \cap (c + U) \neq \emptyset\}$. By the fact that the subsets $C, T(x_U)$ and U are p-convex, we have that $x_U \in \{c \in C : T(x_U) \cap (c + U) \neq \emptyset\}$, which means that $T(x_U) \cap (x_U + U) \neq \emptyset$.

In Case (2), we assume that T is upper semicontinuous: We define $F(b_i) := \{c \in C : T(c) \cap (x_i + \overline{V}) = \emptyset\}$, which is then closed in C (as the subset $x_i + \overline{V}$ is closed) for each $i = 0, 1, \ldots, n$. Then, the argument is similar to the proof for the case of T being USC. By applying Lemmas 5.1.6 and 5.1.7 again, it follows that there exists $x_U \in A$ such that $T(x_U) \cap (x_U + U) \neq \emptyset$. This completes the proof. $\qquad \square$

By Theorem 5.2.2, we have the following Fan–Glicksberg fixed point theorems (Fan [81]) in locally p-convex vector spaces for $p \in (0, 1]$, which also improve or generalize the corresponding results given by Yuan [284], Xiao and Lu [268], and Xiao and Zhu [270] in locally p-convex vector spaces.

Theorem 5.2.3. *Let A be a p-convex compact subset of a locally p-convex vector space X, where $p \in (0, 1]$. Suppose that $T : A \to 2^A$ is USC with nonempty p-convex closed values. Then, T has at least one fixed point.*

Proof. Assume that $\mathfrak{U}$ is the family of neighborhoods of 0 in X, and $U \in \mathfrak{U}$. By Theorem 5.2.2, there exists $x_U \in A$ such that $T(x_U) \cap (x_U + U) \neq \emptyset$. Then, there exists $a_U, b_U \in A$ for which $b_U \in T(a_U)$ and $b_U \in a_U + U$. Now, consider two nets $\{a_U\}$ and $\{b_U\}$ in $\mathrm{Graph}(T)$, which is a compact graph of mapping T as A is compact and T is semicontinuous. We may assume that a_U has a subnet converging to a and $\{b_U\}$ has a subnet converging to b. As $\mathfrak{U}$ is the family of neighborhoods for 0, we should have $a = b$ (e.g., by the Hausdorff separation property) and $a = b \in T(b)$ due to the fact that Graph(T) is close (e.g., see Lemma 1.1 of Yuan [283]), and thus the proof is compete. □

5.3 Graph Approximation for QUSC Set-Valued Mappings

Throughout this section, without loss of generality and unless specified otherwise, for a given p-vector space E, where $p \in (0,1]$, we always denote by $\mathfrak{U}$ the base of the p-vector space E's topology structure, which is the family of its θ-neighborhoods, and we assume that all p-vector spaces E are Hausdorff unless specified for $p \in (0,1]$.

Now, we are going to discuss how to establish the main results regarding the existence of a fixed point theorem for USC set-valued mappings defined on s-convex subsets under the framework of Hausdorff locally p-convex spaces, where $s, p \in (0,1]$.

By following Repovš *et al.* [223] (see also Ewert and Neubrunn [80], Neubrunn [174], and Holá and Mirmostafaee [108]), we recall the following definition for quasi upper semicontinuous (in short, QUSC) mappings, which are a generalization of USC mappings.

Definition 5.3.1 (QUSC Mappings). Let X and Y be two topological spaces and $T : X \longrightarrow 2^Y$ be a set-valued mapping. The mapping T is said to be quasi upper semicontinuous (QUSC) at $x \in X$ if, for each of its neighborhood $W(x)$ and for each neighborhood V of the origin in Y, there exists a point $q(x) \in W(x)$ such that $x \in int T_{-1}(T(q(x)) + V))$, where $T_{-1}(T(q(x)) + V)) = \{z \in X : T(z) \subset T(q(x)) + V\}$, and the notation $int T_{-1}(T(q(x)) + V))$ denotes the (topological) interior of the set $T_{-1}(T(q(x)) + V))$ in X. The mapping T is said to be QUSC if it is QUSC at each point of its domain.

Remark 5.3.2. It is clear that in Definition 5.3.1 for QUSC mappings, for each $x \in X$, by taking $q(x)$ simply as x itself, then it is the same as the definition for USC mappings given by Definition 5.2.1. Therefore, a USC mapping is QUSC, but a QUSC mapping may not be USC, as shown by the following example (which is the example given by Repovš *et al.* [223, p. 1094]).

Example 5.3.3. A QUSC mapping may not be USC, as shown in this example. Let X and Y be two topological spaces. Let A be a dense subset of a space X with $X \setminus A \neq \emptyset$, and for a fixed point $y_0 \in Y$, we define the following set-valued mapping $F : X \to 2^Y$ by

$$F(x) = \begin{cases} \{y_0\} & \text{if } x \in X \setminus A; \\ Y & \text{if } x \in A. \end{cases} \qquad (5.1)$$

Then, F is USC at all points in A, and F is QUSC (but not USC) at points of $X \setminus A$.

However, in this book, we focus on the study of fixed point theorem for USC set-valued mappings in locally p-convex spaces, where $p \in (0, 1]$. In addition, interested readers can find more details from Ewert and Neubrunn [80], Holá and Mirmostafaee [108], and Neubrunn [174] for a comprehensive study on the quasi continuity for both single and set-valued mappings and related application, as well as related reference therein.

By following the idea used by Repovš *et al.* [223] for the graph approximation of QUSC set-valued mappings, along with the concept of "p-convexity" used in locally p-convex spaces to replace the usual concept of "convexity" used in topological vector spaces (see also Ben-El-Mechaieh and Mechaiekh [17], Ben-El-Mechaiekh and Saidi [18], Cellina [45], Kryszewski [147], Repovš *et al.* [223], and related references), we have the following Lemma 3.1, which is then used as a tool to establish a general fixed point theorem for USC set-valued mappings in Hausdorff locally p-convex spaces for $p \in (0, 1]$.

Here, we also recall that if X and Y are two topological spaces and $F : X \to 2^Y$ is a set-valued mapping, and we denote by either $Graph(F)$ or Γ_F the graph of F in $X \times Y$, and α is a given open cover of Γ_F in $X \times Y$, then a (single- or set-valued) mapping $G : X \to Y$ is said to be an α-approximation (also called α-graph approximation) of F if for each point $p \in \Gamma_G$, there exists a point $q \in \Gamma_F$ such that p and q lie in some common element of the cover α. In the case where Y is a topological vector space, if Ω is the open cover of X and V is an open neighborhood of its origin in Y, then $\Omega \times \{y + V\}_{y \in Y}$ is one open cover of $X \times Y$, which is denoted by $\Omega \times V$ in this section. We also refer the readers to the reference books by Dugungji [73] and Kelly [126] for the corresponding notations and concepts used in general topology.

Lemma 5.3.4. *Let X be a paracompact space, Y be a topological vector space, and $p \in (0, 1]$. If $F : X \to 2^Y$ is a QUSC mapping with p-convex values, then for each open cover Ω of X and each p-convex open neighborhood V of the origin in Y, there exists a continuous single-valued $(\Omega \times V)$-approximation for the set-valued mapping F. In particular, the conclusion holds if V is any convex open neighborhood of the origin in Y.*

Proof.   Let Ω be an open covering of X, and let V be a p-convex open neighborhood of the origin in Y. For each $x \in X$, fix an arbitrary element $W(x) \in \Omega$ such that $x \in W(x)$. Then, we first claim the following statements:

(1) By the QUSC of the mapping F, for each $x \in X$, there exists a point $q(x) \in W(x)$ and an open neighborhood $U(x) \subset W(x)$ such that $F(z) \subset F(q(x)) + V$ for all $z \in U(x)$.

(2) As X is paracompact, by Theorem 3.5 of Dugundji [73] (see also Theorem 28 of Kelly [126, Chapter 5]), without loss of generality, let the family $\{G(x)\}_{x \in X}$ be a covering which is a star refinement of the covering $\{U(x)\}_{x \in X}$ of X (and see also the discussion given by Dugundji [73, pp. 167–168] for the concept of star refinement for a given covering).

(3) Using the quasi upper semicontinuity property again for the mapping F, for each $x \in X$, there exists $q'(x) \in G(x)$ and a neighborhood $U'(x) \subset G(x)$ such that $F(z) \subset F(q'(x)) + V$ for all $z \in U'(x)$;

(4) Let $\{e_\alpha\}_{\alpha \in A}$ be a locally finite continuous partition of unity inscribed into the covering $\{U'(x)\}_{x \in X}$ of X, where A is the index set, with $\Sigma_{\alpha \in A}\, e_{\alpha(x)} = 1$ for each $x \in X$; and for each $\alpha \in A$, we can choose $x_\alpha \in X$ such that $supp\, e_\alpha \subset U'(x_\alpha)$ and choose one point $y_\alpha \in F(q'(x_\alpha))$, where $supp\, e_\alpha$ is the support of e_α (defined by $supp\, e_\alpha := \overline{\{x \in X : e_\alpha(x) \neq 0\}}$).

(5) Finally, define a mapping $f : X \to Y$ by $f(x) := \Sigma_{\alpha \in A} e_\alpha^{\frac{1}{p}}(x) y_\alpha$ for each $x \in X$, where $y_\alpha \in F(q'(x_\alpha))$, as given by (4) above. Then, f is well defined, where the sum is taken over all $\alpha \in A$ with $e_\alpha(x) > 0$. By (3), it follows that $\Sigma_{\alpha \in A}(e_\alpha^{\frac{1}{p}}(x))^p = \Sigma_{\alpha \in A} e_\alpha(x) = 1$.

Now, we show that f is indeed the desired single-valued continuous mapping, which is the $(\Omega \times V)$-approximation for the mapping F. Indeed, for any given $x_0 \in X$, we have that

$$x_0 \in St\{x_0, \{supp\ e_\alpha\}_{\alpha \in A}\} \subset St\{x_0, \{U'(x)\}_{x \in X}\}$$
$$\subset St\{x_0, \{G(x)\}_{x \in X}\} \subset U(x') \subset W(x')$$

for some $x' \in X$, where $St\{x_0, \{supp\ e_\alpha\}_{\alpha \in A}\}$ denotes the star of the point $\{x_0\}$ with respect to the family $\{supp\ e_\alpha\}_{\alpha \in A}$ and defined by $St\{x_0, \{supp\ e_\alpha\}_{\alpha \in A}\} := \cup\{U : x_0 \in U, U \in \{supp\ e_\alpha\}_{\alpha \in A}\}$ (see also the corresponding discussion for the notation and concept given by Ageev and Repovš [4, p. 349]).

By the definition of quasi upper semicontinuity, we have that $q(x') \in W(x')$. Hence, the points x_0 and $q(x')$ are Ω-close.

Secondly, if $e_\alpha(x_0) > 0$ for $\alpha \in A$, then $x_0 \in G(x_\alpha)$ and $q'(x_\alpha) \in G(x_\alpha)$ by (3) above. Thus, $q'(x_\alpha) \in St\{x_0, \{G(x)\}_{x \in X}\} \subset U(x')$. Therefore, $y_\alpha \in F(q'(x_\alpha)) \subset F(q(x')) + V$, i.e., $y_\alpha - v_\alpha \in V$ for some $v_\alpha \in F(q(x'))$ for $\alpha \in A$. But then, for $v := \Sigma_\alpha e_\alpha^{\frac{1}{p}}(x_0)v_\alpha \in F(q(x'))$, as F is p-convex valued and we know that $\Sigma_{\alpha \in A}(e_\alpha^{\frac{1}{p}}(x))^p = \Sigma_{\alpha \in A}e_\alpha(x) = 1$, as shown by (5) above, and $y_\alpha - v_\alpha \in V$, too, for $\alpha \in A$, thus we have that $f(x_0) - v = \Sigma e_\alpha^{\frac{1}{p}}(x_0)(y_\alpha - v_\alpha) \in V$, as V is p-convex. Hence, the point $(x_0, f(x_0)) \in Graph(f)$ is $(\Omega \times V)$-close to the point $(q(x'), v) \in Graph(F)$.

In particular, as each convex neighborhood of the origin in Y is also p-convex for each $p \in (0, 1]$, the conclusion holds. The proof is complete. $\square$

5.4 Fixed Point Theorems for QUSC Set-Valued Mappings

Now, we have the following main result for QUSC set-valued mappings in Hausdorff locally p-convex spaces.

Theorem 5.4.1. *Let K be a nonempty compact s-convex subset of a Hausdorff locally p-convex space X, where $p, s \in (0, 1]$. If $T : K \to 2^K$ is a QUSC set-valued mapping with nonempty closed p-convex values and the graph of T is closed, then T has a fixed point in K.*

Proof. We give the proof by using the graph approximation approach for USC set-valued mappings established in this section above. Let $\mathfrak{U}$ be a family of absolutely p-convex open neighborhoods of the origin in X. By this fact, the family $\{x + u\}_{x \in K}$ is an open covering of K, and we denote the family $\{x + u\}_{x \in K}$ by Ω. Now, by Lemma 5.3.4, it follows that there exists one (single-valued) continuous mapping $f_u : K \to K$, which is ($\Omega \times u$)-approximation of the mapping T. By Theorem 4.2.7, f_u has a fixed point $x_u = f_u(x_u)$ in K for each $u \in \mathfrak{U}$. Note that $(x_u, f_u(x_u)) = (x_u, x_u) \in Graph(f_u)$, which is the ($\Omega \times u$)-approximation of $Graph(T)$, and the graph of T is closed due to the assumption. We now prove that T has a fixed point x^*, which is indeed the limit of some sub-net of the family $\{x_u\}_{u \in \mathcal{U}}$ in K, i.e., $x^* \in T(x^*)$, by using notations of the language in general topology (for related references, see Cellina [45], Ben-El-Mechaieh and Mechaiekh [17], and Fan [81]).

Indeed, for any given open p-convex member u in $\mathfrak{U}$, as the set $\{x + u\}_{x \in K} \times \{y + u\}_{y \in K}$ is an open cover of $K \times K$, by Lemma 5.3.4, there exists a single-valued continuous mapping $f_u : K \to K$ which is the ($\Omega \times u$)-approximation of $Graph(T)$, where $\Omega := \{x + u\}_{x \in K}$, as mentioned above. By Theorem 4.2.7, f_u has a fixed point $x_u = f_u(x_u)$ in K for each $u \in \mathfrak{U}$. Now, for $x_u \in K$, by following the proof of Lemma 5.3.4, we observe that, firstly, there exist $x'_u \in K$ and $q(x'_u) \in K$ such that $x_u \in x'_u + u$, and also $q(x'_u) \in x'_u + u$; and secondly, there also exists $v_u \in F(q(x'_u))$ such that $f_u(x_u) - v_u \in u$, which means that $f_u(x_u) \in v_u + u$.

In summary, for any given $u \in \mathfrak{U}$, there exists a continuous mapping $f_u : K \to K$, which has at least one fixed point $x_u \in K$ such that $x_u = f_u(x_u)$ with $(x_u, x_u) = (x_u, f_u(x_u)) \in Graph(f_u)$, and we also have the following statements:

(1) there exists $x'_u \in K$ and $q(x'_u) \in K$ such that $q(x'_u) \in x'_u + u$ and $x_u \in x'_u + u$; and
(2) there exists $v_u \in F(q(x'_u))$ such that $f_u(x_u) - v_u \in u$, which means $f_u(x_u) \in v_u + u$.

Since K is compact, without loss of the generality, we may assume that there exists a sub-net $(x_{u_i})_{u_i \in \mathfrak{U}}$ that converges to x^* in K. Now, we show that x^* is the fixed point of T, i.e., $x^* \in T(x^*)$.

As K is compact, without loss of generality, we may assume that three nets $\{x_u\}_{u \in \mathfrak{U}}$, $\{x'_u\}_{u \in \mathfrak{U}}$, and $\{q(x'_u)\}_{u \in \mathfrak{U}}$ in K have three sub-nets $\{x_{u_i}\}_{u_i \in \mathfrak{U}}$ converging to x^*, $\{x'_{u_i}\}_{u_i \in \mathfrak{U}}$ converging to x'^*, and $\{q(x'_{u_i})\}_{u_i \in \mathfrak{U}}$ converging to $q(x'^*)$ in K, respectively, in K. By the statement in (1) above, it is clear that we must have $x^* = x'^* = q(x'^*)$, as the family $\mathfrak{U}$ is the base of absolutely p-convex open neighborhoods for the origin in X; otherwise, by (1), we would have a contradiction, and thus our claim that $x^* = x'^* = q(x'^*)$ is true in locally p-convex space X.

Now, we prove that x^* is a fixed point of T by using the statement in (2) for all $u \in \mathfrak{U}$. As the net $\{v_u\}_{u \in \mathfrak{U}} \subset K$, we may also assume its sub-net $\{v_{u_i}\}_{u_i \in \mathfrak{U}}$ converges to v^*. Then, by the statement given by (2), it is clear that we have $\lim_{u_i \in \mathfrak{U}} v_{u_i} = v^* = \lim_{u_i \in \mathfrak{U}} f_{u_i}(x_{u_i}) = \lim_{u_i \in \mathfrak{U}} x_{u_i} = x^*$. By the fact that $(v_{u_i}, q(x'_{u_i})) \in Graph(T)$ and the graph of T is closed by the assumption, it follows that $x^* = v^* \in T(x^*)$, which means x^* is a fixed point of T. The proof is complete. $\square$

Remark 5.4.2. Here, we may ask if the assumption "**T(x) is with nonempty closed p-convex values**" could be replaced by the condition "**T(x) is with nonempty closed s-convex values**" in Theorem 5.4.1. In fact, the good news is that the answer is "**YES**"! As we will see in Section 5.6 of this chapter, this leads us to establishing fixed point theorems for USC set-valued mappings in topological vector spaces and p-vector spaces (see Theorem 5.6.6.) to unify fixed point theorems of both single-valued continuous and USC mappings in Hausdorff topological vector spaces (and p-vector spaces). In turn, this provides a unified positive answer to the Schauder conjecture, with the support of the classic fact that all p-norms in finite-dimensional spaces are equivalent, by applying a proof method originally developed by Fan [81] in 1952 and Kalton's embedding theorem [119] in 1977 in embedding a compact convex subset from a TVS into a locally p-convex space for $p \in (0, 1)$.

By following the same idea used in the proof of Theorem 5.4.1, the conclusion of Theorem 4.2.7 still holds for compact QUSC set-valued mappings, as stated by Theorem 5.4.3, and thus we omit its proof here.

Theorem 5.4.3. *If K is a nonempty closed s-convex subset of a Hausdorff locally p-convex space X, where $s, p \in (0, 1]$, then any compact QUSC set-valued mapping $T : K \to 2^K$ with nonempty* *closed p-convex values and the graph of T being closed, has at least one fixed point.*

As an immediate consequence of Theorem 5.4.3, we have the following fixed point result for USC set-valued mappings in Hausdorff locally convex spaces for compact s-convex subsets, which include the common compact convex sets as a special class.

Corollary 5.4.4. *If K is a nonempty closed s-convex subset of a Hausdorff locally convex space X, where $s \in (0, 1]$, then any compact USC set-valued mapping $T : K \to 2^K$ with nonempty closed convex values has at least one fixed point.*

Proof. Let $p = 1$ in Theorem 5.4.3. Then, the conclusion follows by Theorem 5.4.3. This completes the proof. □

As a special case of Theorem 5.4.3 or Corollary 5.4.4, we also have the following corollary.

Corollary 5.4.5. *If K is a nonempty compact s-convex subset of a Hausdorff locally convex space X, where $s \in (0, 1]$, then any USC set-valued mapping $T : K \to 2^K$ with nonempty closed convex values has at least one fixed point.*

Remark 5.4.6. We note that Theorem 5.4.3 improves or unifies the corresponding results given by Cauty [42, 43], Dobrowolski [67], Nhu [175], Park [189], Reich [222], Smart [250], Xiao and Lu [268], Xiao and Zhu [270], and Yuan [284, 285] under the framework of compact single-valued or USC set-valued mappings.

We would also like to mention that by comparing with the topological degree approach or other related o methods used or developed by Cauty [42, 43], Nhu [175] and others, the arguments used in this section actually provides an accessible way for the study of nonlinear analysis of p-convex vector spaces, where $p \in (0, 1]$. The results given in this book are new and perhaps easily understandable for general readers in the mathematical community. For more details, see the works by Yuan [286–291] and related references on the study of nonlinear analysis and its applications in both p-vector and locally p-convex spaces for $p \in (0, 1]$.

5.5 Fixed Point Theorems for 1-Set Contractive and Condensing Mappings

In order to establish fixed point theorems for the classes of 1-set contractive and condensing mappings in p-vector spaces by using the concept of the measure of noncompactness (or the noncompactness measures), which were introduced and made well known in the mathematical community by Kuratowski [148], Darbo [63], and related references therein, we first need to briefly introduce the concept of noncompactness measures for the so-called Kuratowski or Hausdorff measures of noncompactness in normed spaces (see Alghamdi *et al.* [5], Machrafi and Oubbi [162], Nussbaum [176], Sadovskii [235], Silva *et al.* [249], and Xiao and Lu [268] for the general concepts under the framework of p-seminorm or only for locally convex p-convex settings, where $p \in (0, 1]$, which will also be discussed here).

For a given metric space (X, d) (or a p-normed space $(X, \| \cdot \|_p)$) for $p \in (0, 1]$, we recall the notions of completeness, boundedness, relatively compactness, and compactness as follows.

Let (X, d) and (Y, d) be two metric spaces and $T : X \to Y$ be a mapping (or, alternatively, an operator). Then, we recall that:

(1) T is said to be bounded if for each bounded set $A \subset X$, $T(A)$ is a bounded set of Y;
(2) T is said to be continuous if for every $x \in X$, $\lim_{n \to \infty} x_n = x$ implies that $\lim_{n \to \infty} T(x_n) = T$; and
(3) T is said to be completely continuous if T is continuous and $T(A)$ is relatively compact for each bounded subset A of X.

Now, let A_1 and $A_2 \subset X$ be bounded in a metric space (X, d). We also recall that the Hausdorff metric $d_H(A_1, A_2)$ between A_1 and A_2 is defined by

$$d_H(A_1, A_2) := \max\{ \sup_{x \in A_1} \inf_{y \in A_2} d(x, y), \sup_{y \in A_2} \inf_{x \in A_1} d(x, y)\}.$$

The Hausdorff and Kuratowski measures of noncompactness (denoted by β_H and β_K, respectively) for nonempty bounded subset D in X are the nonnegative real numbers $\beta_H(D)$ and $\beta_K(D)$ defined by

$$\beta_H(D) := \inf\{\epsilon > 0 : D \text{ has a finite}\epsilon\text{-net}\},$$

and

$$\beta_K(D) := \inf\{\epsilon > 0 : D \subset \cup_{i=1}^n D_i,$$
$$D_i \text{ is bounded and } diam D_i \leq \epsilon, \, n \in \mathbb{N}\},$$

where $diam D_i$ means the diameter of the set D_i, and it is well known that $\beta_H \leq \beta_K \leq 2\beta_H$.

Moreover, we would like to point out that the notions above have been well defined under the framework of p-seminorm spaces $(E, \|\cdot\|_p)_{p \in \mathfrak{P}}$ by following a similar idea and method used by Chen and Singh [57], Ko and Tasi [143], Kozlov *et al.* [144], and the references therein.

For a given space X, we suppose that T is a mapping from $D \subset X$ to X. Then, we recall that:

(1) T is said to be a k-set contraction with respect to β_K (or β_H) if there is a number $k \in [0, 1)$ such that $\beta_K(T(A)) \leq k\beta_K(A)$ (or $\beta_H(T(A)) \leq k\beta_H(A)$) for all bounded sets A in D; and
(2) T is said to be β_K-condensing (or β_H-condensing) if $(\beta_K(T(A)) < \beta_K(A))$ (or $\beta_H(T(A)) < \beta_H(A)$) for all bounded sets A in D with $\beta_K(A) > 0$ (or $\beta_H(A) > 0$).

For the convenience of our discussion, throughout the remainder of this book, if a mapping "is β_K-condensing (or β_H-condensing)", we simply say that it is "a condensing mapping" unless specified otherwise. Moreover, it is easy to see that: (1) if T is a compact operator, then T is a k-set contraction; and (2) if T is a k-set contraction for $k \in (0, 1)$, then T is condensing.

In order to establish the fixed points of set-valued condensing mappings in p-vector spaces for $p \in (0, 1]$, we need to recall some notions introduced by Machrafi and Oubbi [162] for the measure of noncompactness in locally p-convex vector spaces. This also satisfies some necessary (common) properties of the classical measures of noncompactness, such as β_K and β_H mentioned above, as introduced by Kuratowski [148] and Sadovskii [235] (see also related discussion by Alghamdi *et al.* [5], Ayerbe Toledano *et al.* [12], Nussbaum [176], Silva *et al.* [249], Xiao and Lu [268], and the references therein). In particular, the noncompactness measures in locally p-vector spaces (for $p \in (0,1]$) should possess a stable property, which means that the measure of noncompactness A remains the same upon transition to the (closure) for the p-convex hull of subset A.

For the convenience of discussion, we use α and β to denote the Kuratowski and Hausdorff measures of noncompactness in topological vector spaces, respectively (as used by Machrafi and Oubbi [162]), unless otherwise stated. E is used to denote a Hausdorff topological vector space over the field $\mathbb{K} \in \{\mathbb{R}, \mathbb{Q}\}$, where $\mathbb{R}$ denotes for all real numbers, $\mathbb{Q}$ denotes for all complex numbers, and $p \in (0,1]$. Here, the base set of the family of all balanced zero neighborhoods in E is denoted by $\mathfrak{V}_0$.

We recall that $U \in \mathfrak{V}_0$ is said to be shrinkable, if it is absorbing, balanced, and $rU \subset U$ for all $r \in (0,1)$, and we know that any topological vector space admits a local base at zero consisting of shrinkable sets (see Klee [141] or Jarchow [116] for details).

Recall that a topological vector space E is said to be a locally p-convex space if E has a local base at zero consisting of p-convex sets. The topology of a locally p-convex space is always given by an upward-directed family P of p-seminorms, where a p-seminorm on E is any nonnegative real-valued and subadditive functional $\| \cdot \|_p$ on E such that $\|\lambda x\|_p = |\lambda|^p \|x\|_p$, for each $x \in E$ and $\lambda \in \mathbb{R}$ (i.e., the real number line). When E is Hausdorff, then for every $x \neq 0$, there is some $p \in P$ such that $P(x) \neq 0$. Whenever the family P is reduced to a singleton, one says that $(E, \| \cdot \|)$ is a p-seminormed space. A p-normed space is a Hausdorff p-seminormed space, and when $p = 1$, which is the usual locally convex case. Furthermore, a p-normed space is a metric vector space with the translation-invariant metric $d_p(x, y) := \|x-y\|_p$ for all $x, y \in E$, which is the same notation used above.

By Remark 2.2.3, if P is a continuous p-seminorm on E, then the ball $B_p(0, s) := \{x \in E : P(x) < s\}$ is shrinkable for each $s > 0$. Indeed, if $r \in (0,1)$ and $x \in r\overline{B_p(0, s)}$, then there exists a net $(x_i)_{i \in I} \subset B_p(0, s)$ such that rx_i converges to x. By the continuity of P, we get $P(x) \leq r^p s < s$, which means that $r\overline{B_p(0, s)} \subset B_P(0, s)$. In general, it can be shown that every p-convex $U \in \mathfrak{V}_0$ is shrinkable.

We recall that a given such neighborhood U, a subset $A \subset E$ is said to be U-small if $A - A \subset U$ (or, in other words, small of order U, as stated by Robertson [225]). Now, by following the idea of Kaniok [125] in the setting of a topological vector space E, we use zero neighborhoods in E instead of seminorms to define the measure of noncompactness in (local convex) p-vector spaces, where $p \in (0, 1]$, as follows.

For each $A \subset E$, the U-measures of noncompactness $\alpha_U(A)$ and $\beta_U(A)$ for A are defined as follows:

$$\alpha_U(A) := \inf \{r > 0 : A \text{ is covered by a finite number of } rU-$$
$$\text{small sets } A_i, \text{ for } i = 1, 2, \ldots, n\},$$

and

$$\beta_U(A) := \inf\{r > 0 : \exists x_1, \ldots, x_n \in E \text{ such that } A \subset \cup_{i=1}^n (x_i + rU)\},$$

where we set $\inf \emptyset := \infty$.

By the definition above, it is clear that when E is a normed space and U is the closed unit ball of E, α_U and β_U are nothing but the Kuratowski measure β_K and Hausdorff measure β_H of noncompactness, respectively. Thus, if $\mathfrak{U}$ denotes a fundamental system of balanced and closed zero neighborhoods in E and $\mathfrak{F}_\mathfrak{U}$ is the space of all functions $\phi : \mathfrak{U} \to R$, endowed with pointwise ordering, then the α_U (resp., β_U) measures for noncompactness introduced by Kaniok [125] can be expressed by the Kuratowski (resp., the Hausdorff) measure of noncompact $\alpha(A)$ (resp., $\beta(A)$) for a subset A of E as a function defined from $\mathfrak{U}$ into $[0, \infty)$ by

$$\alpha(A)(U) := \alpha_U(A) \text{ (resp., } \beta(A)(U) := \beta_U(A)).$$

On the other hand, by following Machrafi and Oubbi [162], in order to define the measure of noncompactness in general p-vector space E, we need the following notions, which are called "basic (and sufficient) collections of zero neighborhoods" (in short, BCZN(SCZN)) in a topological vector space. To do this, let us introduce an equivalence relation on V_0 by saying that U is related to V, denoted by "$U\mathfrak{R}V$", if and only if there exist $r, s > 0$ such that $rU \subset V \subset sU$. We now have the following definition.

Definition 5.5.1 (BCZN and SCZN). We say that $\mathfrak{B} \subset \mathfrak{V}_0$ is a basic collection of zero neighborhoods (in short, BCZN) if it contains at most one representative member from each equivalence class with respect to $\mathfrak{R}$. It is said to be sufficient (in short, SCZN) if it is basic and, for every $V \in \mathfrak{V}_0$, there exists some $U \in \mathfrak{B}$ and some $r > 0$ such that $rU \subset V$.

Remark 5.5.2. By Remark 2.2.3, it follows that for a locally p-convex space E, its base set $\mathfrak{U}$, the family of all open p-convex subsets for 0, is BCZB. We also note that: (1) In the case where E is a normed space, if f is a continuous functional on E, $U := \{x \in E : |f(x)| < 1\}$, and V is the open unit ball of E, then $\{U\}$ is basic but not sufficient, but $\{V\}$ is sufficient; (2) if (E, τ) is a locally convex space whose topology is given by an upward-directed family P of seminorms, so that no two of them are equivalent, the collection $(B_p)_{p \in \mathbb{P}}$ is a SCZN, where B_p is the open unit ball of p. Further, if $\mathfrak{W}$ is a fundamental system of zero neighborhoods in a topological vector space E, then there exists an SCZN consisting of $\mathfrak{W}$ members; and (3) by following Oubbi [183], we recall that a subset A of E is called uniformly bounded with respect to a sufficient collection $\mathfrak{B}$ of zero neighborhoods if there exists $r > 0$ such that $A \subset rV$ for all $V \in \mathfrak{B}$. Note that in the locally convex space $C_c(X) := C_c(X, \mathbb{K})$, the set $B_\infty := \{f \in C(X) : \|f\|_\infty \leq 1\}$ is uniformly bounded with respect to the SCZN $\{B_k, k \in \mathbb{K}\}$, where B_k is the (closed or) open unit ball of the seminorm P_k, where $k \in \mathbb{K}$.

Now, we are ready to give the definition for the measure of noncompactness in general p-vector space E as follows.

Definition 5.5.3. Let $\mathfrak{B}$ be an SCZN in E. For each $A \subset E$, we define the measure of noncompactness of A with respect to $\mathfrak{B}$ by $\alpha_{\mathfrak{B}}(A) := \sup_{U \in \mathfrak{B}} \alpha_U(A)$.

By the definition above, it is clear that: (1) The measure of noncompactness α_B holds the property of semi-additivity, i.e., $\alpha_B(A \cup B) = \max\{\alpha_B(A), \alpha_B(B)\}$; and 2) $\alpha_B(A) = 0$ if and only if A is a precompact subset of E (for details on more properties, see Proposition 1 and related discussion by Machraf and Oubbi [183]).

As we know, under the normed spaces (and even seminormed spaces), Kuratowski [148], Darbo [63], and Sadovskii [235] introduced the notions of k-set-contractions for $k \in (0, 1)$ and condensing mappings to establish fixed point theorems in the setting of Banach spaces, normed or seminormed. By following the same idea, if E is a

Hausdorff p-vector (or locally p-convex) space, we have the following definition for general (nonlinear) mappings.

Definition 5.5.4. Let C be a subset of E. A mapping $T : C \to 2^C$ is said to be a k-set contraction (resp., condensing), if there is some SCZN $\mathfrak{B}$ in E consisting of p-convex sets, such that (resp., condensing) for any $U \in \mathfrak{B}$, there exists $k \in (0,1)$ (resp., condensing) such that $\alpha_U(T(A)) \leq k\alpha_U(A)$ for $A \subset C$ (resp., $\alpha_U(T(A)) < \alpha_U(A)$ for each $A \subset C$ with $\alpha_U(A) > 0$).

It is clear that a contraction mapping on C is a k-set contraction mapping (where we always mean $k \in (0,1)$), and a k-set contraction mapping on C is condensing; and they all reduce to the usual cases by the definitions of β_K and β_H, which are the Kuratowski and Hausdorff measure of noncompactness, respectively, in normed spaces (see Kuratowski [148]).

Hereinafter, we denote by $\mathfrak{V}_0$ the set of all shrinkable zero neighborhoods in E. We have the following result, which is Theorem 1 of Machrafi and Oubbi [162], stating that in the general setting of locally p-convex spaces, the measure of noncompactness α for U given by Definition 5.5.3 above is stable from U to its p-convex hull $C_p(A)$ of the subset A in E, which is key for us to establish the fixed points for condensing mappings in locally p-convex spaces for $p \in (0,1]$. This also means that the key property for the measures due to the Kuratowski and Hausdorff measures of noncompactness in normed (or p-seminorm) spaces, which also holds for the measure of noncompactness by Definition 5.5.3 above in the setting of locally p-convex spaces with $p \in (0,1]$ (see more similar and related discussion given in detail by Alghamdi *et al.* [5] and Silva *et al.* [249]).

Lemma 5.5.5. *We have the following conclusions:*

(1) *If $U \in \mathfrak{V}_0$ is p-convex for some $p \in (0,1]$, then $\alpha(C_p(A)) = \alpha(A)$ for every $A \subset E$.*
(2) *If $U \in \mathfrak{V}_0$ is p-convex for some $p \in (0,1]$, then $\alpha_U(C_s(A)) = \alpha_U(A)$ for every $A \subset E$ and every $0 < s \leq p \leq 1$.*

Proof. Following the proof of Theorem 1 given by Machrafi and Oubbi [162], let $A \subset E$ and $s \in (0,1]$ be such that $0 < s \leq p \leq 1$.

Since $A \subset co_s(A)$, we clearly have $\alpha_U(co_s(A)) \geq \alpha_U(A)$. Now, we prove the conclusion by using the contradiction method. Assume $\alpha_U(co_s(A)) > \alpha_U(A)$, and choose $r > 0$, so that $\alpha_U(co_s(A)) > r > \alpha_U(A)$. Then, there exist rU-small sets $A_1, \ldots, A_n$, $n \geq 1$ such that $A \subset \cup_{i=1}^n A_i$. Then, for each $i = 1, \ldots, n$, $A_i \subset A_i + rU \subset co_s(A_i) + rU$. Since $co_s(A_i) + rU$ is s-convex, it follows that $co_s(A_i) \subset co_s(A_i) + rU$. Hence, $co_s(A_i) - co_s(A_i) \subset rU$. Without loss of the generality, we may then assume that each A_i is s-convex. Now, choose some $a_i \in A_i$, then $A_i \subset a_i + rU$, and since U is a neighborhood of 0, there exists $M > 0$, such that $a_i \in MU$ for every $i = 1, \ldots, n$. Therefore, we have that

$$\cup_{i=1}^n A_i \subset MU + rU \subset \sqrt[s]{M^s + r^s}U.$$

If $x \in co_s(A)$, then there are $n \in \mathbb{N}$, and for every i with $1 \leq i \leq n$, $\mu_i > 0$, and $x_i \in A_i$, such that $\Sigma_{i=1}^n \mu_i^s = 1$ and $x = \Sigma_{i=1}^n \mu_i x_i$. For arbitrary $\delta > 0$, since the set $P := \{(\lambda_1, \ldots, \lambda_n) \in \mathbb{R}^n, \lambda_i \geq 0 \text{ and } \Sigma_i \lambda_i^s = 1\}$ is precompact, there exist $m \in \mathbb{N}$ and, for every $j = 1, \ldots, m$, $\mu^j := (\mu_1^j, \ldots, \mu_n^j) \in P$ such that $P \subset \Sigma_{j=1}^m (\mu^j + \Delta(\delta))$, with $\Delta(\delta) := \{(\lambda_1, \ldots, \lambda_n) \in \mathbb{R}^n : \max_{i=1}^n |\lambda_i| < \delta\}$. Set, for every $j \in \{1, \ldots, m\}$, $S^j := \mu_1^j A_1 + \cdots, \mu_n^j A_n$. Then, clearly we have $S^j - S^j \subset rU$, and there exists $j \in \{1, \ldots, m\}$ such that $\max_{i=1}^n |\mu_i - \mu_i^j| < \delta$. Therefore, for $y := \Sigma_{i=1}^n \mu_i^j x_i \in S^j$, we have

$$x - y = \Sigma_{i=1}^n (\mu_i - \mu_i^j)x_i \in \delta \sqrt[s]{n(M^s + r^s)}U.$$

Choosing δ small enough, so that $\delta \sqrt[s]{n(M^s + r^s)}U \leq \epsilon$, we conclude that $co_s(A)$ is covered by the sets $S_j + \epsilon U$. But for every $a, b \in S^j$ and $y, z \in U$, we have

$$(a + \epsilon y) - (b + \epsilon z) = (a - b) + \epsilon(x - y) \in rU + \epsilon \sqrt[s]{2}U$$

$$\subset r \sqrt[s]{1 + 2\frac{\epsilon^s}{r^s}}U.$$

Since ϵ is arbitrary, $S^j + \epsilon U$ is rU-small. By the fact that $\alpha_U = \alpha_{\overline{U}}$, $\alpha_U(co_s(A)) \leq r$. This contradicts our assumption on r, and thus we must have $\alpha_U(C_s(A)) = \alpha_U(A)$. The proof is complete. $\square$

Now, based on the definition for the measure of noncompactness given by Definition 5.5.3 (originally from Machrafi and Oubbi [162]), we have the following general extension version of Schauder, Darbo, and Sadovskii type fixed point theorems in the context of locally p-convex vector spaces for condensing mappings.

Theorem 5.5.6 (Schauder fixed point theorem for condensing mappings). *Let $C \subset E$ be a complete p-convex subset of a Hausdorff locally p-convex space E, with $0 < p \leq 1$. If $T : C \to 2^C$ is an upper semicontinuous and (α) condensing set-valued mapping with nonempty p-convex closed values, then T has a fixed point in C, and the set of fixed points of T is compact.*

Proof. Let $\mathfrak{B}$ be a sufficient collection of p-convex zero neighborhoods in E with respect to which T is condensing and for any given $U \in \mathfrak{B}$. We choose some $x_0 \in C$, and let $\mathfrak{F}$ be the family of all closed p-convex subsets A of C with $x_0 \in A$ and $T(A) \subset A$. Note that $\mathfrak{F}$ is not empty since $C \in \mathfrak{F}$. Let $A_0 = \cap_{A \in \mathfrak{F}} A$. Then, A_0 is a nonempty closed p-convex subset of C, such that $T(A_0) \subset A_0$. We shall show that A_0 is compact. Let $A_1 = \overline{C_p(T(A_0) \cup \{x_0\})}$. Since $T(A_0) \subset A_0$ and A_0 is closed and p-convex, $A_1 \subset A_0$. Hence, $T(A_1) \subset T(A_0) \subset A_1$. It follows that $A_1 \in \mathfrak{F}$ and therefore $A_1 = A_0$. Now, by Lemma 5.5.5 above (and also see Proposition 1 of Machrafi and Oubbi [162]), we get $\alpha_U(T(A_0)) = \alpha_U(A)$. The assumption on T shows that $\alpha_U(A_0) = 0$ since T is condensing. As U is arbitrary from the family $\mathfrak{B}$, A_0 is p-convex and compact (see Proposition 4 in [162]). Now, the conclusion follows by Theorem 5.4.3 above. Secondly, let C_0 be the set of fixed points of T in C. Then, it follows that $C_0 \subset T(C_0)$ and the upper semicontinuity of T implies that its graph is closed, and so is the set C_0. As T is condensing, we have $\alpha_U(T(C_0)) \leq \alpha_U(C_0)$, which implies that $\alpha_U(C_0) = 0$. As U is arbitrary from the family $\mathfrak{B}$, which implies that C_0 is compact (by Proposition 4 in [162] again). The proof is complete. $\square$

As applications of Theorem 5.5.6, we have in the following a few fixed point theorems for condensing mappings in locally p-convex spaces.

Corollary 5.5.7 (Darbo-type fixed point theorem). *Let C be a complete p-convex subset of a Hausdorff locally p-convex space E*

with $0 < p \leq 1$. *If* $T : C \to 2^C$ *is a (k)-set-contraction (where* $k \in (0,1)$*) with closed and p-convex values, then* T *has a fixed point.*

Corollary 5.5.8 (Sadovskii-type fixed point theorem). *Let* $(E, \| \cdot \|)$ *be a complete p-normed space and* C *be a bounded, closed, and p-convex subset of* E, *where* $0 < p \leq 1$. *Then, every continuous and condensing mapping* $T : C \to 2^C$ *with closed and p-convex values has a fixed point.*

Proof. In Theorem 5.5.6, let $\mathfrak{B} := \{B_p(0,1)\}$, where $B_p(0,1)$ stands for the closed unit ball of E, and by the fact that it is clear that $\alpha(A) = (\alpha_{\mathfrak{B}}(A))^p$ for each $A \subset E$. Then, T satisfies all conditions of Theorem 5.5.6. This completes the proof. $\square$

Corollary 5.5.9 (Darbo type). *Let* $(E, \| \cdot \|)$ *be a complete p-normed space and* C *be a bounded, closed, and p-convex subset of* E, *where* $p \in (0,1]$. *Then, each mapping* $T : C \to C$ *which is continuous and a set-contraction has a fixed point.*

Remark 5.5.10. In this chapter, based on the Schauder fixed point theorems in p-vector spaces for continuous single-valued mappings as a fundamental tool, combining with the KKM principle and the graph approximation method for QUSC set-valued mappings, we develop the general fixed point theory for QUSC set-valued mappings in locally p-convex spaces, where $p \in (0,1]$. These results will be used in the following three chapters to establish the best approximation, fixed point theorems, and nonlinear alternative principles for nonlinear functional analysis under the general locally p-convex spaces by unifying the corresponding results in the existing literature.

Actually, our Theorem 5.5.6 above improves Theorem 5 of Machrafi and Oubbi [162] for general condensing mappings, which are general USC mappings with closed p-convex values. It also unifies the corresponding results in the existing literature, e.g., see Alghamdi et al. [5], Ayerbe Toledano *et al.* [12], Górniewicz [100], Górniewicz *et al.* [101], Nussbaum [176], Silva *et al.* [249], Xiao and Lu [268], Xiao and Zhu [270], and the references therein.

Secondly, as an application of the KKM principle for abstract convex spaces discussed in Section 5.1 for compact p-convex sets in topological vector spaces (TVSs), we also establish fixed point theorems for USC set-valued mappings under the framework of locally p-convex spaces for $p \in (0, 1]$.

5.6 Fixed Point Theorems of USC Set-Valued Mappings for Affirmatively Solving the Schauder Conjecture in p-Vector Spaces

The goal of this section is to establish fixed point theorems of USC set-valued mappings defined on s-convex subsets in Hausdorff p-TVSs by using the functional analysis method without requiring the metric on the underlying p-vector spaces for $s, p \in (0, 1]$. These results include or unify the corresponding results in the existing literature and also provide a positive answer to Schauder's open question. We emphasize that the method established in this section is mainly based on tools from functional analysis and topology, unlike Cauty's approach [42]–[44] (and the related references therein), by introducing a new concept called "algebraic absolute neighborhood retracts" (ANRs), which comes under the category of algebraic topology in mathematics.

As pointed out in Remark 5.4.2 in Section 5.4 above, we would like to establish fixed point theorems for USC set-valued mappings by taking s-convex values instead of p-convex values in locally p-convex and p-vector spaces for $s, p \in (0, 1]$, which would allow us to provide a complete affirmative solution for the Schauder conjecture in general Hausdorff TVSs. In order to do so, we need to first discuss the equivalence of s- and p-norms in $l^s(n)$ and $l^p(n)$ spaces under the framework of a standard n-dimensional Euclidean space $\mathbb{R}^n$, with its standard norm $\| \cdot \|$, where $s, p \in (0, 1]$ (see their definitions and further explanation in detail in the following).

Indeed, we first recall that for a finite n-dimensional TVS V_n, where $n \in \mathbb{N}^+$, Tychonoff [260] in 1935 proved that each n-dimensional TVS V is isomorphic to $\mathbb{R}^n$ with its usual topology. More precisely, if $\{e_1, \ldots, e_n\}$ is a basis of V_n, then the mapping $\mathbb{R}^n \to V_n$ given by $(a_1, \ldots, a_n) \to a_1 e_1 + \cdots + a_n e_n$ is an isomorphism (see also Theorem 1.21 of Rudin [232, p. 16] or Theorem 3.5.6 of Jarchow [116, p. 66]), which means that all s- and p-norms on a finite-dimensional vector space over $\mathbb{R}$ (or $\mathbb{C}$) are equivalent for

$s, p \in (0, 1]$. Here, we recall that the "equivalence" between the p-norm of the n-dimensional space $l^p(n)$ and the standard Euclidean norm $\| \cdot \|$ for the Euclidean space $\mathbb{R}^n$ means that both the p-norm and $\| \cdot \|$ generate the same topology structures (i.e., they generate all the same open subsets) for $p \in (0, 1]$. Now, we recall the following proposition, which is actually Theorem 3.1.6 in Chapter 3, restated here for the convenience of readers.

Proposition 5.6.1. *Let Y be an n-dimensional subspace of a topological vector space X with real $\mathbb{R}$ (or complex $\mathbb{C}$) scalars, where $n \in \mathbb{N}^+$. We have that:*

(a) *every isomorphism of $\mathbb{R}^n$ (or $\mathbb{C}^n$) onto Y is a homeomorphism;*
(b) *Y is closed; and*
(c) *the s-normed space $(Y, \| \cdot \|_s)$ (including $l^s(n)$, $\mathbb{R}^n$, any other finite n-dimensional vector space) is equivalent to any p-normed space $(Y, \| \cdot \|_p)$ for $s, p \in (0, 1]$, and all finite n-dimensional spaces are locally convex.*

Proof. Its proof is given by Theorem 3.1.6. □

Actually, based on a topological property in functional analysis for a continuous functional on a compact set to obtain its minimum and maximum values, we have the following Proposition 5.6.2, which explains result (c) of Proposition 5.6.1 more rigorously in terms of p- and s-norms for both $l^p(n)$ and $l^s(n)$, respectively, where $s, p \in (0, 1]$ and $n \in \mathbb{N}^+$. The proposition states that the family of s-convex subsets generated by an s-norm in $l^s(n)$ is equivalent to the family of p-convex subsets generated by a p-norm in any n-dimensional vector space (such as $l^s(n)$, $l^s(n)$, and $\mathbb{R}^n$).

Proposition 5.6.2. *For any given two n-dimensional spaces V_n (e.g., $l^s(n)$ and $l^p(n)$), the s-norm $\| \cdot \|_s$ of $l^s(n)$ is equivalent to the p-norm $\| \cdot \|_p$ of $l^p(n)$ for any $s, p \in (0, 1]$, where $n \in \mathbb{N}^+$.*

Proof. For any given standard n-dimensional Euclidean space $\mathbb{R}^n$, we use $\| \cdot \|$ to denote the standard Euclidean norm defined by $\|x\| := (\sum_{i=1}^{n} |x_i|^2)^{\frac{1}{2}}$ for each $x = (x_i)_{i=1}^{n} \in \mathbb{R}^n$. For any $p \in (0, 1]$, the p-norm space $l^p(n)$ has a p-norm $\| \cdot \|_p$ defined by $\|x\|_p := \sum_{i=1}^{n} |x_i|^p$

for $x = (x_i)_{i=1}^n \in l^p(n)$. Then, we first have that

$$\|x\|_p = \sum_{i=1}^n |x_i|^p \leq \sum_{i=1}^n ((|x_i|^2)^{\frac{1}{2}})^p \leq \sum_{i=1}^n ((\sum_{j=1}^n |x_j|^2)^{\frac{1}{2}})^p \leq n\|x\|^p,$$

for each $x \in \mathbb{R}^n$, which implies $\|x - y\|_p \leq n\|x - y\|^p$ for each $x, y \in \mathbb{R}^n$. This means the p-norm of $l^p(n)$ from the Euclidean space $(\mathbb{R}^n, \|\cdot\|)$ to $l^p(n)$ is continuous. Let $S := \{x \in \mathbb{R}^n : \|x\| = 1\}$. Then, S is compact in $\mathbb{R}^n$, so the p-norm $\|\cdot\|_p$ attains its minimum and maximum on S, which means there exist x_{p0} and x_{p1} in $S \subset \mathbb{R}^n$, such that for all $x \in S$, we have $0 \neq \|x_{p0}\|_p \leq \|x\|_p \leq \|x_{p1}\|_p \neq 0$. Now, for any $x \in \mathbb{R}^n$ with $x \neq 0$, we have $\frac{x}{\|x\|} \in S$, which implies that $\|x_{p0}\|_p \leq \|\frac{x}{\|x\|}\|_p \leq \|x_{p1}\|_p$. Thus, we have

$$\|x_{p0}\|_p \cdot \|x\|^p \leq \|x\|_p \leq \|x_{p1}\|_p \cdot \|x\|^p,$$

for each $x \in \mathbb{R}^n$ (which is also true for $x = 0$). This means the p-norm $\|\cdot\|_p$ of the space $l^p(n)$ is equivalent to the standard Euclidean norm $\|\cdot\|$ of the Euclidean space $\mathbb{R}^n$.

Next, for any $s \in (0, 1]$, with the s-norm of space $l^s(n)$, by following the same approach as above and assuming that s-norm $\|\cdot\|_s$ attains its minimum and maximum on S at x_{s0} and x_{s1} in S, respectively, then we have the following:

$$\|x_{s0}\|_s \cdot \|x\|^s \leq \|x\|_s \leq \|x_{s1}\|_s \cdot \|x\|^s,$$

for each $x \in \mathbb{R}^n$. By putting it all together, we have that for $s, p \in (0, 1]$, the following inequality holds:

$$\|x_{s0}\|_s \cdot (\frac{\|x\|_p}{\|x_{p1}\|_p})^{\frac{s}{p}} \leq \|x\|_s \leq \|x_{s1}\|_s \cdot (\frac{\|x\|_p}{\|x_{p0}\|_p})^{\frac{s}{p}}$$

for each $x \in l^s(n)$ and $l^p(n)$. The above inequality states that for any $s, p \in (0, 1]$, the s-norm $\|\cdot\|_s$ in $l^s(n)$ is equivalent to the p-norm $\|\cdot\|_p$ in $l^p(n)$. This completes the proof. $\square$

Now, as an application of either Proposition 5.6.1 or Proposition 5.6.2, and using the approximation method of finite-dimensional spaces employed to prove the existence of fixed points for USC mappings in p-normed spaces, we first recall the following fixed point theorem for USC set-valued mappings in p-normed spaces. This was actually first given as Theorem 4.1 by Ennassik *et al.* [78] (which includes Theorem 2.15 of Xiao and Zhu [270] as a special case). To ensure that the text is self-contained, we give its proof in detail here by following the argument method developed and used by Xiao and Zhu [270], Ennassik *et al.* [78], and the references therein.

Theorem 5.6.3. *Let C be a compact s-convex subset of a p-normed space $(X, \| \cdot \|_p)$ for $s, p \in (0, 1]$. If $T : C \to 2^C$ is USC and $T(x)$ is nonempty closed and s-convex for each $x \in C$, then T has a fixed point in C, i.e., there exists $x_0 \in C$ such that $x_0 \in T(x_0)$.*

Proof. Since C is compact, it is totally bounded. Then, for any integer $n \in \mathbb{N}^+$, $C \subset \cup_{x \in C} B_p(x, \frac{1}{n})$, where $B_p(x, \frac{1}{n}) := \{z \in C : \|x - z\|_p < \frac{1}{n}\}$, which is open in C, there are a finite number of points $x_{1,n}, \ldots, x_{k_n,n}$ in C such that $C \subset_{i=1}^{k_n} B_p(x_{i,n}, \frac{1}{n})$. Now, for $i = 1, \ldots, k_n$, we define a mapping $\phi_{i,n} : C \to \mathbb{R}$ by $\phi_{i,n}(x) := \max\{\frac{1}{n} - \|x - x_{i,n}\|_p, 0\}$, for each $x \in C$. Then clearly, $\phi_{i,n}$ is a continuous functional on C and $\sum_{i=1}^{k_n} \phi_{i,n}(x) > 0$ for all $x \in C$. Now, taking $y_{i,n} \in T(x_{i,n})$, we define a mapping $T_n : C \to C$ by

$$T_n(x) := \sum_{i=1}^{k_n} \left(\frac{\phi_{i,n}(x)}{\phi_n(x)} \right)^{\frac{1}{s}} y_{i,n},$$

where $\phi_n(x) = \sum_{i=1}^{k_n} \phi_{i,n}(x)$ for each $x \in C$. As T_n is continuous from C to C, by Theorem 4.1.6, T_n has a fixed point $x_n \in C$ such that $x_n = T_n(x_n)$. Since $(x_n)_{n \in \mathbb{N}^+} \subset C$ and C is compact, without loss of generality, we may assume that $(x_n)_{n \in \mathbb{N}^+}$ converges to some $x^* \in C$. We now want to prove that x^* is a fixed point of T.

Indeed, for each $\epsilon > 0$, the subset $U_\epsilon = T(x^*) + B_p(0, \epsilon)$ is an open neighborhood of the closed subset $T(x^*)$, hence we have $\cap_{\epsilon > 0} U_\epsilon = \overline{T(x^*)}$. Now, by the upper semi-continuity of T, there exists $\delta > 0$ such that $T(B_p(x^*, \delta)) \subset U_\epsilon$. As $x_n \to x^*$, there is a positive integer $N > \frac{2}{\delta}$ such that $x_n \in B_p(x^*, \frac{\delta}{2})$ for all $n \geq N$. Let $n \geq N$ and $1 \leq i \leq k_n$. Then, $\|x_n - x_{i,n}\|_p < \frac{1}{n}$ whenever $\phi_{i,n}(x_n) > 0$. Thus, $x_{i,n} \in B_p(x^*, \delta)$. Hence, $y_{i,n} \in T(x_{i,n}) \subset T(B_p(x^*, \delta)) \subset U_\epsilon$ and therefore $y_{i,n} \in U_\epsilon$. Since $T(x^*)$ and $B_p(0, \epsilon))$ are s-convex, we know that U_ϵ is also s-convex. Consequently,

$$x_n = T_n(x_n) = \sum_{i=1}^{k_n} \left(\frac{\phi_{i,n}(x)}{\phi_n(x)} \right)^{\frac{1}{s}} y_{i,n} \in U_\epsilon.$$

Now, letting $n \to \infty$, we obtain $x^* \in \overline{U_\epsilon} \subset U_{2\epsilon}$, and so, we have $x^* \in \cap_{\epsilon>0} U_\epsilon = \overline{T(x^*)} = T(x^*)$. This completes the proof. $\qquad\square$

Now, we can establish a fixed point theorem for USC set-valued mappings with closed s-convex values defined on a closed s-convex subset of Hausdorff locally p-convex spaces for $s, p \in (0, 1]$ by applying an idea originally introduced by Fan [81] in 1952 (also used by Ennassik *et al.* [78]), which is supported by the fact that "each finite-dimensional space V_n is linearly homeomorphic to an Euclidean space $\mathbb{R}^n$, and each s-norm of $(V_n, \| \cdot \|_s)$ is equivalent to the p-norm of $(V_n, \| \cdot \|_p)$ for any $s, p \in (0, 1]$", as given by Proposition 5.6.1 or Proposition 5.6.2 above.

Theorem 5.6.4. *If K is a nonempty compact s-convex subset of a Hausdorff locally p-convex space X, where $s, p \in (0, 1]$. Then any USC set-valued mapping $T : K \to 2^K$ with nonempty closed s-convex values has at least one fixed point.*

Proof. By following the idea developed by Fan [81] (see also Ennassik *et al.* [78] for the proof of their Theorem 4.3), let $\mathcal{U}$ be a basis of absolutely p-convex open neighborhoods of the null element zero that generates the locally p-convex topology of X. Let $U \in \mathcal{U}$ be fixed, and let $V \in \mathcal{U}$ be such that $\overline{V} \subset U$. Since C is compact, there are $x_1, \ldots, x_n \in C$ such that $C \subset \{x_1, \ldots, x_n\} + V$. Now, let $K := co_s\{x_1, \ldots, x_n\}$, the s-convex hull of $\{x_1, \ldots, x_n\}$. It is easily seen that $K \subset C$ and K is also a compact s-convex subset of a finite-dimensional topological vector space, denoted by $V_n := span\{x_1, \ldots, x_n\}$ being n-dimensional. Now, we define a set-valued mapping $T_U : K \to 2^K$ by $T_U(x) := (T(x) + \overline{V}) \cap K$ for each $x \in K$.

Then, we first claim that for each $x \in K$, $T_U(x)$ is a nonempty closed s-convex subset K in the n-dimensional space V_n. Indeed, for each $x \in K$, as $T(x) \subset C \subset \{x_1, \ldots, x_n\} + V$, so for any $y \in T(x)$, there exist $x_y \in \{x_1, \ldots, x_n\} \subset K$ and $v_y \in V$ such that $y = x_y + v_y$. Then, we have $y + (-v_y) = x_y \in K$, and also $y + (-v_y) \in T(x) + V$, as V is absolute. This implies that $y + (-v_y) \in (T(x) + \overline{V}) \cap K$, which means that $T_U(x) = (T(x) + \overline{V}) \cap K \neq \emptyset$ for each $x \in K$. Second, we claim that, indeed, $T_U(x)$ is s-convex with the s-norm in V_n, as both the s- and p-norms are equivalent (in the sense of homeomorphism to $\mathbb{R}^n$, with the Euclidean norm defined by $\| \cdot \|$ above in Proposition 5.6.2). Indeed, by the assumption, it is clear that $T_U(x)$ is p-convex in V_n for $p \in (0, 1]$; but, on the other hand, as $T_U(x) \subset K$ is nonempty in a finite n-dimensional space V_n, by Proposition 5.6.1 or Proposition 5.6.2 above, we can now claim that $T_U(x)$ is s-convex, as both the s- and p-norms are equivalent in the n-dimensional space V_n. Third, we can also show that the mapping T_U is USC. To see this, let $x_0 \in K$ and V_0 be an open set such that $T_U(x_0) \subset V_0$. Since $T_U(x_0)$ is compact, there is $W_0 \in \mathcal{U}$ such that $T_U(x_0) + W_0 \subset V_0$ by Theorem 1.10 of Rudin [232, p. 10]. On the other hand, keeping in mind the closedness of $T(x_0) + \overline{V}$, by Lemma 1 of Fan [81], there exists $W_1 \in \mathcal{U}$ such that

$$((T(x_0) + \overline{V}) + W_1) \cap (K + W_1) \subset T_U(x_0) + W_0.$$

Since T is USC, there exists an open neighborhood U_0 of x_0 such that $T(x) \subset T(x_0) + W$ for all $x \in C \cap U_0$. Thus, we have $T(x) + \overline{V} \subset (T(x_0) + \overline{V} + W_1$ for all $x \in K \cap U_0$. Therefore, we have

$$T_U(x) \subset ((T(x_0) + \overline{V}) + W_1) \cap K \subset ((T(x_0) + \overline{V}) + W_1)$$
$$\cap (K + W_1) \subset T_U(x_0) + W_0.$$

Thus, $T_U(x) \subset T_U(x_0) + W_0$. Since $T_U(x_0) + W_0 \subset V_0$, $T_U(x) \subset V_0$. This proves that T_U is USC. Then, by Theorem 5.6.3 above, there is $x_U \in K$ such that $x_U \in T_U(x_U) \subset T(x_U) + \overline{V}$, which means $x_U \in T(x_U) + U$.

Now, we have the net $\{x_U\}_{U \in \mathcal{U}} \subset C$ with the property that $x_U \in T_U(x_U)$ for each $U \in \mathcal{U}$. Note that C is compact, and we may assume that the family $\{x_U\}_{U \in \mathcal{U}}$ has a sub-net $(x_{U_i})_{i \in I}$ which converges to some element $x^* \in C$. By the upper semi-continuity of T_U with nonempty closed values, we have that the graph of T is closed, and thus we have $x^* \in \cap_{U \in \mathcal{U}}(T(x^*) + U) = T(x^*)$, which means x^* is a fixed point of T.

(Here, we also share another way to prove the existence of T as follows: If we define $F_U := \{x \in C : x \in T(x) + \overline{U}\}$ for each $U \in \mathcal{U}$, then F_U is nonempty, closed, and compact (as a subset of the compact C), and the family $\{F_U\}_{U \in \mathcal{U}}$ has the finite intersection property. Thus, $\cap_{U \in \mathcal{U}} F_U \neq \emptyset$, and any point $x \in \cap_{U \in \mathcal{U}} F_U$ is a fixed point of T.)

This completes the proof. □

As a special case of Theorem 5.6.4 above by taking $s = 1$, we have the following result.

Corollary 5.6.5. *If K is a nonempty compact convex subset of a Hausdorff locally p-convex space X, where $p \in (0, 1]$, then any USC set-valued mapping $T : K \to 2^K$ with nonempty closed convex values has at least one fixed point.*

By applying Corollary 5.6.5 as a tool, we can establish the fixed point theorem for USC set-valued mappings with nonempty closed and convex values defined on a nonempty convex compact subset of (Hausdorff) topological vector spaces. This provides a positive answer for the Schauder conjecture [237], which has remained open for a long time. Theorem 5.6.6 includes both continuous single-valued and USC set-valued mappings in a Hausdorff topological vector space (or, in general, a p-vector space) as its special classes.

Theorem 5.6.6. *If K is a nonempty compact convex subset of a Hausdorff topological vector space (or, in general, a p-vector space) X, then any USC set-valued mapping $T : K \to 2^K$ with nonempty closed convex values has at least one fixed point.*

Proof. As K is compact and convex in X, taking any point $x_0 \in K$, let $K_0 := K - \{x_0\}$. Then, K_0 is still convex and compact and contains a zero element; thus, K_0 is p-convex for $p \in (0, 1)$ by Lemma 2.2.4(ii). We now define a mapping $T_0 : K_0 \to 2^{K_0}$ by $T_0(x - x_0) := T(x) - x_0$ for each $x \in K$. Then T_0 is still a USC set-valued mapping with non-empty closed and convex values.

As K_0 is p-convex for $p \in (0, 1)$, by the Kalton embedding theorem (Theorem 2.3.6 in Chapter 2), it follows that K_0 can be linearly embedded in a locally p-convex space Y, which means that there exists a linear mapping $L : \text{lin}(K_0) \to Y$ whose restriction to K_0 is a homeomorphism. We define another mapping $S : L(K_0) \to 2^{L(K_0)}$ by $S(L(z)) := L(T_0(z))$ for each $z \in K_0$. Then, this mapping can be easily checked to be well defined. The mapping S is also a USC set-valued mapping with nonempty closed and convex values since L is a (linear and continuous) homeomorphism and T_0 is a USC set-valued mapping with nonempty closed and convex values on K_0. Furthermore, the set $L(K_0)$ is compact, being the image of a compact set under a continuous mapping L, and $L(K_0)$ is also convex (and p-convex) since it is the image of a convex (and p-convex) subset under a linear mapping. Then, by Corollary 5.6.5, S has a fixed point $y \in L(K_0)$, which means that there exists $x \in K$ such that $y = L(x - x_0) \in S(L(x - x_0)) = L(T_0(x - x_0)) = L(T(x) - x_0)$, which implies that $L(x - x_0) \in L(T(x) - x_0)$. As L is a homeomorphism on K_0, meaning $x - x_0 \in T(x) - x_0$, we have that $x \in T(x)$ is a fixed point of T. This completes the proof. $\square$

Remark 5.6.7. Theorem 5.6.6 states that any USC set-valued mapping with nonempty closed convex values defined on a nonempty s-convex compact subset of a Hausdorff topological vector space (or, in general, a p-vector space) has a fixed point. This result is significant, as it not only provides a positive answer to the Schauder conjecture for USC set-valued mappings but also unifies or improves the corresponding results in the existing literature for both single-valued continuous and USC set-valued mappings as a special class. In addition, we can prove that Theorem 5.6.6 is also true for a compact USC mapping defined on a closed s-convex subset of a Hausdorff topological vector space (p-vector spaces), which includes or improves most of the available results for fixed point theorems in the existing literature as special cases (e.g., see the discussion given by Yuan [292]–[293], and associated comprehensive discussion given by Khan *et al.* [127]–[128], He and Yannelis [106]–[107] and references wherein), including, among others, Agarwal *et al.* [1], Ben-El-Mechaiekh and Mechaiekh [17], Ben-El-Mechaiekh and Saidi [18], Ennassik and Taoudi [79], Fan [81]–[82], Guo *et al.* [104], Mauldin [167], Granas and Dugundji [102], Park [189], and the references therein). In particular, we note that an affirmative answer to the Schauder conjecture for a single-valued continuous mapping was recently obtained by Ennaassik and Taoudi [79], defined on a nonempty compact p-convex subset in Hausdorff topological vector spaces, where $p \in (0, 1]$.

Secondly, we note that without assuming the underlying topological vector space or p-vector space with any metric, Theorem 5.6.6 improves or unifies the corresponding results established by Cauty [42]–[43] by applying the algebraic ANRs concept under the category of algebraic topology, (see also comments and discussion given by Dobrowolski [67], Nhu [175], Park [189], and the comprehensive references therein), Xiao and Lu [268], Xiao and Zhu [270], and Yuan [284]–[291] under the framework of locally p-convex spaces for compact USC set-valued mappings defined on s-convex subsets which may not be compact for $s, p \in (0, 1]$.

Moreover, we would like to mention that by comparing with the algebraic topology approach or other related topological approaches developed by Cauty [42]–[43] (see also Nhu [175] and the related references therein), the proof obtained by following the functional analysis method in this paper actually provides an accessible and understandable way for the study of nonlinear analysis on p-convex vector spaces for $p \in (0, 1]$. The argument and approach used in this paper are also easily understandable for interested readers in the mathematical community. Readers may refer to Yuan [286]–[291] and related references for more details on the development of nonlinear analysis and its related applications in p-vector spaces $p \in (0, 1]$.

In addition, we would also like to share with readers that a proof for the Schauder conjecture using the Dugundji extension method in p-vector spaces was recently given by Yuan [291] for continuous single-valued mappings in topological vector spaces for $p \in (0, 1]$. Thus, the fixed point theorems established in this paper for USC set-valued mappings include those given by Yuan [291] as special cases (see also some recent results given by Yuan and Xiao [294] by using the Dugundji extension method in p-vector spaces as a tool for the study of fixed point theorems in p-normed spaces).

Finally, before ending this chapter, we remark that by comparing with the topological method or related arguments used by Askoura *et al.* [8] and Cauty [42]–[43] (see also Dobrowolski [67], Nhu [175], and Reich [222]), the fixed point results given in this chapter improve or unify the corresponding ones given by Alghamdi *et al.* [5], Darbo [63], Liu [160], Machrafi and Oubbi [162], Sadovskii [235], Park [189], Silva *et al.* [249], Xiao and Lu [268], and Yuan [286], as well as those from the references therein.

Last but not least, hereinafter, all the discussions in the remaining part of this book, i.e., in Chapters 6–8, focus on locally p-convex spaces for set-valued mappings. In fact, they can also be studied under the general framework of topological vector spaces or p-vector spaces by using Theorem 5.6.6 as a basic tool, but we do not include these discussions in this book.

Chapter 6

Best Approximation in Locally
p-Convex Spaces

The goal of this chapter is to establish general best approximation results for the classes of 1-set continuous and hemicompact (see its definition as follows) non-self mappings, which are then used as a tool to derive the general principles for the existence of solutions to problems defined by Birkhoff and Kellogg [23] in 1922 and fixed points for non-self 1-set contractive mappings. This chapter consists of two sections, which are introduced briefly as follows.

The first section focuses on the framework of the best approximations for 1-set contractive single-valued mappings in locally p-convex spaces; in the second section, we study the best approximations for 1-set contractive set-valued mappings in locally p-convex spaces, where $p \in (0, 1]$.

Here, we also recall that the Birkhoff–Kellogg theorem was first introduced and proved by Birkhoff and Kellogg [23] while discussing the existence of solutions for the equation $x = \lambda F(x)$, where λ is a real parameter and F is a general nonlinear non-self mapping defined on an open convex subset U of a topological vector space E. Therefore, the general form of the Birkhoff–Kellogg problem now involves finding the so-called invariant direction for the nonlinear single-valued or set-valued mappings F, i.e., to find $x_0 \in \overline{U}$ (or $x_0 \in \partial \overline{U}$) and $\lambda > 0$ such that $\lambda x_0 = F(x_0)$, or $\lambda x_0 \in F(x_0)$ when F is a set-valued mapping. In particular, this chapter focuses on the study of single-valued mappings in locally p-convex spaces for $p \in (0, 1]$.

Since Birkhoff and Kellogg proposed their theorem, the study of Birkhoff–Kellogg problems received significant attention from scholars. For example, one of the fundamental results in nonlinear functional analysis, called the Leray–Schauder alternative, developed by Leray and Schauder [154] in 1934, was established via the topological degree. Thereafter, certain other types of Leray–Schauder alternatives were proved using different techniques other than the topological degree; see the works by Granas and Dugundji [102] and Furi and Pera [92] in the Banach space setting, applications to the boundary value problems for ordinary differential equations, and a general class of mappings for nonlinear alternative of Leray–Schauder type in normal topological spaces. In addition, for Birkhoff–Kellogg-type theorems for general class mappings in TVS, see the works of Agarwal *et al.* [1], Agarwal and O'Regan [2, 3], and Park [186]. In particular, O'Regan [180] recently used the Leray–Schauder-type coincidence theory to establish some Birkhoff–Kellogg problems and also Furi–Pera-type results for a general class of single-valued or set-valued mappings.

In this chapter, the best approximation results for 1-set contractive mappings in locally p-convex spaces are first established, which are then used to establish the existence of solutions for Birkhoff–Kellogg problems and related nonlinear alternatives. These new results allow us to derive a general principle for the Leray–Schauder principle and related fixed point theorems of non-self mappings in locally p-convex spaces for $p \in (0, 1]$. The new results given in this part not only include the corresponding results in the existing literature as special cases but are also expected to play a fundamental role in the development of nonlinear problems arising from theory to practice for 1-set contractive mappings under the framework of p-vector or locally p-convex spaces, which include the general vector topological spaces and locally convex spaces as special classes.

We also note that the general principles for nonlinear alternative related to the Leray–Schauder alternative and other types under the framework of locally p-convex spaces for $p \in (0, 1]$ given in this section would be useful tools for the study of nonlinear problems. In addition, we note that the corresponding results in the existing literature for Birkhoff–Kellogg problems and the Leray–Schauder alternatives have been studied comprehensively by Granas and Dugundji [102], Isac [118], Kim *et al.* [129], Park [187–189],

Carbone and Conti [38], Chang *et al.* [54], [53], Chang and Yen [55], Shahzad [238,240], and Singh [248]. In particular, many general forms have been recently obtained by O'Regan [181], Yuan [286–291], and the references therein.

6.1 Best Approximation and Fixed Point for 1-Set Contractive Single-Valued Mappings

The goal of this section is to establish the framework for the best approximation and related fixed point results for non-self 1-set contractive single-valued mappings in locally p-convex spaces, where $p \in (0, 1]$. In order to study the general existence of fixed points for non-self mappings in locally p-convex spaces, we need some definitions and notions, as follows.

Definition 6.1.1 (Inward and outward sets in p-vector spaces). Let C be a subset of a p-vector space E and $x \in E$ for $0 < p \leq 1$. Then, the p-inward set $I_C^p(x)$ and p-outward set $O_C^p(x)$ are defined, respectively, by
$$I_C^p(x) := \{x + r(y - x) : y \in C, \text{ for any } r \geq 0 \text{ (1) if } 0 \leq r \leq 1,$$
with $(1 - r)^p + r^p = 1$; or (2) if $r \geq 1$, with $(\frac{1}{r})^p + (1 - \frac{1}{r})^p = 1\}$; and $O_C^p(x) := \{x + r(y - x) : y \in C, \text{ for any } r \leq 0 \text{ (1) if } 0 \leq |r| \leq 1$, with $(1 - |r|)^p + |r|^p = 1$; or (2) if $|r| \geq 1$, with $(\frac{1}{|r|})^p + (1 - \frac{1}{|r|})^p = 1\}$.

   From the definitions, it is obvious that when $p = 1$, both inward and outward sets, $I_C^p(x)$ and $O_C^p(x)$, are reduced to the definition for the inward set $I_C(x)$ and the outward set $O_C(x)$, respectively, in topological vector spaces introduced by Halpern and Bergman [105] and used for the study of non-self mappings related to nonlinear functional analysis in the literature. In this book, we mainly focus on the study of the p-inward set $I_U^p(x)$ for the best approximation related to the boundary condition for the existence of fixed points in locally p-convex spaces. By the special property of the p-convex concept for $p \in (0, 1)$ with $p = 1$, we have the following fact.

Lemma 6.1.2. *Let C be a subset of a p-vector space E and $x \in E$, where for $0 < p \leq 1$. Then, for both p-inward and -outward sets, $I_C^p(x)$ and $O_C^p(x)$, defined above, we have the following:*

(1) When $p \in (0,1)$, $I_C^p(x) = [\{x\} \cup C]$, and $O_C^p(x) = [\{x\} \cup \{2x\} \cup -C]$.

(2) When $p = 1$, $[\{x\} \cup C] \subset I_C^p(x)$, and $[\{x\} \cup \{2x\} \cup -C] \subset O_C^p(x)$.

Proof. First, when $p \in (0,1)$, by the definition of $I_C^p(x)$, the only real number $r \geq 0$ satisfying the equation $(1-r)^p + r^p = 1$ for $r \in [0,1]$ is $r = 0$ or $r = 1$, and when $r \geq 1$, the equation $(\frac{1}{r})^p + (1 - \frac{1}{r})^p = 1$ implies that $r = 1$. Applying the same reason for $O_C^p(x)$, it follows that $r = 0$ and $r = -1$.

Secondly, when $p = 1$, all $r \geq 0$ and all $r \leq 0$ satisfy the requirement of the definitions for $I_C^p(x)$ and $O_C^p(x)$, respectively; thus, the proof is compete. $\square$

By following the original idea given by Tan and Yuan [258] for hemicompact mappings in metric spaces, we introduce the following definition for a mapping being hemicompact in p-seminorm spaces for $p \in (0,1]$, which is indeed the "**(H) Condition**" used in Theorem 6.1.5 below to prove the existence of the best approximation results for 1-set contractive mappings in locally p-convex spaces for $p \in (0,1]$.

Definition 6.1.3 (The "(H) Condition" for hemicompact mappings). Let E be a locally p-convex space for $p \in (0,1]$. For a given bonded (closed) subset D in E, a mapping $F : D \to 2^E$ is said to be hemicompact if each sequence $\{x_n\}_{n \in N}$ in D has a convergent subsequence with limit x_0 such that $x_0 \in F(x_0)$, whenever $\lim_{n \to \infty} d_{P_U} P(x_n, F(x_n)) = 0$ for each $U \in \mathfrak{U}$, where $d_{P_U} P(x, C) := \inf\{P_U(x - y) : y \in C\}$ is the distance of a single point x with the subset C in E based on P_U. P_U is the Minkowski p-functional in E for $U \in \mathfrak{U}$, which is the base of the family consisting of all open p-convex subsets for θ-neighborhoods in E.

Remark 6.1.4 (On the Meaning of (H) and (H1) Conditions). First, we may initially explain that **Definition 6.1.3** above for the "**(H) Condition**" is indeed an extension of a "**Hemicompact mapping**" by Tan and Yuan [258], defined from a metric space to a locally p-convex space with the p-seminorm, which is in turn an extension of the concept of compact mappings widely used in functional analysis, where $p \in (0,1]$. In addition, by the monotonicity of Minkowski p-functionals, i.e., the larger the θ-neighborhoods, the smaller the values of the Minkowski p-functionals (see also

Balachandran [13, pp.178]), Definition 6.1.3 thus describes the convergence for the distance between x_n and $F(x_n)$ by using the language of seminorms in terms of Minkowski p-functionals for each θ-neighborhood in $\mathfrak{U}$ (the base), which is the family consisting of its open p-convex θ-neighborhoods in p-vector space E.

Secondly, on the other hand, for the "**(H1) Condition**" that appears and is used in Theorem 6.1.5, we may see it is an extension of the so-called **demiclosedness principle**, first given by Browder [32] in 1968 for nonexpansive mappings defined in normed spaces which are either uniformly convex or satisfy the Opial condition given by Opial [179] in 1967. Indeed, we will see that the "**(H1) Condition**" is also automatically satisfied by single-valued or set-valued nonexpansive mappings, as shown by Lemma 7.2.3 (and Lemma 7.5.2) defined in uniformly convex spaces or by Lemma 7.5.9 defined in normed spaces which satisfy the Opial condition in Chapter 7, and by Theorem 8.6.11 in Chapter 8 for nonexpansive mappings defined in locally convex spaces satisfying the "*P*-**Opial condition**", which is an extension given by Chen and Singh [57] in 1992 based on the original definition of the Opial condition for Banach spaces [179].

Now, we have the following Schauder fixed point theorem for 1-set contractive mappings in locally p-convex spaces for $p \in (0,1]$.

Theorem 6.1.5 (Schauder fixed point theorem). *Let U be a nonempty bounded open subset of a (Hausdorff) locally p-convex space E, where $p \in (0,1]$ and its zero element $\theta \in U$, and let $C \subset E$ be a closed p-convex subset of E such that $\theta \in C$. Let $F : C \cap \overline{U} \to C \cap \overline{U}$ be a continuous 1-set contractive single-valued mapping and satisfying the following "(H) Condition" or "(H1) Condition":*

(H) Condition: *The sequence $\{x_n\}_{n \in \mathbb{N}}$ in $\overline{U}$ has a convergent subsequence with limit $x_0 \in \overline{U}$ such that $x_0 \in F(x_0)$, whenever $\lim_{n \to \infty} d_{P_U}(x_n, F(x_n)) = 0$, where $d_{P_U}(x_n, F(x_n)) := P_U(x_n - F(x_n)\}$, with P_U being the Minkowski p-functional for any $U \in \mathfrak{U}$, which is the family of all nonempty open p-convex subsets of zero in E.*

(H1) Condition: *There exists x_0 in $\overline{U}$ with $x_0 = F(x_0)$ if there exists $\{x_n\}_{n \in \mathbb{N}}$ in $\overline{U}$ such that $\lim_{n \to \infty} d_{P_U}(x_n, F(x_n)) = 0$, where P_U is the Minkowski p-functional for any $U \in \mathfrak{U}$, which is the family of all nonempty open p-convex subsets of zero in E.*

Then, F has at least one fixed point in $C \cap \overline{U}$.

Proof. Let U be any element in $\mathfrak{U}$, which is the family of all nonempty open p-convex subsets for zero in E. As the mapping T is 1-set contractive, consider an increasing sequence $\{\lambda_n\}$ such that $0 < \lambda_n < 1$ and $\lim_{n \to \infty} \lambda_n = 1$, where $n \in \mathbb{N}$. Now, we define a mapping $F_n : C \to C$ by $F_n(x) := \lambda_n F(x)$ for each $x \in C$ and $n \in \mathbb{N}$. Then, it follows that F_n is a λ_n-set contractive mapping with $0 < \lambda_n < 1$. By Theorem 5.5.6 on the condensing mapping F_n in p-vector space with p-seminorm P_U for each $n \in \mathbb{N}$, there exists $x_n \in C$ such that $x_n \in F_n(x_n) = \lambda_n F(x_n)$. As P_U is the Minkowski p-functional of U in E, it follows that P_U is continuous, as $0 \in int(U) = U$. Note that for each $n \in \mathbb{N}$, $\lambda_n x_n \in \overline{U} \cap C$, which implies that $x_n = r(\lambda_n F(x_n)) = \lambda_n F(x_n)$, and thus $P_U(\lambda_n F(x_n)) \leq 1$ by Lemma 2.2.10. We note that

$$P_U(F(x_n) - x_n) = P_U(F(x_n) - \lambda_n F(x_n)) = P_U\left(\frac{(1 - \lambda_n)\lambda_n F(x_n)}{\lambda_n}\right)$$

$$\leq \left(\frac{1 - \lambda_n}{\lambda_n}\right)^p P_U(\lambda_n F(x_n)) \leq \left(\frac{1 - \lambda_n}{\lambda_n}\right)^p,$$

which implies that $\lim_{n \to \infty} P_U(F(x_n) - x_n) = 0$ for all $U \in \mathfrak{U}$.

Now, we prove the conclusion by considering the following two conditions (H) and (H1):

(i) If F satisfies the (H) condition, it implies that the consequence $\{x_n\}_{n \in \mathbb{N}}$ has a convergent subsequence which converges to x_0 such that $x_0 = F(x_0)$. Without loss of generality, we assume that $\lim_{n \to \infty} x_n = x_0$ is with $x_n = \lambda_n F(x_n)$, and $\lim_{n \to \infty} \lambda_n = 1$, which implies that $x_0 = \lim_{n \to \infty}(\lambda_n F(x_n))$ and consequently $\lim_{n \to \infty} F(x_n) = x_0$.

(ii) If F satisfies the (H1) condition, then by that condition, it follows that there exists x_0 in $\overline{U}$ such that $x_0 = F(x_0)$, which is a fixed point of F. We complete the proof. $\square$

If the mapping in Theorem 6.1.5 is not a self-mapping, we have the following best approximation result for non-self mappings under the framework of locally p-convex spaces, where $p \in (0, 1]$.

Theorem 6.1.6 (Best approximation). *Let U be a bounded open p-convex subset of a locally p-convex space E, where $p \in (0, 1]$ and the zero element $\theta \in U$, and let C be a (bounded) closed p-convex subset* of E also with zero element $\theta \in C$. Assume that $F : \overline{U} \cap C \to C$ is a continuous 1-set contractive single-valued mapping, and for each $x \in \partial_C U$ with $F(x) \in C \backslash \overline{U}$, $(P_U^{\frac{1}{p}}(F(x)) - 1)^p \leq P_U(F(x) - x)$ for $0 < p \leq 1$ (which is trivial when $p = 1$). In addition, we consider if F satisfies the following conditions, (H) or (H1):

(H) Condition: The sequence $\{x_n\}_{n \in \mathbb{N}}$ in $\overline{U}$ has a convergent subsequence with limit $x_0 \in \overline{U}$ such that $x_0 = F(x_0)$, whenever $\lim_{n \to \infty} d_{P_U}(x_n, F(x_n)) = 0$, where $d_{P_U}(x_n, F(x_n)) := \inf\{P_U(x_n - F(x_n))\}$, with P_U being the Minkowski p-functional for any $U \in \mathfrak{U}$, which is the family of all nonempty open p-convex subsets containing the zero in E.

(H1) Condition: There exists x_0 in $\overline{U}$ with $x_0 = F(x_0)$ if there exists $\{x_n\}_{n \in \mathbb{N}}$ in $\overline{U}$ such that $\lim_{n \to \infty} d_{P_U}(x_n, F(x_n)) = 0$, where P_U is the Minkowski p-functional for any $U \in \mathfrak{U}$, which is the family of all nonempty open p-convex subsets containing the zero in E.

Then, we have that there exists $x_0 \in C \cap \overline{U}$ such that

$$P_U(F(x_0) - x_0) = d_P(y_0, \overline{U} \cap C) = d_p(F(x_0), \overline{I_U^p(x_0)} \cap C),$$

where P_U is the Minkowski p-functional of U. More precisely, we have either (I) or (II) holding in the following:

(I) F has a fixed point $x_0 \in \overline{U} \cap C$, i.e., $0 = P_U(F(x_0) - x_0) = d_P(F(x_0), \overline{U} \cap C) = d_p(F(x_0), \overline{I_U^p(x_0)} \cap C)$.

(II) There exists $x_0 \in \partial_C(U)$ and $F(x_0) \notin \overline{U}$ with

$$P_U(F(x_0) - x_0) = d_P(F(x_0), \overline{U} \cap C) = d_p(F(x_0),$$

$$\overline{I_U^p(x_0)} \cap C) = (P_U^{\frac{1}{p}}(F(x_0)) - 1)^p > 0.$$

Proof. As E is a locally p-convex space E, it suffices to prove that for each open p-convex subset U in $\mathfrak{U}$ (which is the family of all nonempty open p-convex subsets containing the zero in E), there exists a sequence $(x_n)_{n \in \mathbb{N}}$ in $\overline{U}$ such that $\lim_{n \to \infty} P_U(F(x_n) - x_n) = 0$, and the conclusion follows by applying the (H) condition.

Let $r : E \to U$ be a retraction mapping defined by

$$r(x) := \frac{x}{\max\{1, (P_U(x))^{\frac{1}{p}}\}}$$

for each $x \in E$, where P_U is the Minkowski p-functional of U. Since the zero element of the space E, $\theta \in U (= intU$, as U is open), it follows that r is continuous by Lemma 2.2.10. As the mapping F is 1-set contractive, consider an increasing sequence $\{\lambda_n\}$ such that $0 < \lambda_n < 1$ and $\lim_{n \to \infty} \lambda_n = 1$, where $n \in \mathbb{N}$. Now, for each $n \in \mathbb{N}$, we define a mapping $F_n : C \cap \overline{U} \to C$ by $F_n(x) := \lambda_n F \circ r(x)$ for each $x \in C \cap \overline{U}$. By the fact that C and $\overline{U}$ are p-convex, it follows that $r(C) \subset C$ and $r(\overline{U}) \subset \overline{U}$, and thus $r(C \cap \overline{U}) \subset C \cap \overline{U}$. Therefore, F_n is a mapping from $\overline{U} \cap C$ to itself. For each $n \in \mathbb{N}$, by the fact that F_n is a λ_n-set contractive mapping with $0 < \lambda_n < 1$, it follows by Theorem 5.5.6 for the condensing mapping that there exists $z_n \in C \cap \overline{U}$ such that $F_n(z_n) = \lambda_n F \circ r(z_n)$. As $r(C \cap \overline{U}) \subset C \cap \overline{U}$, let $x_n = r(z_n)$. Then, we have that $x_n \in C \cap \overline{U}$ with $x_n = r(\lambda_n F_n(x_n))$ such that either (1) or (2) holds in the following for each $n \in \mathbb{N}$: (1) $\lambda_n F_n(x_n) \in C \cap \overline{U}$; (2) $\lambda_n F_n(x_n) \in C \backslash \overline{U}$.

Now, we prove the conclusion by considering the following two cases under the (H) and (H1) conditions:

- Case (I) For each $n \in N$, $\lambda_n F(x_n) \in C \cap \overline{U}$.
- Case (II) There exists a positive integer n such that $\lambda_n F(x_n) \in C \backslash \overline{U}$.

First, according to case (I), for each $n \in \mathbb{N}$, $\lambda_n F(x_n) \in \overline{U} \cap C$, which implies that $x_n = r(\lambda_n F(x_n)) = \lambda_n F(x_n)$, and thus $P_U(\lambda_n F(x_n)) \le 1$ by Lemma 2.2.10. Note that

$$P_U(F(x_n) - x_n) = P_U(F(x_n) - x_n) = P_U(F(x_n) - \lambda_n F(x_n))$$

$$= P_U\left(\frac{(1 - \lambda_n)\lambda_n F(x_n)}{\lambda_n}\right)$$

$$\le \left(\frac{1 - \lambda_n}{\lambda_n}\right)^p P_U(\lambda_n F(x_n)) \le \left(\frac{1 - \lambda_n}{\lambda_n}\right)^p,$$

which implies that $\lim_{n \to \infty} P_U(F(x_n) - x_n) = 0$. Now, for any $V \in \mathbb{U}$, without loss of generality, let $U_0 = V \cap U$. Then, we have the following conclusion:

$$P_{U_0}(F(x_n) - x_n) = P_{U_0}(F(x_n) - x_n) = P_{U_0}(F(x_n) - \lambda_n F(x_n))$$

$$= P_{U_0}\left(\frac{(1 - \lambda_n)\lambda_n F(x_n)}{\lambda_n}\right)$$

$$\le \left(\frac{1 - \lambda_n}{\lambda_n}\right)^p P_{U_0}(\lambda_n F(x_n)) \le \left(\frac{1 - \lambda_n}{\lambda_n}\right)^p,$$

which implies that $\lim_{n\to\infty} P_{U_0}(F(x_n) - x_n) = 0$, where P_{U_0} is the Minkowski p-functional of U_0 in E.

Now, if F satisfies the (H) condition, if follows that the consequence $\{x_n\}_{n\in\mathbb{N}}$ has a convergent subsequence which converges to x_0 such that $x_0 = F(x_0)$. Without loss of generality, we assume that $\lim_{n\to\infty} x_n = x_0$, $x_n = \lambda_n y_n$, and $\lim_{n\to\infty} \lambda_n = 1$, and as $x_0 = \lim_{n\to\infty}(\lambda_n F(x_n))$, it implies that $F(x_0) = \lim_{n\to\infty} F(x_n) = x_0$. Thus, there exists $x_0 = F(x_0)$, and thus we have $0 = d_p(x_0, F(x_0)) = d(y_0, \overline{U} \cap C) = d_p(F(x_0), \overline{I^p_{\overline{U}}(x_0)} \cap C))$, as, indeed, $x_0 = F(x_0) \in \overline{U} \cap C \subset \overline{I^p_{\overline{U}}(x_0)} \cap C)$.

If F satisfies the (H1) condition, if follows that there exists $x_0 \in \overline{U} \cap C$ with $x_0 = F(x_0)$. Then, we have $0 = P_U(F(x_0) - x_0) = d_P(F(x_0), \overline{U} \cap C) = d_p(F(x_0), \overline{I^p_{\overline{U}}(x_0)} \cap C)$.

Second, according to case (II), there exists a positive integer n such that $\lambda_n F(x_n) \in C \backslash \overline{U}$. Then, we have that $P_U(\lambda_n F(x_n)) > 1$ and also $P_U(F(x_n)) > 1$, as $\lambda_n < 1$. As $x_n = r(\lambda_n F(x_n)) = \dfrac{\lambda_n F(x_n)}{(P_U(\lambda_n F(x_n)))^{\frac{1}{p}}}$, it implies that $P_U(x_n) = 1$, and thus $x_n \in \partial_C(U)$. Note that

$$P_U(F(x_n) - x_n) = P_U\left(\frac{(P_U(F(x_n))^{\frac{1}{p}} - 1)F(x_n)}{P_U(F(x_n))^{\frac{1}{p}}}\right) = (P_U^{\frac{1}{p}}(F(x_n)) - 1)^p.$$

By the assumption, we have $(P_U^{\frac{1}{p}}(F(x_n)) - 1)^p \leq P_U(F(x_n) - x)$ for $x \in C \cap \partial \overline{U}$, which follows that

$$P_U(F(x_n)) - 1 \leq P_U(F(x_n)) - \sup\{P_U(z) : z \in C \cap \overline{U}\}$$

$$\leq \inf\{P_U(F(x_n) - z) : z \in C \cap \overline{U}\}$$

$$= d_p(F(x_n), C \cap \overline{U}).$$

Thus, we have the best approximation:

$$P_U(F(x_n) - x_n) = d_P(y_n, \overline{U} \cap C) = (P_U^{\frac{1}{p}}(F(x_n) - 1)^p > 0.$$

Now, we want to show that

$$P_U(y_n - x_n) = d_P(F(x_n), \overline{U} \cap C) = d_p(F(x_n), \overline{I^p_{\overline{U}}(x_0)} \cap C) > 0.$$

By the fact that $(\overline{U} \cap C) \subset \overline{I^p_{\overline{U}}(x_n)} \cap C$, let $z \in \overline{I^p_{\overline{U}}(x_n)} \cap C \backslash (\overline{U} \cap C)$. We first claim that $P_U(F(x_n) - x_n) \leq P_U(F(x_n) - z)$. If not, we have

$P_U(F(x_n) - x_n) > P_U(F(x_n) - z)$. As $z \in I_{\overline{U}}^p(x_n) \cap C \setminus (\overline{U} \cap C)$, there exists $y \in \overline{U}$ and a non-negative number c (actually $c \geq 1$ as will be soon shown), with $z = x_n + c(y - x_n)$. Since $z \in C$ but $z \notin \overline{U} \cap C$, it implies that $z \notin \overline{U}$. By the fact that $x_n \in U$ and $y \in \overline{U}$, we must have the constant $c \geq 1$; otherwise, it implies that $z(= (1-c)x_n + cy) \in \overline{U}$, which is impossible by our assumption, i.e., $z \notin \overline{U}$. Thus, we have that $c \geq 1$, which implies that $y = \frac{1}{c}z + (1 - \frac{1}{c})x_n \in C$ (as both $x_n \in C$ and $z \in C$). On the other hand, as $z \in I_{\overline{U}}^p(x_n) \cap C \setminus (\overline{U} \cap C)$, and $c \geq 1$ with $(\frac{1}{c})^p + (1 - \frac{1}{c})^p = 1$, combining with our assumption that for each $x \in \partial_C \overline{U}$ and $y \in F(x_n) \setminus \overline{U}$, $P_U^{\frac{1}{p}}(y) - 1 \leq P_U^{\frac{1}{p}}(y - x)$ for $0 < p \leq 1$, it then follows that

$$P_U(F(x_n) - y) = P_U \left[\frac{1}{c}(F(x_n) - z) + \left(1 - \frac{1}{c}\right)(F(x_n) - x_n) \right]$$
$$\leq \left[\left(\frac{1}{c}\right)^p P_U(F(x_n) - z) \right.$$
$$\left. + \left(1 - \frac{1}{c}\right)^p P_U(F(x_n) - x_n) \right]$$
$$< P_U(F(x_n) - x_n),$$

which contradicts that $P_U(F(x_n) - x_n) = d_P(F(x_n), \overline{U} \cap C)$. As shown above, we know that $y \in \overline{U} \cap C$, and we should have $P_U(F(x_n) - x_n) \leq P_U(F(x_n) - y)$! This helps us complete the claim $P_U(F(x_n) - x_n) \leq P_U(F(x_n) - z)$ for any $z \in I_{\overline{U}}^p(x_n) \cap C \setminus (\overline{U} \cap C)$, which means that the following best approximation of Fan type (see [84], [85]) holds:

$$0 < d_P(F(x_n), \overline{U} \cap C) = P_U(F(x_n) - x_n) = d_p(F(x_n), I_{\overline{U}}^p(x_n) \cap C).$$

Now, by the continuity of P_U, it follows that the following best approximation of Fan type is also true:

$$0 < P_U(F(x_n) - x_n) = d_P(F(x_n), \overline{U} \cap C) = d_p(F(x_n), I_{\overline{U}}^p(x_n) \cap C)$$
$$r = d_p(F(x_n), \overline{I_{\overline{U}}^p(x_n)} \cap C);$$

and we have the following conclusion due to the fact that $\lim_{n\to\infty} x_n = x_0$ and the continuity of F (actually $x_0 \neq F(x_0)$):

$$P_U(F(x_0) - x_0) = d_P(F(x_0), \overline{U} \cap C) = d_p(F(x_0), I_{\overline{U}}^p(x_0) \cap C)$$

$$= d_p(F(x_0), \overline{I_{\overline{U}}^p(x_0)} \cap C)$$

$$= (P_U^{\frac{1}{p}}(F(x_0)) - 1)^p > 0.$$

This completes the proof. □

Remark 6.1.7. We note that Theorems 6.1.5 and 6.1.6 also improve the corresponding best approximations for 1-set contractive mappings given by Li *et al.* [156], Liu [160], Xu [278], Xu *et al.* [279], and other results from the references therein. When $p = 1$, we have a similar best approximation result for the mapping F in the locally convex spaces with outward set boundary condition, which is discussed later.

Though the main focus of this chapter is on the study of the best approximations and fixed point theorems for single-valued mappings, when a p-vector space E (for $p = 1$) is a locally convex space, we can also have the following best approximation for upper semicontinuous (USC) set-valued mappings by applying Theorem 5.5.6 with arguments used in Theorems 6.1.5 and 6.1.6 above (see also more discussion given by Yuan [286–291] and the references therein).

Theorem 6.1.8 (Best approximation for USC set-valued mappings). *Let U be a bounded open convex subset of a locally convex space E (i.e., $p = 1$) with zero element $\theta \in intU = U$ (the interior $intU = U$, as U is open), and let C be a closed p-convex subset of E also with zero element $\theta \in C$. Assume that $F : \overline{U} \cap C \to 2^C$ is a USC 1-set-contractive set-valued mapping with nonempty closed convex values and satisfying condition (H) or (H1) above (in Theorem 6.1.6). Then, there exists $x_0 \in \overline{U} \cap X$ and $y_0 \in F(x_0)$ such that $P_U(y_0 - x_0) = d_P(y_0, \overline{U} \cap C) = d_p(y_0, \overline{I_U(x_0)} \cap C)$, where P_U is the Minkowski p-functional of U. More precisely, we have either (I) or (II) holding in the following:*

(I) *F has a fixed point $x_0 \in U \cap C$, i.e., $x_0 \in F(x_0)$ (so that $P_U(y_0 - x_0) = P_U(y_0 - x_0) = d_P(y_0, \overline{U} \cap C) = d_p(y_0, \overline{I_U(x_0)} \cap C)) = 0$).*

(II) *There exist $x_0 \in \partial_C(U)$ and $y_0 \in F(x_0)$, with $y_0 \notin \overline{U}$, such that*

$$P_U(y_0 - x_0) = d_P(y_0, \overline{U} \cap C) = d_p(y_0, I_{\overline{U}}(x_0) \cap C)$$
$$= d_p(y_0, \overline{I_{\overline{U}}(x_0)} \cap C) > 0.$$

Proof. By following the proof used in Theorems 6.1.5 and 6.1.6 and then applying Theorem 5.5.6 for $p = 1$, the conclusion follows. This completes the proof. $\qquad\square$

Now, by the application of Theorem 6.1.6 with Remark 6.1.7 and the argument used in Theorem 6.1.6, we have the following general principle for the existence of solutions for Birkhoff–Kellogg problems in p-seminorm spaces for locally p-convex spaces, where $0 < p \le 1$.

Theorem 6.1.9 (Principle of Birkhoff–Kellogg alternative). *Let U be a bounded open p-convex subset of a locally p-convex space E, where $p \in (0, 1]$, with zero element $\theta \in intU = (U)$ (the interior $intU$ as U is open), and let C be a closed p-convex subset of E also with zero element $\theta \in C$. Assume that $F : \overline{U} \cap C \to C$ is a continuous 1-set-contractive mapping and satisfying condition (H) or (H1) above. Then, F has at least one of the following two properties:*

(I) *F has a fixed point $x_0 \in U \cap C$ such that $x_0 = F(x_0)$;*
(II) *there exist $x_0 \in \partial_C(U)$, $F(x_0) \notin \overline{U}$, and $\lambda = \dfrac{1}{(P_U(F(x_0))^{\frac{1}{p}}} \in$*
    $(0, 1)$ such that $x_0 = \lambda F(x_0)$; In addition, if for each $x \in \partial_C U$,
    $P_U^{\frac{1}{p}}(F(x)) - 1 \le P_U^{\frac{1}{p}}(F(x) - x)$ for $0 < p \le 1$ (which is trivial when $p = 1$), then the best approximation between $\{x_0\}$ and $F(x_0)$ is given by

$$P_U(F(x_0) - x_0) = d_p(F(x_0), \overline{I_{\overline{U}}^p(x_0)} \cap C) = (P_U^{\frac{1}{p}}(F(x_0)) - 1)^p.$$

Proof. If (I) is not the case, then (II) is proved by following the proof in Theorem 6.1.6 for case (ii): $F(x_0) \notin \overline{U}$ with $F(x_0) = f(x_0)$, where f is the restriction of the continuous retraction r with respect to the set U in E defined in the proof of Theorem 6.1.6 above. Indeed, as $F(x_0) \notin \overline{U}$, it follows that $P_U(F(x_0)) > 1$, and $x_0 = f(F(x_0)) = F(x_0)\dfrac{1}{(P_U(F(x_0))^{\frac{1}{p}}}$. Now, let $\lambda = \dfrac{1}{(P_U(F(x_0))^{\frac{1}{p}}}$. Then, we have $\lambda < 1$ and $x_0 = \lambda F(x_0)$. Finally, the additional assumption in (II) allows us to obtain the best approximation between x_0 and $F(x_0)$ by following the proof of Theorem 6.1.6, as $P_U(F(x_0) - x_0) = d_P(F(x_0), \overline{U} \cap C) = d_p(F(x_0), \overline{I_{\overline{U}}^p(x_0)} \cap C) > 0$. This completes the proof. $\qquad\square$

As an application of Theorem 6.1.8 for the non-self USC set-valued mappings discussed in Theorem 6.1.9, we have the following general principle of Birkhoff–Kellogg alternative in locally convex spaces.

Theorem 6.1.10 (Principle of Birkhoff–Kellogg alternative). *Let U be a bounded open p-convex subset of a locally convex space E with the zero element $\theta \in U$, and let C be a closed convex subset of E also with zero element $\theta \in C$. Assume that $F : \overline{U} \cap C \to 2^C$ is a USC 1-set contractive set-valued mapping with nonempty closed convex values and satisfying condition (H) or (H1) above. Then, it has at least one of the following two properties:*

(I) *F has a fixed point $x_0 \in U \cap C$ such that $x_0 \in F(x_0)$.*
(II) *There exist $x_0 \in \partial_C(U)$ and $y_0 \in F(x_0)$, with $y_0 \notin \overline{U}$ and $\lambda \in (0,1)$, such that $x_0 = \lambda y_0$, and the best approximation between $\{x_0\}$ and $F(x_0)$ is given by $P_U(y_0 - x_0) = d_P(y_0, \overline{U} \cap C) = d_p(y_0, \overline{I_{\overline{U}}^p(x_0)} \cap C) > 0$.*

On the other hand, by the proof of Theorems 6.1.6, we note that for case (II) of Theorem 6.1.6, the assumption "each $x \in \partial_C U$ with $P_U^{\frac{1}{p}}(F(x) - 1 \leq P_U^{\frac{1}{p}}(F(x) - x)$" is only used to guarantee the best approximation "$P_U(F(x_0) - x_0) = d_P(F(x_0), \overline{U} \cap C) = d_p(F(x_0), \overline{I_{\overline{U}}^p(x_0)} \cap C) > 0$". Thus, we have the following Leray–Schauder alternative in p-vector spaces, which, of course, includes the corresponding results in locally convex spaces as special cases.

Theorem 6.1.11 (Leray–Schauder nonlinear alternative). *Let C a closed p-convex subset of the p-seminorm space E with $p \in (0,1]$ and the zero element $\theta \in C$. Assume the $F : C \to C$ is a continuous 1-set contractive mapping and satisfying condition (H) or (H1) above. Let $\varepsilon(F) := \{x \in C : x = \lambda F(x), \text{ for some } 0 < \lambda < 1\}$. Then, either F has a fixed point in C or the set $\varepsilon(F)$ is unbounded.*

Proof. We prove the conclusion by assuming that F has no fixed point, and then we claim that the set $\varepsilon(F)$ is unbounded. Otherwise, assume that the set $\varepsilon(F)$ is bounded and that P is the continuous p-seminorm for E. Then, there exists $r > 0$ such that the set $B(0, r) := \{x \in E : P(x) < r\}$, which contains the set $\varepsilon(F)$, i.e., $\varepsilon(F) \subset B(0, r)$, which means for any $x \in \varepsilon(F)$, $P(x) < r$. Then, $B(0.r)$ is an open p-convex subset of E and the zero element $\theta \in B(0, r)$ by Lemma 2.2.10 and Remark 2.2.12. Now, let $U := B(0, r)$ in Theorem 6.1.9; it follows that the mapping $F : B(0, r) \cap C \rightarrow C$ satisfies all the general conditions of Theorem 6.1.9, and we have that for any $x_0 \in \partial_C B(0, r)$, no $\lambda \in (0, 1)$ such that $x_0 = \lambda F(x_0)$. Indeed, for any $x \in \varepsilon(F)$, it follows that $P(x) < r$, as $\varepsilon(F) \subset B(0, r)$, but for any $x_0 \in \partial_C B(0, r)$, we have $P(x_0) = r$. Thus, conclusion (II) of Theorem 6.1.9 does not hold. By Theorem 6.1.9 again, F must have a fixed point, but this contradicts our assumption that F is fixed point free. This completes the proof. $\qquad\square$

Now, assume a given p-vector space E equipped with the P-seminorm (by assuming that it is continuous at zero) for $p \in (0, 1]$. Then, we know that $P : E \rightarrow \mathbb{R}^+$, $P^{-1}(0) = 0$, $P(\lambda x) = |\lambda|^p P(x)$ for any $x \in E$ and $\lambda \in \mathbb{R}$. Then, we have the following useful result for fixed points due to Rothe and Altman in p-vector spaces, in particular, for locally p-convex spaces, which plays an important role in optimization problems, variational inequality, and complementarity problems (see Isac [118] or Yuan [284] and the references therein for details of related studies).

Corollary 6.1.12. *Let U be a bounded open p-convex subset of a locally p-convex space E and zero element $\theta \in U$, and let C be a closed p-convex subset of E with $U \subset C$, where $p \in (0, 1]$. Assume that $F : \overline{U} \rightarrow C$ is a continuous 1-set contractive mapping and satisfying condition (H) or (H1) above. We consider if one of the following is satisfied:*

(1) *Rothe-type condition: $P_U(F(x)) \leq P_U(x)$ for $x \in \partial U$.*
(2) *Petryshyn-type condition: $P_U(F(x)) \leq P_U(F(x) - x)$ for $x \in \partial U$.*
(3) *Altman-type condition: $|P_U(F(x))|^{\frac{2}{p}} \leq [P_U(F(x)) - x)]^{\frac{2}{p}} + [P_U(x)]^{\frac{2}{p}}$ for $x \in \partial U$.*

Then, F has at least one fixed point.

Proof. By conditions (1), (2), and (3), it follows that the conclusion of (II) in Theorem 6.1.9, "there exist $x_0 \in \partial_C(U)$ and $\lambda \in (0,1)$ such that $x_0 \neq \lambda F(x_0)$", does not hold; therefore, by the alternative of Theorem 6.1.9, F has a fixed point. This completes the proof. $\square$

For $p = 1$, when a p-vector space is a locally convex space, we have the following classical Fan's best approximation [84], which is a powerful tool for nonlinear functional analysis, helpful in the study of optimization, mathematical programming, game theory, mathematical economics, and others related topics in applied mathematics.

Corollary 6.1.13 (Fan's best approximation). *Let U be a bounded open convex subset of a locally convex space E with the zero element $\theta \in U$, and let C be a closed convex subset of E also with zero element $\theta \in C$. Assume that $F : \overline{U} \cap C \to C$ is a continuous 1-set contractive mapping and satisfying condition (H) or (H1) above. Assuming that P_U is the Minkowski p-functional of U in E, then there exists $x_0 \in \overline{U} \cap X$ such that $P_U(F(x_0) - x_0) = d_P(F(x_0), \overline{U} \cap C) = d_p(F(x_0), \overline{I_{\overline{U}}(x_0)} \cap C)$.*

More precisely, we have either (I) or (II) holding in the following:

(I) *F has a fixed point $x_0 \in U \cap C$, i.e., $x_0 = F(x_0)$ (so that $0 = P_U(F(x_0) - x_0) = d_P(F(x_0), \overline{U} \cap C) = d_p(F(x_0), \overline{I_{\overline{U}}(x_0)} \cap C)$).*

(II) *There exists $x_0 \in \partial_C(U)$ and $F(x_0) \notin \overline{U}$ with*

$$P_U(F(x_0) - x_0) = d_p(F(x_0), \overline{I_{\overline{U}}(x_0)} \cap C) = P_U(F(x_0)) - 1 > 0.$$

Proof. When $p = 1$, it automatically satisfies the inequality $P_U^{\frac{1}{p}}((x)) - 1 \leq P_U^{\frac{1}{p}}(F(x) - x)$. Now, if F has no fixed points, by Theorem 6.1.9, indeed, we have that for $x_0 \in \partial_C(U)$, $P_U(F(x_0) - x_0) = d_P(F(x_0), \overline{U} \cap C) = d_p(F(x_0), \overline{I_{\overline{U}}(x_0)} \cap C) = P_U(F(x_0)) - 1$. The conclusions are given by Theorem 6.1.6 (or Theorem 6.1.10). The proof is complete. $\square$

Before concluding this section, we summarize that the best approximation and fixed point theorems for 1-set contractive single-valued mappings in locally p-convex spaces have been established. But more results will be derived in the following section under the framework of locally p-convex spaces for $p \in (0, 1]$.

We would like to point out that similar results on Rothe and Leray–Schauder alternatives have been developed by Isac [118], Park [185], Potter [213], Shahzad [240], [238], Xiao and Zhu [270], and Yuan [286–291], which could be used as tools for nonlinear analysis in p-vector spaces.

6.2 Best Approximation and Fixed Point for 1-Set Contractive Set-Valued Mappings

The goal of this section is to establish best approximations and fixed points for non-self 1-set contractive set-valued mappings in locally p-convex spaces, where $p \in (0, 1]$.

By following Definition 6.1.3 above, we first recall a key notion called "hemicompact mapping" in p-seminorm spaces.

Hemicompact Mapping: Let E be a p-vector space with p-seminorm for $p \in (0, 1]$. For a given bonded (closed) subset D in E, a mapping $F : D \to 2^E$ is said to be hemicompact if each sequence $\{x_n\}_{n \in N}$ in D has a convergent subsequence with limit x_0 such that $x_0 \in F(x_0)$, whenever $\lim_{n \to \infty} d_{P_U} P(x_n, F(x_n)) = 0$ for each $U \in \mathfrak{U}$, where $d_{P_U} P(x, C) := \inf\{P_U(x - y) : y \in C\}$ is the distance of a single point x, with the subset C in E based on P_U, and P_U is the Minkowski p-functional in E for $U \in \mathfrak{U}$, which is the base of the family consisting of all subsets of θ-neighborhoods in E.

Remark 6.2.1. We would like to point that the concept of "hemicompact mapping" for a set-valued mapping used in this chapter is an extension from a metric space to a p-vector space with the p-seminorm, where $p \in (0, 1]$, which was first introduced by Tan and Yuan [258]. By the monotonicity of Minkowski p-functionals, i.e., the larger the θ-neighborhoods, the smaller the values of the Minkowski p-functionals (see also p. 178 of Balachandran [13]), the definition of hemicompact mapping describes the converge for the distance between x_n and $F(x_n)$ by using the language of seminorms in terms of Minkowski p-functionals for each θ-neighborhood in $\mathfrak{U}$ (the base), which is the family consisting of its θ-neighborhoods in p-vector space E.

Now, we have the following Schauder fixed point theorem for 1-set contractive mappings in locally p-convex spaces, where $p \in (0, 1]$.

Theorem 6.2.2 (Schauder fixed point theorem). *Let U be a nonempty bounded open subset of a (Hausdorff) locally p-convex space E and its zero element $\theta \in U$, and let $C \subset E$ be a closed p-convex subset of E such that $\theta \in C$, with $p \in (0,1]$. We consider if $F : C \cap \overline{U} \to 2^{C \cap \overline{U}}$ is a USC 1-set contractive set-valued mappings with nonempty closed p-convex values and satisfying the following condition (H) or (H1):*

  (H) condition: *The sequence $\{x_n\}_{n \in \mathbb{N}}$ in $\overline{U}$ has a convergent subsequence with limit $x_0 \in \overline{U}$ such that $x_0 \in F(x_0)$, whenever $\lim_{n \to \infty} d_{P_U}(x_n, F(x_n)) = 0$, where $d_{P_U}(x_n, F(x_n)) := \inf\{P_U(x_n - z) : z \in F(x_n)\}$, with P_U being the Minkowski p-functional for any $U \in \mathfrak{U}$, which is the family of all nonempty open p-convex subsets containing the zero in E;*

  (H1) condition: *There exists x_0 in $\overline{U}$ with $x_0 \in F(x_0)$ if there exists $\{x_n\}_{n \in \mathbb{N}}$ in $\overline{U}$ such that $\lim_{n \to \infty} d_{P_U}(x_n, F(x_n)) = 0$, where, P_U is the Minkowski p-functional for any $U \in \mathfrak{U}$, which is the family of all nonempty open p-convex subsets containing the zero in E.*

  Then, *F has at least one fixed point in $C \cap \overline{U}$.*

Proof. Let $\mathfrak{U}$ be the family of all nonempty open p-convex subsets containing the zero in E and U be any element in $\mathfrak{U}$. As the mapping T is 1-set contractive, consider an increasing sequence $\{\lambda_n\}$ such that $0 < \lambda_n < 1$ and $\lim_{n \to \infty} \lambda_n = 1$, where $n \in \mathbb{N}$. Now, we define a mapping $F_n : C \to 2^C$ by $F_n(x) := \lambda_n F(x)$ for each $x \in C$ and $n \in \mathbb{N}$. Then, it follows that F_n is a λ_n-set contractive mapping with $0 < \lambda_n < 1$. By Theorem 5.5.6 on the condensing mapping F_n in p-vector space with p-seminorm P_U for each $n \in \mathbb{N}$, there exists $x_n \in C$ such that $x_n \in F_n(x_n) = \lambda_n F(x_n)$. Thus, there exists $y_n \in F(x_n)$ such that $x_n = \lambda_n y_n$. As P_U is the Minkowski p-functional of U in E, it follows that P_U is continuous, as $\theta \in int(U) = U$. Note that for each $n \in \mathbb{N}$, $\lambda_n x_n \in \overline{U} \cap C$, which implies that $x_n = r(\lambda_n y_n) = \lambda_n y_n$, and thus $P_U(\lambda_n y_n) \le 1$ by Lemma 2.2.10. Note that

$$P_U(y_n - x_n) = P_U(y_n - x_n) = P_U(y_n - \lambda_n y_n) = P_U\left(\frac{(1 - \lambda_n)\lambda_n y_n}{\lambda_n}\right)$$

$$\le \left(\frac{1 - \lambda_n}{\lambda_n}\right)^p P_U(\lambda_n y_n) \le \left(\frac{1 - \lambda_n}{\lambda_n}\right)^p,$$

which implies that $\lim_{n \to \infty} P_U(y_n - x_n) = 0$ for all $U \in \mathfrak{U}$.

Now,

(i) If F satisfies the (H) condition, it implies that the consequence $\{x_n\}_{n\in\mathbb{N}}$ has a convergent subsequence which converges to x_0 such that $x_0 \in F(x_0)$. Without loss of generality, we assume that $\lim_{n\to\infty} x_n = x_0$, where $y_n \in F(x_n)$ is with $x_n = \lambda_n y_n$, and $\lim_{n\to\infty} \lambda_n = 1$, implying that $x_0 = \lim_{n\to\infty}(\lambda_n y_n)$, which means $y_0 := \lim_{n\to\infty} y_n = x_0$. There exists $y_0(=x_0) \in F(x_0)$.

(ii) If F satisfies the (H1) condition, then by that condition, it follows that there exists x_0 in $\overline{U}$ such that $x_0 \in F(x_0)$, which is a fixed point of F. We complete the proof. $\square$

If the set-valued mapping in Theorem 6.2.2 is not self-mapping, then we have the following best approximation result under the framework of locally p-convex spaces for $p \in (0.1]$.

Theorem 6.2.3 (Best approximation). *Let U be a bounded open p-convex subset of a locally p-convex space E, where $p \in (0, 1]$, the zero element $\theta \in U$, and let C be a (bounded) closed p-convex subset of E also with zero element $\theta \in C$. Assume that $F : \overline{U} \cap C \to 2^C$ is a USC 1-set contractive set-valued mapping with nonempty closed p-convex values, and for each $x \in \partial_C U$ with $y \in F(x) \cap (C\backslash\overline{U}))$, $(P_U^{\frac{1}{p}}(y) - 1)^p \leq P_U(y - x)$ for $p \in (0, 1]$ (which is trivial when $p = 1$). In addition, we consider F satisfies the following condition (H) or (H1):*

(H) condition: *The sequence $\{x_n\}_{n\in\mathbb{N}}$ in $\overline{U}$ has a convergent subsequence with limit $x_0 \in \overline{U}$ such that $x_0 \in F(x_0)$, whenever $\lim_{n\to\infty} d_{P_U}(x_n, F(x_n)) = 0$, where, $d_{P_U}(x_n, F(x_n)) := \inf\{P_U(x_n - z) : z \in F(x_n)\}$, where P_U is the Minkowski p-functional for any $U \in \mathfrak{U}$, which is the family of all nonempty open p-convex subsets containing the zero in E;*

(H1) condition: *There exists x_0 in $\overline{U}$ with $x_0 \in F(x_0)$ if there exists $\{x_n\}_{n\in\mathbb{N}}$ in $\overline{U}$ such that $\lim_{n\to\infty} d_{P_U}(x_n, F(x_n)) = 0$, where P_U is the Minkowski p-functional for any $U \in \mathfrak{U}$, which is the family of all nonempty open p-convex subsets containing the zero in E.*

Then, there exist $x_0 \in C \cap \overline{U}$ and $y_0 \in F(x_0)$ such that
$$P_U(y_0 - x_0) = d_P(y_0, \overline{U} \cap C) = d_p(y_0, \overline{I_U^p(x_0)} \cap C),$$

where P_U is the Minkowski p-functional of U. More precisely, we have either (I) or (II) holding in the following:

(I) F has a fixed point $x_0 \in \overline{U} \cap C$, $x_0 \in F(x_0)$ such that $0 = P_U(x_0 - x_0) = d_P(x_0, \overline{U} \cap C) = d_p(x_0, \overline{I_{\overline{U}}^p(x_0)} \cap C)$.

(II) There exists $x_0 \in \partial_C(U)$ and $y_0 \in F(x_0) \backslash \overline{U}$ with

$$P_U(y_0 - x_0)* = d_P(y_0, \overline{U} \cap C) = d_p(y_0, \overline{I_{\overline{U}}^p(x_0)} \cap C)$$

$$= (P_{\overline{U}}^{\frac{1}{p}}(y_0) - 1)^p > 0.$$

Proof. As E is p-convex space and U is a bounded open p-convex subset of E, it suffices to prove that there exists a sequence $(x_n)_{n \in \mathbb{N}}$ in $\overline{U}$ and $y_n \in F(x_n)$ such that $\lim_{n \to \infty} P_U(y_n - x_n) = 0$, and the conclusion follows by applying the (H) condition.

Let $r : E \to U$ be a retraction mapping defined by $r(x) := \dfrac{x}{\max\{1, (P_U(x))^{\frac{1}{p}}\}}$ for each $x \in E$, where P_U is the Minkowski p-functional of U. Since the space E's zero $0 \in U (= intU$, as U is open), it follows that r is continuous by Lemma 2.2.10. As the mapping F is 1-set contractive, consider an increasing sequence $\{\lambda_n\}$ such that $0 < \lambda_n < 1$ and $\lim_{n \to \infty} \lambda_n = 1$, where $n \in \mathbb{N}$. Now, for each $n \in \mathbb{N}$, we define a mapping $F_n : C \cap \overline{U} \to 2^C$ by $F_n(x) := \lambda_n F \circ r(x)$ for each $x \in C \cap \overline{U}$. By the fact that C and $\overline{U}$ are p-convex, it follows that $r(C) \subset C$ and $r(\overline{U}) \subset \overline{U}$, and thus $r(C \cap \overline{U}) \subset C \cap \overline{U}$. Therefore, F_n is a mapping from $\overline{U} \cap C$ to itself. For each $n \in \mathbb{N}$, by the fact that F_n is a λ_n-set-contractive mapping with $0 < \lambda_n < 1$, it follows by Theorem 5.5.6 for the condensing mapping that there exists $z_n \in C \cap \overline{U}$ such that $z_n \in F_n(z_n) = \lambda_n F \circ r(z_n)$. As $r(C \cap \overline{U}) \subset C \cap \overline{U}$, let $x_n = r(z_n)$. Then, we have that $x_n \in C \cap \overline{U}$, and there exists $y_n \in F(x_n)$ with $x_n = r(\lambda_n y_n)$ such that (1) or (2) holds in the following for each $n \in \mathbb{N}$: (1) $\lambda_n y_n \in C \cap \overline{U}$; (2) $\lambda_n y_n \in C \backslash \overline{U}$.

Now, we prove the conclusion by considering the following two cases under conditions (H) and (H1):

- Case (I) For each $n \in N$, $\lambda_n y_n \in C \cap \overline{U}$.
- Case (II) There exists a positive integer n such that $\lambda_n y_n \in C \backslash \overline{U}$.

First, according to case (I), for each $n \in \mathbb{N}$, $\lambda_n y_n \in \overline{U} \cap C$, which implies that $x_n = r(\lambda_n y_n) = \lambda_n y_n$, and thus $P_U(\lambda_n y_n) \leq 1$ by Lemma 2.2.10. Note that

$$P_U(y_n - x_n) = P_U(y_n - x_n) = P_U(y_n - \lambda_n y_n)$$
$$= P_U\left(\frac{(1 - \lambda_n)\lambda_n y_n}{\lambda_n}\right)$$
$$\leq \left(\frac{1 - \lambda_n}{\lambda_n}\right)^p P_U(\lambda_n y_n) \leq \left(\frac{1 - \lambda_n}{\lambda_n}\right)^p,$$

which implies that $\lim_{n \to \infty} P_U(y_n - x_n) = 0$. Now, for any $V \in \mathbb{U}$, without loss of generality, let $U_0 = V \cap U$. Then, we have the following conclusion:

$$P_{U_0}(y_n - x_n) = P_{U_0}(y_n - x_n)$$
$$= P_{U_0}(y_n - \lambda_n y_n) = P_{U_0}\left(\frac{(1 - \lambda_n)\lambda_n y_n}{\lambda_n}\right)$$
$$\leq \left(\frac{1 - \lambda_n}{\lambda_n}\right)^p P_{U_0}(\lambda_n y_n) \leq \left(\frac{1 - \lambda_n}{\lambda_n}\right)^p,$$

which implies that $\lim_{n \to \infty} P_{U_0}(y_n - x_n) = 0$, where P_{U_0} is the Minkowski p-functional of U_0 in E.

Now, if F satisfies the (H) condition, if follows that the consequence $\{x_n\}_{n \in \mathbb{N}}$ has a convergent subsequence which converges to x_0 such that $x_0 \in F(x_0)$. Without loss of the generality, we assume that $\lim_{n \to \infty} x_n = x_0$, where $y_n \in F(x_n)$ is with $x_n = \lambda_n y_n$, and $\lim_{n \to \infty} \lambda_n = 1$, and as $x_0 = \lim_{n \to \infty} (\lambda_n y_n)$, it implies that $y_0 = \lim_{n \to \infty} y_n = x_0$. Thus, there exists $y_0 (= x_0) \in F(x_0)$, and we have $0 = d_p(x_0, F(x_0)) = d(y_0, \overline{U} \cap C) = d_p(y_0, \overline{I_U^p(x_0)} \cap C))$, as, indeed, $x_0 = y_0 \in F(x_0) \in \overline{U} \cap C \subset \overline{I_U^p(x_0)} \cap C)$.

If F satisfies the (H1) condition, if follows that there exists $x_0 \in \overline{U} \cap C$ with $x_0 \in F(x_0)$. Then, we have $0 = P_U(y_0 - x_0) = d_P(y_0, \overline{U} \cap C) = d_p(y_0, \overline{I_U^p(x_0)} \cap C)$.

Second, according to case (II), there exists a positive integer n such that $\lambda_n y_n \in C \setminus \overline{U}$. Then, we have that $P_U(\lambda_n y_n) > 1$, and also $P_U(y_n) > 1$, as $\lambda_n < 1$. As $x_n = r(\lambda_n y_n) = \dfrac{\lambda_n y_n}{(P_U(\lambda_n y_n))^{\frac{1}{p}}}$, it implies that $P_U(x_n) = 1$, and thus $x_n \in \partial_C(U)$. Note that

$$P_U(y_n - x_n) = P_U\left(\frac{(P_U(y_n)^{\frac{1}{p}} - 1)y_n}{P_U(y_n)^{\frac{1}{p}}}\right) = \left(P_U^{\frac{1}{p}}(y_n) - 1\right)^p.$$

By the assumption, we have $(P_U^{\frac{1}{p}}(y_n) - 1)^p \leq P_U(y_n - x)$ for $x \in C \cap \partial \overline{U}$, and it follows that

$$P_U(y_n) - 1 \leq P_U(y_n) - \sup\{P_U(z) : z \in C \cap \overline{U}\}$$
$$\leq \inf\{P_U(y_n - z) : z \in C \cap \overline{U}\} = d_p(y_n, C \cap \overline{U}).$$

Thus, we have the best approximation: $P_U(y_n - x_n) = d_P(y_n, \overline{U} \cap C) = (P_U^{\frac{1}{p}}(y_n) - 1)^p > 0$.

Now, we want to show that $P_U(y_n - x_n) = d_P(y_n, \overline{U} \cap C) = d_p(y_n, \overline{I_{\overline{U}}^p(x_0)} \cap C) > 0$.

By the fact that $(\overline{U} \cap C) \subset I_{\overline{U}}^p(x_n) \cap C$, let $z \in I_{\overline{U}}^p(x_n) \cap C \backslash (\overline{U} \cap C)$. We first claim that $P_U(y_n - x_n) \leq P_U(y_n - z)$. If not, we have $P_U(y_n - x_n) > P_U(y_n - z)$. As $z \in I_{\overline{U}}^p(x_n) \cap C \backslash (\overline{U} \cap C)$, there exists $y \in \overline{U}$ and a non-negative number c (actually $c \geq 1$, as will be soon shown) with $z = x_n + c(y - x_n)$. Since $z \in C$ but $z \notin \overline{U} \cap C$, it implies that $z \notin \overline{U}$. By the fact that $x_n \in \overline{U}$ and $y \in \overline{U}$, we must have the constant $c \geq 1$; otherwise, it implies that $z(= (1 - c)x_n + cy) \in \overline{U}$, which is impossible by our assumption, i.e., $z \notin \overline{U}$. Thus, we have that $c \geq 1$, which implies that $y = \frac{1}{c}z + (1 - \frac{1}{c})x_n \in C$ (as both $x_n \in C$ and $z \in C$). On the other hand, as $z \in I_{\overline{U}}^p(x_n) \cap C \backslash (\overline{U} \cap C)$, and $c \geq 1$ with $(\frac{1}{c})^p + (1 - \frac{1}{c})^p = 1$, combining with our assumption that for each $x \in \partial_C \overline{U}$ and $y \in F(x_n) \backslash \overline{U}$, $P_U^{\frac{1}{p}}(y) - 1 \leq P_U^{\frac{1}{p}}(y - x)$ for $0 < p \leq 1$, it then follows that

$$P_U(y_n - y) = P_U\left[\frac{1}{c}(y_n - z) + \left(1 - \frac{1}{c}\right)(y_n - x_n)\right]$$
$$\leq \left[\left(\frac{1}{c}\right)^p P_U(y_n - z) + \left(1 - \frac{1}{c}\right)^p P_U(y_n - x_n)\right]$$
$$< P_U(y_n - x_n),$$

which contradicts that $P_U(y_n - x_n) = d_P(y_n, \overline{U} \cap C)$. As shown above, we know that $y \in \overline{U} \cap C$, and we should have $P_U(y_n - x_n) \leq P_U(y_n - y)$! This helps us complete the claim $P_U(y_n - x_n) \leq P_U(y_n - z)$

for any $z \in I_{\overline{U}}^p(x_n) \cap C \setminus (\overline{U} \cap C)$, which means that the following best approximation of Fan type [84], [85] holds:

$$0 < d_P(y_n, \overline{U} \cap C) = P_U(y_n - x_n) = d_p(y_n, I_{\overline{U}}^p(x_n) \cap C).$$

Now, by the continuity of P_U, it follows that the following best approximation of Fan type is also true:

$$0 < P_U(y_n - x_n) = d_P(y_n, \overline{U} \cap C) = d_p(y_n, I_{\overline{U}}^p(x_n) \cap C)$$

$$= d_p(y_n, \overline{I_{\overline{U}}^p(x_n)} \cap C).$$

The proof is complete. □

Remark 6.2.4. Based on the Proof of Theorem 6.2.3, we have the following: (1) For the condition "$x \in \partial_C U$ with $y \in F(x)$, $P_{\overline{U}}^{\frac{1}{p}}(y) - 1 \le P_U^{\frac{1}{p}}(y - x)$ for $0 < p \le 1$", indeed we only need that for "$x \in \partial_C U$ with $y \in F(x) \cap (C \setminus \overline{U}))$, $P_U^{\frac{1}{p}}(y) - 1 \le P_U^{\frac{1}{p}}(y - x)$ for $p \in (0, 1]$". (2) Theorem 6.2.3 also improves the corresponding best approximation for 1-set contractive mappings given by Li *et al.* [156], Liu [160], Xu [278], Xu *et al.* [279], and results from the references therein. (3) When $p = 1$, we have a similar best approximation result for the mapping F in the locally convex spaces with outward set boundary condition as follows (see Theorem 3 of Park [185] and related discussion in the references therein).

Let $p = 1$ in Theorem 6.2.4. Then, we have the following best approximation result for USC 1-ontractive mappings with nonempty closed convex values in locally convex spaces for the outward set $\overline{O_{\overline{U}}(x_0)}$ based on the point $\{x_0\}$ with respect to the convex subset $\overline{U}$ in E.

Theorem 6.2.5 (Best approximation for outward sets). *Let U be a bounded open convex subset of a locally convex space E with zero element $\theta \in \text{int} U = U$ (the interior $\text{int} U = U$ as U is open), and let C be a closed p-convex subset of E also with zero element $\theta \in C$. Assume that $F : \overline{U} \cap C \to 2^C$ is a USC 1-set contractive set-valued mapping with nonempty closed convex values and satisfying condition (H) or (H1) above. Then, there exist $x_0 \in \overline{U} \cap X$ and $y_0 \in F(x_0)$ such that $P_U(y_0 - x_0) = d_P(y_0, \overline{U} \cap C) = d_p(y_0, \overline{O_{\overline{U}}(x_0)} \cap C)$, where P_U is*

the Minkowski p-functional of U. More precisely, we have either (I)
or (II) holding in the following:

(I) *F has a fixed point $x_0 \in U \cap C$, i.e., $0 = x_0 \in F(x_0)$ (such that*
$P_U(x_0 - x_0) = d_P(x_0, \overline{U} \cap C) = d_p(x_0, \overline{O_{\overline{U}}(x_0)} \cap C))$.
(II) *There exists $x_0 \in \partial_C(U)$ and $y_0 \in F(x_0) \backslash \overline{U}$ with*

$$P_U(y_0 - x_0) = d_P(y_0, \overline{U} \cap C) = d_p(y_0, O_{\overline{U}}(x_0) \cap C)$$

$$= d_p(y_0, \overline{O_{\overline{U}}(x_0)} \cap C) > 0.$$

Proof. We define a new mapping $F_1 : \overline{U} \cap C \to 2^C$ by $F_1(x) :=$
$\{2x\} - F(x)$ for each $x \in \overline{U} \cap C$. Then, F_1 is also a USC mapping
with nonempty closed convex values, and F_1 satisfies all hypotheses
of Theorem 6.2.3 with $p = 1$. Now, by Theorem 6.2.3, there exist
$x_0 \in \overline{U} \cap X$ and $y_1 \in F_1(x_0)$ such that $P_U(y_1 - x_0) = d_P(y_1, \overline{U} \cap C) =$
$d_p(y_1, \overline{I_{\overline{U}}(x_0)} \cap C)$. More precisely, we have either (I) or (II) holding
in the following:

(I) *F_1 has a fixed point $x_0 \in U \cap C$ and $y_0 \in F(x_0)$ (so $0 = P_U(y_1 -$*
$x_0) = P_U(y_1 - x_0) = d_P(y_1, \overline{U} \cap C) = d_p(y_1, \overline{I_{\overline{U}}(x_0)} \cap C))$.
(II) *There exists $x_0 \in \partial_C(U)$ and $y_1 \in F_1(x_0) \backslash \overline{U}$ with*

$$P_U(y_1 - x_0) = d_P(y_1, \overline{U} \cap C) = d_p(y_1, \overline{O_{\overline{U}}(x_0)} \cap C) > 0.$$

Now, for any $x \in O_{\overline{U}}(x_0)$, there exists $r < 0, u \in \overline{U}$ such that
$x = x_0 + r(u - x_0)$. Let $x_1 = 2x_0 - x$. Then, $x_1 = 2x_0 - x_0 - r(u - x_0) =$
$x_0 + (-r)(u - x_0) \in I_{\overline{U}}(x_0)$. Let $y_1 = 2x_0 - y_0$, for some $y_0 \in F(x_0)$.
As we have $P_U(y_1 - x_0) = d_P(y_1, \overline{U} \cap C) = d_p(y_1, \overline{I_{\overline{U}}(x_0)} \cap C)$, it
follows that $P_U(y_1 - x_0) \leq P_U(y_1 - x_1)$, which implies that

$$P_U(x_0 - y_0) = P_U(y_1 - x_0) \leq P_U(y_1 - x_1)$$

$$= P_U(2x_0 - y_0 - (2x_0 - x)) = P_U(y_0 - x)$$

for all $x \in O_{\overline{U}}(x_0)$. Thus, we have $P_U(y_0 - x_0) = d_P(y_0, \overline{U} \cap C) =$
$d_p(y_0, O_{\overline{U}}(x_0) \cap C)$, and by the continuity of P_U, it follows that

$$P_U(y_0 - x_0) = d_P(y_0, \overline{U} \cap C) = d_p(y_0, \overline{O_{\overline{U}}(x_0)} \cap C)(P_U^{\frac{1}{p}}(y_0) - 1)^p > 0.$$

This completes the proof. $\qquad \square$

Now, by the application of Theorem 6.2.3 and Remark 6.2.4 and the argument used in Theorem 6.2.3 (Theorem 6.1.9), we have the following general principle for the existence of solutions for Birkhoff–Kellogg Problems in p-seminorm spaces, where $p \in (0, 1]$.

Theorem 6.2.6 (Principle of the Birkhoff–Kellogg alternative). *Let U be a bounded open p-convex subset of a locally p-convex space E, where $p \in (0, 1]$, with zero element $\theta \in intU = (U)$ (the interior $intU$ as U is open), and let C be a closed p-convex subset of E also with zero element $\theta \in C$. Assume that $F : \overline{U} \cap C \to 2^C$ is a USC 1-set-contractive set-valued mapping with nonempty closed p-convex values and satisfying condition (H) or (H1) above. Then, F has at least one of the following two properties:*

(I) *F has a fixed point $x_0 \in U \cap C$ such that $x_0 \in F(x_0)$.*
(II) *There exist $x_0 \in \partial_C(U)$, $y_0 \in F(x_0) \backslash \overline{U}$, and $\lambda = \dfrac{1}{(P_U(y_0))^{\frac{1}{p}}} \in$*
    $(0, 1)$ such that $x_0 = \lambda y_0 \in \lambda F(x_0)$; in addition, if for each $x \in \partial_C U$, $P_U^{\frac{1}{p}}(y) - 1 \le P_U^{\frac{1}{p}}(y - x)$ for $0 < p \le 1$ (which is trivial when $p = 1$), then the best approximation between x_0 and y_0 is given by

$$P_U(y_0 - x_0) = d_P(y_0, \overline{U} \cap C) = d_p(y_0, \overline{I_U^p(x_0)} \cap C)$$

$$= (P_U^{\frac{1}{p}}(y_0) - 1)^p > 0.$$

Proof. If (I) is not the case, then (II) is proved by Remark 6.2.4 and by following the proof in Theorem 6.2.3 (and for the notation f used in the proof of Theorem 6.1.9), for the case (II): $y_0 \in C \backslash \overline{U}$, with $y_0 = f(x_0) \in F(x_0)$. Indeed, as $y_0 \notin \overline{U}$, and it follows that $P_U(y_0) > 1$, and $x_0 = f(y_0) = \dfrac{y_0}{(P_U(y_0))^{\frac{1}{p}}}$. Now, let $\lambda = \dfrac{1}{(P_U(y_0))^{\frac{1}{p}}}$, and we have $\lambda < 1$ and $x_0 = \lambda y_0$ with $y_0 \in F(x_0)$. Finally, the additional assumption in (II) allows us to obtain the best approximation between x_0 and y_0 by following the proof of Theorem 6.2.3, as $P_U(y_0 - x_0) = d_P(y_0, \overline{U} \cap C) = d_p(y_0, \overline{I_U^p(x_0)} \cap C) > 0$. This completes the proof. $\square$

As an application of Theorem 6.2.3 for the non-self set-valued mappings discussed in Theorem 6.2.6 with outward set conditions, we have the following general principle of Birkhoff–Kellogg alternative in locally convex spaces.

Theorem 6.2.7 (Principle of the Birkhoff–Kellogg alternative in locally convex spaces). *Let U be a bounded open convex subset of a locally convex space E with the zero element $\theta \in U$, and let C be a closed convex subset of E also with zero element $\theta \in C$. Assume that $F : \overline{U} \cap C \to 2^C$ is a USC 1-set contractive set-valued mapping with nonempty closed convex values and satisfying condition (H) or (H1) above. Then, it has at least one of the following two properties:*

(I) *F has a fixed point $x_0 \in U \cap C$ such that $x_0 \in F(x_0)$.*
(II) *There exists $x_0 \in \partial_C(U)$ and $y_0 \in F(x_0) \backslash \overline{U}$ and $\lambda \in (0,1)$ such that $x_0 = \lambda y_0$, and the best approximation between x_0 and y_0 is given by $P_U(y_0 - x_0) = d_P(y_0, \overline{U} \cap C) = d_p(y_0, \overline{I_{\overline{U}}^p(x_0)} \cap C) > 0$.*

Here, we note that by the proof of Theorems 6.2.3, it is easy to see that for case (II) of Theorem 6.2.3, the assumption "each $x \in \partial_C U$ with $y \in F(x)$, $P_U^{\frac{1}{p}}(y) - 1 \le P_U^{\frac{1}{p}}(y - x)$" is only used to guarantee the best approximation "$P_U(y_0 - x_0) = d_P(y_0, \overline{U} \cap C) = d_p(y_0, \overline{I_{\overline{U}}^p(x_0)} \cap C) > 0$". Thus, we have the following Leray–Schauder alternative in p-vector spaces, which, of course, includes the corresponding results in locally convex spaces as special cases.

Theorem 6.2.8 (Leray–Schauder nonlinear alternative). *Let C a closed p-convex subset of the p-seminorm space E, where $p \in (0,1]$ and the zero element $\theta \in C$. Assume that $F : C \to 2^C$ is a USC 1-set contractive set-valued mapping with nonempty closed p-convex values and satisfying condition (H) or (H1) above. Let $\varepsilon(F) := \{x \in C : x \in \lambda F(x), \text{ for some } 0 < \lambda < 1\}$. Then, either F has a fixed point in C or the set $\varepsilon(F)$ is unbounded.*

Proof. We prove the conclusion by assuming that F has no fixed point, and then we claim that the set $\varepsilon(F)$ is unbounded. Otherwise, assume that the set $\varepsilon(F)$ is bounded and that P is the continuous p-seminorm for E. Then, there exists $r > 0$ such that the set $B(0, r) := \{x \in E : P(x) < r\}$, which contains the set $\varepsilon(F)$, i.e., $\varepsilon(F) \subset B(0, r)$, which means for any $x \in \varepsilon(F)$, $P(x) < r$. Then, $B(0.r)$ is an open p-convex subset of E and the zero $0 \in B(0, r)$ by Lemma

2.2.10 and Remark 2.2.12. Now, let $U := B(0, r)$ in Theorem 6.2.6, and it follows that for the mapping $F : B(0, r) \cap C \to 2^C$ satisfies all general conditions of Theorem 6.2.6, and we have that for any $x_0 \in \partial_C B(0, r)$, no $\lambda \in (0, 1)$ such that $x_0 = \lambda y_0$, where $y_0 \in F(x_0)$. Indeed, for any $x \in \varepsilon(F)$, it follows that $P(x) < r$, as $\varepsilon(F) \subset B(0, r)$, but for any $x_0 \in \partial_C B(0, r)$, we have $P(x_0) = r$, and thus conclusion (II) of Theorem 6.2.6 does not hold. By Theorem 6.2.6 again, F must have a fixed point, but this contradicts our assumption that F is fixed point free. This completes the proof. $\square$

Now, assume a given p-vector space E equipped with the P-seminorm (by assuming that it is continuous at zero) for $p \in (0, 1]$. We know that $P : E \to \mathbb{R}^+$, $P^{-1}(0) = 0$, $P(\lambda x) = |\lambda|^p P(x)$ for any $x \in E$ and $\lambda \in \mathbb{R}$. Then, we have the following useful result for fixed points due to Rothe and Altman in p-vector spaces, which plays an important role in optimization problems, variational inequality, and complementarity problems (see Isac [118], or Yuan [284], and the references therein for details on related studies).

Corollary 6.2.9. *Let U be a bounded open p-convex subset of a locally p-convex space E and zero element $\theta \in U$, and let C be a closed p-convex subset of E with $U \subset C$, where $p \in (0, 1]$. Assume that $F : \overline{U} \to 2^C$ is a USC 1-set contractive set-valued mapping with nonempty closed p-convex values and satisfying condition (H) or (H1) above. We consider if one of the following is satisfied:*

(1) *(Rothe-type condition)*: $P_U(y) \leq P_U(x)$ *for* $y \in F(x)$, *where* $x \in \partial U$.
(2) *(Petryshyn-type condition)*: $P_U(y) \leq P_U(y - x)$ *for* $y \in F(x)$, *where* $x \in \partial U$.
(3) *(Altman-type condition)*: $|P_U(y)|^{\frac{2}{p}} \leq [P_U(y) - x)]^{\frac{2}{p}} + [P_U(x)]^{\frac{2}{p}}$ *for* $y \in F(x)$, *where* $x \in \partial U$.

Then, F has at least one fixed point.

Proof. By the conditions (1), (2), and (3), it follows that the conclusion of (II) in Theorem 6.2.6, "there exist $x_0 \in \partial_C(U)$ and $\lambda \in (0, 1)$ such that $x_0 \notin \lambda F(x_0)$", does not hold. Thus, by the alternative of Theorem 6.2.6, F has a fixed point. This completes the proof. $\square$

When $p = 1$, each locally p-convex space is a locally convex space, and thus we have the following classical Fan's best approximation [84] as a powerful tool for the study of optimization, mathematical programming, game theory, mathematical economics, and others related topics in applied mathematics.

Corollary 6.2.10 (Fan's best approximation). *Let U be a bounded open convex subset of a locally convex space E with the zero element $\theta \in U$, and let C be a closed convex subset of E also with zero element $\theta \in C$. Assume that $F : \overline{U} \cap C \to 2^C$ is a USC 1-set contractive set-valued mapping with nonempty closed convex values and satisfying condition (H) or (H1) above. If we assume that P_U is the Minkowski p-functional of U in E, then there exist $x_0 \in \overline{U} \cap X$ and $y_0 \in T(x_0)$ such that $P_U(y_0 - x_0) = d_P(y_0, \overline{U} \cap C) = d_p(y_0, \overline{I_{\overline{U}}(x_0)} \cap C)$. More precisely, we have either (I) or (II) holding in the following, where $W_{\overline{U}}(x_0)$ is either the inward set $I_{\overline{U}}(x_0)$ or the outward set $O_{\overline{U}}(x_0)$:*

(I) *F has a fixed point $x_0 \in U \cap C$, $0 = P_U(y_0 - x_0) = P_U(y_0 - x_0) = d_P(y_0, \overline{U} \cap C) = d_p(y_0, \overline{W_{\overline{U}}(x_0)} \cap C))$.*

(II) *There exists $x_0 \in \partial_C(U)$ and $y_0 \in F(x_0) \backslash \overline{U}$ with*

$$P_U(y_0 - x_0) = d_P(y_0, \overline{U} \cap C) = d_p(y_0, \overline{W_{\overline{U}}(x_0)} \cap C)$$
$$= P_U(y_0) - 1 > 0.$$

Proof. When $p = 1$, then it automatically satisfies the inequality $P_U^{\frac{1}{p}}(y) - 1 \leq P_U^{\frac{1}{p}}(y - x)$, and indeed we have that for $x_0 \in \partial_C(U)$, with $y_0 \in F(x_0)$, $P_U(y_0 - x_0) = d_P(y_0, \overline{U} \cap C) = d_p(y_0, \overline{W_{\overline{U}}(x_0)} \cap C) = P_U(y_0) - 1$. The conclusions are given by Theorem 6.2.3 (or Theorem 6.2.6). The proof is complete. $\qquad\square$

We would like to point out that similar results on Rothe and Leray–Schauder alternatives have been developed by Isac [118], Park [185], Potter [213], Shahzad [238, 240], Xiao and Zhu [270], and related references therein as tools for nonlinear analysis in topological vector spaces. As mentioned above, when $p = 1$ and taking F as a continuous mapping, we obtain the version of Leray–Schauder in TVS, but we omit its detailed statement.

Remark 6.2.11. In concluding this chapter, we summarize that the general best approximation and fixed point theorems for non-self 1-set contractive single-valued and set-valued mappings are established in locally p-convex spaces. These results will be used in the following two chapters to establish some useful tools for nonlinear analysis under the framework of locally p-convex spaces, where $p \in (0, 1]$.

Chapter 7

Nonlinear Analysis of Single-Valued Mappings in Locally p-Convex Spaces

The goal of this chapter is to develop some new results and tools that can be used in nonlinear functional analysis under the framework of locally p-convex spaces for single-valued mappings which may be 1-set contractive, semiclosed 1-set contractive, or nonexpansive. This chapter consists of five sections, which are briefly introduced as follows.

The first section discusses the general principle of Birkhoff–Kellogg-type alternatives and related fixed point results for non-self mappings of single-valued 1-set contractive mappings. Section 7.2 is concerned with the study of fixed point theorems for non-self mappings by combining with the demiclosedness principle under the framework of spaces which are either uniformly convex or satisfy the so-called Opial condition. In Section 7.3, we study fixed point theorems for non-self semiclosed 1-set contractive mappings under the general conditions. Section 7.4 discusses the alternative principle and related fixed point theorems for non-self semiclosed 1-set contractive mappings associated with various boundary conditions in locally p-convex spaces, where $p \in (0, 1]$. In the final section, we focus on the study of fixed point theorems for non-self semiclosed 1-set contractive and nonexpansive mappings.

The general results established in this chapter unify the corresponding results in the existing literature on the study of fixed points and related nonlinear alternative principles in nonlinear functional analysis.

7.1 Principle of Alternative in Locally p-Convex Spaces

As applications of the results established in Chapter 6, the goal of this section is to establish the general existence of solutions for Birkhoff–Kellogg problems and the principle of Leray–Schauder alternatives in locally p-convex spaces, where $p \in (0, 1]$.

For the convenience of our discussion, starting with this section and in the following chapters, for a given vector space E, we always denote by $CB(E)$, $K(E)$, and $KC(E)$ the family of nonempty closed bounded subsets, nonempty compact subsets, and nonempty compact convex subsets of E, respectively, unless specified otherwise.

We now first discuss the following result related to the general principle of alternative for non-self mappings in locally p-convex spaces. We note that though Theorem 7.1.1 is a special case of Theorem 6.2.6, here we give its direct proof in detail for the convenience of the reader.

Theorem 7.1.1 (Birkhoff–Kellogg alternative in locally p-convex spaces). *Let U be a bounded open p-convex subset of a locally p-convex space E, where $p \in (0, 1]$, with the zero element $\theta \in U$, and let C be a closed p-convex subset of E also with zero element $\theta \in C$, and assume that $F : \overline{U} \cap C \to C$ is a continuous 1-set contractive single-valued mapping and satisfying condition (H) or (H1) above. In addition, for each $x \in \partial_C(U)$,*
$$P_U^{\frac{1}{p}}(F(x)) - 1 \le P_U^{\frac{1}{p}}(F(x) - x) \text{ for } 0 < p \le 1 \text{ (which is trivial when}$$
$p = 1$), where P_U is the Minkowski p-functional of U. Then, we have that either (I) or (II) holding in the following:

(I) *there exists $x_0 \in \overline{U} \cap C$ such that $F(x_0) = x_0$;*
(II) *there exists $x_0 \in \partial_C(U)$ with $F(x_0) \notin \overline{U}$ and $\lambda > 1$ such that $\lambda x_0 = F(x_0)$, i.e., $F(x_0) \in \{\lambda x_0 : \lambda > 1\} \neq \emptyset$.*

Proof. By following the argument and symbols used in the proof of Theorem 6.2.3 (and the notation for f used in the proof of Theorem 6.1.9), we have that either:

(1) F has a fixed point $x_0 \in \overline{U} \cap C$; or
(2) there exists $x_0 \in \partial_C(U)$ and $x_0 = f(F(x_0))$ such that

$$P_U(F(x_0) - x_0) = d_p(F(x_0), \overline{I_{\overline{U}}(x_0)} \cap C) = P_U(F(x_0)) - 1 > 0,$$

where $\partial_C(U)$ denotes the boundary of U relative to C in E and f is the restriction of the continuous retraction r with respect to the set U in E defined in the proof of Theorem 6.2.3 above.

If F has no fixed point, then (2) above holds and $x_0 \neq F(x_0)$. As given by the proof of Theorem 6.2.3, we have that $F(x_0) \notin \overline{U}$, and thus $P_U(F(x_0)) > 1$ and $x_0 = f(F(x_0)) = \dfrac{F(x_0)}{(P_U(F(x_0))^{\frac{1}{p}}}$, which means $F(x_0) = (P_U(F(x_0))^{\frac{1}{p}} x_0$. Let $\lambda = (P_U(F(x_0)))^{\frac{1}{p}}$. Then, $\lambda > 1$, and we have $\lambda x_0 = F(x_0)$. This completes the proof. $\qquad \square$

Theorem 7.1.2 (Birkhoff–Kellogg alternative in LCS). *Let U be a bounded open convex subset of a locally convex space (LCS) E with the zero element $\theta \in U$, and let C be a closed convex subset of E also with zero element $\theta \in C$. Assume that $F : \overline{U} \cap C \to C$ is a continuous 1-set contractive single-valued mapping and satisfying condition (H) or (H1) above. Then, we have either (I) or (II) holding in the following:*

(I) *There exists $x_0 \in \overline{U} \cap C$ such that $x_0 = F(x_0)$.*
(II) *There exists $x_0 \in \partial_C(U)$ with $F(x_0) \notin \overline{U}$ and $\lambda > 1$ such that $\lambda x_0 = F(x_0)$, i.e., $F(x_0) \in \{\lambda x_0 : \lambda > 1\} \neq \emptyset$.*

Proof. When $p = 1$, then it automatically satisfies the inequality $P_U(F(x)) - 1 \leq P_U(F(x_0) - x)$, and indeed we have that for $x_0 \in \partial_C(U)$, $P_U(F(x_0) - x_0) = d_P(F(x_0), \overline{U} \cap C) = d_p(F(x_0), \overline{W_{\overline{U}}(x_0)} \cap C) = P_U(F(x_0)) - 1$. The conclusions are given by Theorem 7.1.1 or Theorem 6.2.6. The proof is complete. $\qquad \square$

Indeed, we have the following fixed points for non-self mappings in p-vector spaces for $p \in (0, 1]$ under various boundary conditions in locally p-convex spaces.

Theorem 7.1.3 (Fixed points for non-self mappings in locally p-convex space). *Let U be a bounded open p-convex subset of a locally p-convex space E, where $p \in (0, 1]$ with the zero element $\theta \in U$, and let C be a closed p-convex subset of E also with zero element $\theta \in C$. Assume that $F : \overline{U} \cap C \to C$ is a continuous 1-set contractive single-valued mapping and satisfying condition (H) or (H1) above. In* *addition, for each* $x \in \partial_C(U)$, $(P_U^{\frac{1}{p}}(F(x)) - 1)^p \leq P_U^{\frac{1}{p}}(F(x) - x)$ *for* $0 < p \leq 1$ *(which is trivial when $p = 1$), where P_U is the Minkowski p-functional of U. We consider if F satisfies any one of the following conditions for any $x \in \partial_C(U) \backslash F(x)$:*

(i) $P_U(F(x) - z) < P_U(F(x) - x)$ *for some* $z \in \overline{I_{\overline{U}}(x)} \cap C$;

(ii) *there exists* λ *with* $|\lambda| < 1$ *such that* $\lambda x + (1-\lambda)F(x) \in \overline{I_{\overline{U}}(x)} \cap C$;

(iii) $F(x) \in \overline{I_{\overline{U}}(x)} \cap C$;

(iv) $F(x) \in \{\lambda x : \lambda > 1\} = \emptyset$ *(i.e., Leray–Schauder boundary condition)*;

(v) $F(\partial U) \subset \overline{U} \cap C$;

(vi) $P_U(F(x) - x) \neq ((P_U(F(x))^{\frac{1}{p}} - 1)^p$.

Then, F must has a fixed point.

Proof. By following the argument and symbols used in the proof of Theorem 6.2.3 (see also Theorem 6.2.6), we have that either:

(1) F has a fixed point $x_0 \in \overline{U} \cap C$; or

(2) there exists $x_0 \in \partial_C(U)$, with $x_0 = f(F(x_0))$, such that

$$P_U(F(x_0) - x_0) = d_P(F(x_0), \overline{U} \cap C) = d_p(F(x_0), \overline{I_{\overline{U}}(x_0)} \cap C)$$
$$= P_U(F(x_0)) - 1 > 0,$$

where $\partial_C(U)$ denotes the boundary of U relative to C in E and f is the restriction of the continuous retraction r with respect to the set U in E.

First, suppose that F satisfies condition (i). If F has no fixed point, then (2) holds and $x_0 \neq F(x_0)$. Then, by condition (i), it follows that $P_U(F(x_0) - z) < P_U(F(x_0) - x_0)$ for some $z \in \overline{I_{\overline{U}}(x)} \cap C$, which contradicts the best approximation equation given by (2) above, and thus F must have a fixed pint.

Second, suppose that F satisfies condition (ii). If F has no fixed point, then (2) holds and $x_0 \neq F(x_0)$. Then, by condition (ii), there exists $|\lambda| < 1$ such that $\lambda x_0 + (1 - \lambda)F(x_0) \in \overline{I_{\overline{U}}(x)} \cap C$. It follows that

$$P_U(F(x_0) - x_0) \leq P_U(F(x_0) - (\lambda x_0 + (1 - \lambda F(x_0)))$$
$$= P_U(\lambda(F(x_0) - x_0))$$
$$= |\lambda|^p P_U(F(x_0) - x_0) < P_U(F(x_0) - x_0),$$

which is impossible, and thus F must have a fixed point in $\overline{U} \cap C$.

Third, suppose that F satisfies condition (iii), i.e., $F(x) \in I_{\overline{U}}(x) \cap C$. Then, by (2), we have that $P_U(F(x_0) - x_0) = 0$, and thus $x_0 = F(x_0)$, which means F has a fixed point.

Fourth, suppose that F satisfies condition (iv). If F has no fixed point, then (2) holds and $x_0 \neq F(x_0)$. As given by the proof of Theorem 6.2.3, we have that $F(x_0) \notin \overline{U}$, thus $P_U(F(x_0)) > 1$ and $x_0 = f(F(x_0)) = \dfrac{F(x_0)}{(P_U(F(x_0)))^{\frac{1}{p}}}$, which means $F(x_0) = (P_U(F(x_0)))^{\frac{1}{p}} x_0$, where $(P_U(F(x_0)))^{\frac{1}{p}} > 1$, which contradicts assumption (iv), and thus F must have a fixed point in $\overline{U} \cap C$.

Fifth, suppose that F satisfies condition (v). Then, $x_0 \neq F(x_0)$. As $x_0 \in \partial_C U$, now by condition (v), we have that $F(\partial U) \subset \overline{U} \cap C$, and it follows that for $F(x_0)$, we have $F(x_0) \in \overline{U} \cap C$. Thus, $F(x_0) \notin \overline{U} \backslash C$, which implies that $0 < P_U(F(x_0) - x_0) = d_P(F(x_0), \overline{U} \cap C) = 0$, which is impossible, and thus F must have a fixed point. Here, as mentioned in Remark 6.2.4, based on condition (v), applying $F(\partial U) \subset \overline{U} \cap C$ is enough to know that the mapping F has a fixed point, making the following general hypothesis unnecessary: "for each $x \in \partial_C(U)$, $P_U^{\frac{1}{p}}(F(x)) - 1 \leq P_U^{\frac{1}{p}}(F(x) - x)$ for $0 < p \leq 1$".

Finally, suppose that F satisfies condition (vi). If F has no fixed point, then (2) holds and $x_0 \neq F(x_0)$. Then, condition (v) implies that $P_U(F(x_0) - x_0) \neq ((P_U(F(x_0))^{\frac{1}{p}} - 1)^p$; however, our proof of Theorem 6.2.3 shows that $P_U(F(x_0) - x_0) = ((P_U(F(x_0)))^{\frac{1}{p}} - 1)^p$, which is impossible, and thus F must have a fixed point. Then, the proof is complete. $\square$

Now, by taking the set C in Theorem 7.1.1 as the whole locally p-convex space E itself, we have the following general results for non-self continuous mappings, which include results of the fixed point types given by Rothe, Petryshyn, Altman, and Leray–Schauder as special cases in locally convex spaces.

Taking $p = 1$ and $C = E$ in Theorem 7.1.3, we have the following fixed points for non-self single-valued mappings in locally convex spaces, and the corresponding results for upper semicontinuous set-valued mappings are discussed by Yuan [286–291] and related references therein.

Theorem 7.1.4 (Fixed point for non-self mappings with boundary conditions). *Let U be a bounded open convex subset of a locally convex space E with the zero element $\theta \in U$, and assume that $F : \overline{U} \to E$ is a continuous 1-set contractive single-valued mapping and satisfying condition (H) or (H1) above. We consider if F satisfies any one of the following conditions for any $x \in \partial(U)\backslash F(x)$:*

(i) $P_U(F(x) - z) < P_U(F(x) - x)$ *for some* $z \in \overline{I_{\overline{U}}(x)}$;
(ii) *there exists λ with $|\lambda| < 1$ such that* $\lambda x + (1 - \lambda)F(x) \in \overline{I_{\overline{U}}(x)}$;
(iii) $F(x) \in \overline{I_{\overline{U}}(x)}$;
(iv) $F(x) \in \{\lambda x : \lambda > 1\} = \emptyset$;
(v) $F(\partial(U) \subset \overline{U}$;
(vi) $P_U(F(x) - x) \neq P_U(F(x)) - 1$.

Then, F must have a fixed point.

In what follows, based on the best approximation theorem in p-seminorm space, we also give some fixed point theorems for non-self mappings with various boundary conditions which are related to the study of the existence of solutions for partial differential equations (PDEs) and differential equations with boundary problems (see Browder [32], Petryshyn [205, 206], and Reich [222]), which play a key role in nonlinear analysis for p-seminorm space, as shown in the following.

First, as discussed in Remark 6.2.4, the proof of Theorem 6.2.3, with only the strong boundary condition "$F(\partial(U)) \subset \overline{U} \cap C$", we can prove that F has a fixed point, and thus we have the following fixed point theorem of Rothe type in p-vector spaces.

Theorem 7.1.5 (Rothe type). *Let U be a bounded open p-convex subset of a locally p-convex space E, where $p \in (0, 1]$ with the zero element $\theta \in U$. Assume that $F : \overline{U} \to E$ is a continuous 1-set contractive single-valued mapping, satisfying condition (H) or (H1) above, and such that $F(\partial(U)) \subset \overline{U}$. Then, F must has a fixed point.*

Now, as applications of Theorem 7.1.5, we give the following Leray–Schauder alternative in locally p-convex spaces for non-self mappings associated with boundary conditions, which often appear in applications (see Isac [118] and the references therein for the study of complementary problems and related topics in optimization).

Theorem 7.1.6 (Leray–Schauder alternative in locally p-convex spaces). *Let E be a locally p-convex space, where $p \in (0,1]$, $B \subset E$ is a bounded closed p-convex such that $\theta \in \mathrm{int}B$. Let $F : [0,1] \times B \to E$ be a continuous 1-set contractive single-valued mapping, satisfying condition (H) or (H1) above, and such that the set $F([0,1] \times B)$ being relatively compact in E. We consider if the following assumptions are satisfied:*

(1) $x \neq F(t,x)$ for all $x \notin \partial B$ and $t \in [0,1]$;
(2) $F(\{0\} \times \partial B) \subset B$.

Then, there is an element $x^* \in B$ such that $x^* = F(1, x^*)$.

Proof. For $n \in \mathbb{N}$, we consider the mapping

$$
F_n(x) = \begin{cases} F(\dfrac{1 - P_B(x)}{\epsilon_n}, \dfrac{x}{P_B(x)}) & \text{if } 1 - \epsilon \leq P_B(x) \leq 1, \\[3mm] F(1, \dfrac{X}{1 - \epsilon_n}) & \text{if } P_B(x) < 1 - \epsilon_n, \end{cases} \tag{7.1}
$$

where P_B is the Minkowski p-functional of B and $\{\epsilon_n\}_{n \in N}$ is a sequence of real numbers such that $\lim_{n \to \infty} \epsilon_n = 0$ and $0 < \epsilon_n < \frac{1}{2}$ for any $n \in \mathbb{N}$, and we also observe that the mapping F_n is 1-set contractive continuous with nonempty closed p-convex values on B. From assumption (2), we have that $F_n(\partial B) \subset B$, and the assumptions of Theorem 7.1.5 are satisfied. Then, for each $n \in \mathbb{N}$, there exists an element $u_n \in B$ such that $u_n = F_n(u_n)$.

We first prove the statement, "It is impossible to have an infinite number of the elements u_n satisfy the inequality: $1 - \epsilon_n \leq P_B(u_n) \leq 1$."

If not, we assume to have an infinite number of the elements u_n that satisfy the following inequality:

$$
1 - \epsilon_n \leq P_B(u_n) \leq 1.
$$

As $F_n(B)$ is relatively compact and by the definition of mappings F_n, we have that $\{u_n\}_{n\in\mathbb{N}}$ is contained in a compact set in E. Without loss of generality (indeed, each compact set is also countably compact), we define the sequence $\{t_n\}_{n\in\mathbb{N}}$ by $t_n := \frac{1-P_B(u_n)}{\epsilon}$ for each $n \in N$. Then, we have that $\{t_n\}_{n\in\mathbb{N}} \subset [0,1]$, and we may assume that $\lim_{n\to\infty} t_n = t \in [0,1]$. The corresponding subsequence of $\{u_n\}_{n\in\mathbb{N}}$ is denoted again by $\{u_n\}_{n\in\mathbb{N}}$, and it also satisfies the inequality $1 - \epsilon_n \le P_B(u_n) \le 1$, which implies that $\lim_{n\to\infty} P_B(u_n) = 1$.

Now, let u^* be an accumulation point of $\{u_n\}_{n\in\mathbb{N}}$, and we thus have $\lim_{n\to\infty}(t_n, \frac{u_n}{P_B(u_n)}, u_n) = (t, u^*, u^*)$. By the fact that F is compact, we can assume that $u_n = F(t_n, \frac{u_n}{P_B(u_n)})$ for each $n \in \mathbb{N}$. It follows that $u^* = F(t, u^*)$, which contradicts assumption (1), as we have $\lim_{n\to\infty} P_B(u_n) = 1$ (which means that $u^* \in \partial B$, which is impossible).

Thus, it is impossible to have that "to have an infinite number of elements u_n satisfy the inequality: $1 - \epsilon_n \le P_B(u_n) \le 1$", which means that there is only a finite number of elements of sequence $\{u_n\}_{n\in N}$ satisfying the inequality: $1-\epsilon_n \le P_B(u_n) \le 1$. Now, without loss of generality, for $n \in \mathbb{N}$, we have the following inequality:

$$P_B(u_n) < 1 - \epsilon_n.$$

By the fact that $\lim_{n\to\infty}(1 - \epsilon_n) = 1$, $u_n \in F(1, \frac{u_n}{1-\epsilon})$ for all $n \in \mathbb{N}$ and assuming that $\lim_{n\to} u_n = u^*$, the continuity of F with nonempty closed values implies that by $u_n = F(1, \frac{u_n}{1-\epsilon})$ for each $n \in \mathbb{N}$, $u^* = F(1, u^*)$. This completes the proof. □

As a special case of Theorem 7.1.6, we have the following principle for the implicit form of Leray–Schauder-type alternative in locally p-convex spaces for $0 < p \le 1$.

Corollary 7.1.7 (Implicit Leray–Schauder alternative). *Let E be a locally p-convex space, where $p \in (0,1]$ and $B \subset E$ is a bounded closed p-convex such that $\theta \in int B$. Let $F : [0,1] \times B \to E$ be 1-set contractive and continuous, satisfying condition (H) or (H1) above, and the set $F([0,1] \times B)$ be relatively compact in E. We consider if the following assumptions are satisfied:*

(1) $F(\{0\} \times \partial B) \subset B$;
(2) $x \ne F(0, x)$ for all $x \in \partial B$.

Then, at least one of the following properties is satisfied:

(i) There exists $x^* \in B$ such that $x^* = F(1, x^*)$.
(ii) There exists $(\lambda^*, x^*) \in (0, 1) \times \partial B$ such that $x^* = F(\lambda^*, x^*)$.

Proof. The result is an immediate consequence of Theorem 7.1.6. This completes the proof. □

We would like to point out that similar results on Rothe and Leray–Schauder alternatives have been developed by Furi and Pera [92], Granas and Dugundji [102], Górniewicz [100], Górniewicz *et al.* [101], Isac [118], Li *et al.* [156], Liu [160], Park [185], Potter [213], Shahzad [238,240], Xu [278], Xu *et al.* [279], and related references therein. These are used as tools for nonlinear analysis in the Banach space setting and applications to boundary value problems for ordinary differential equations in noncompact problems, a general class of mappings for nonlinear alternative of Leray–Schauder type in normal topological spaces. Some Birkhoff–Kellogg-type theorems for general class mappings in topological vector spaces are also established by Agarwal *et al.* [1], Agarwal and O'Regan [2,3], Park [186], and the references therein. In particular, recently, O'Regan [180] used the Leray–Schauder-type coincidence theory to establish some Birkhoff–Kellogg problems and Furi–Pera-type results for a general class of 1-set contractive mappings.

Before concluding this section, we would like to share with readers that as the application of the best approximation result for 1-set contractive mappings, we established some fixed point theorems and the general principle of Leray–Schauder alternative for non-self mappings. These can play important roles for nonlinear analysis under the framework of locally *p*-convex (seminorm) spaces. This can lead us to the further development of nonlinear analysis under a framework of locally topological vector spaces, normed spaces, or Banach spaces.

7.2 Fixed Point Theorems for 1-Set Contractive Mappings

The goal of this section is to establish general fixed point theorems for non-self mappings by combining with the demiclosedness principle under the framework of spaces which are either uniformly convex or satisfy the so-called Opial condition. In this section, based on the best approximation Theorem 6.2.3 for classes of 1-set contractive mappings developed in Chapter 6, we show how it can be used as a useful tool to establish fixed point theorems for non-self upper semicontinuous mappings in locally p-convex (seminorm) spaces for $p \in (0,1]$, which include norm spaces and uniformly convex Banach spaces as special classes.

Following Browder [32], Li [155], Goebel and Kirk [96], Petryshyn [205, 206], Tan and Yuan [258], Xu [278], and the references therein, we first recall the following definition under the framework of p-seminorm spaces for $p \in (0,1]$.

Definition 7.2.1. Let D be a nonempty (bounded) closed subset of locally p-convex spaces $(E, \| \cdot \|_p)$, where $p \in (0,1]$. Suppose $f : D \to X$ is a (single-valued) mapping. Then, we recall that:

(1) f is said to be nonexpansive if for each $x, y \in D$, we have $\|f(x) - f(y)\|_p \le \|x - y\|_p$;

(2) f is said to be demiclosed (actually, the mapping $(I - f)$ is closed at (x_0, y_0) under the product topology $(E, \sigma(E, E^*)) \times (E, \| \cdot \|_p)$, where I denotes the identity on E, $\sigma(E, E^*)$ is the weak topology, and $\| \cdot \|_p$ is the p-seminorm (or strong) topology), (see Browder [32]) at $y \in X$ if for any sequence $\{x_n\}_{n \in \mathbb{N}}$ in D, the conditions $x_n \to x_0 \in D$ weakly and $(I - f)(x_n) \to y_0$ strongly imply that $(I - f)(x_0) = y_0$, where I is the identity mapping;

(3) f is said to be hemicompact (which is Definition 6.1.3 in Chapter 6) if each sequence $\{x_n\}_{n \in \mathbb{N}}$ in D has a convergent subsequence with the limit x_0 such that $x_0 = f(x_0)$, whenever $\lim_{n \to \infty} d_p(x_n, f(x_n)) = 0$, where $d_P(x_n, f(x_n)) := \inf\{P_U(x_n - z) : z \in f(x_n)\}$ and P_U is the Minkowski p-functional for any $U \in \mathfrak{U}$, which is the family of all nonempty open p-convex subsets containing the zero in E;

(4) f is said to be demicompact (by Petryshyn [205]) if each sequence $\{x_n\}_{n \in \mathbb{N}}$ in D has a convergent subsequence whenever $\{x_n - f(x_n)\}_{n \in \mathbb{N}}$ is a convergent sequence in X;

(5) f is said to be a semi-closed 1-set contractive mapping if f is 1-set contractive mapping, and $(I - f)$ is closed, which means the graph of the mapping $U = I - f$ (denoted by $G(U)$) is closed in the product topology $(E, \| \cdot \|_p) \times (E, \| \cdot \|_p)$, where I is the identity mapping (which is Definition 2 of Li [155]); and

(6) f is said to be semi-contractive (see Petryshyn [206] and Browder [32]) if there exists a mapping $V : D \times D \to 2^X$ such that $f(x) = V(x, x)$ for each $x \in D$, with (a) for each fixed $x \in D$, $V(\cdot, x)$ is nonexpansive from D to X, and (b) for each fixed $x \in D$, $V(x, \cdot)$ is completely continuous from D to X, uniformly for u in a bounded subset of D (which means that if v_j converges weakly to v in D and u_j is a bounded sequence in D, then $V(u_j, v_j) - V(u_j, v) \to 0$, converging strongly in D).

From the above, we first observe that definitions (1)–(6) for set-valued mappings can be given in a similar way with the Hausdorff metric H. Secondly, if f is a continuous demicompact mapping, then $(I - f)$ is closed, where I is the identity mapping on X. It is also clear from the definitions that every demicompact mapping is hemi-compact in seminorm spaces, but the converse is not true, as shown by the following example.

Example 7.2.2. Let $X = \mathbb{R}$, and define a function $f : X \to X$ by $f(x) = x + \tan^{-1}(x)$ for each $x \in X$. Then, it is clear that f is hemicompact, but not demicompact.

Actually, it is evident that if f is demicompact, then $I - f$ is demiclosed. It is known that for each condensing mapping f, when D or $f(D)$ is bounded, then f is hemicompact; and also f is demicompact in metric spaces by Lemmas 2.1 and 2.2 of Tan and Yuan [258], respectively. In addition, it is known that every nonexpansive mapping is a 1-set-contractive mapping; and also if f is a hemicompact 1-set-contractive mapping, then f is a 1-set-contractive mapping satisfying the following **condition (H1)** (which is the same as "condition (H1)" but slightly different from condition (H) in Chapter 5):

(**H1**) **condition**: Let D be a nonempty bounded subset of a space E, and assume that $F : \overline{D} \to 2^E$ is a set-valued mapping. If $\{x_n\}_{n \in \mathbb{N}}$ is any sequence in D such that for each x_n, there exists $y_n \in F(x_n)$ with $\lim_{n \to \infty}(x_n - y_n) = 0$, then there exists a point $x \in \overline{D}$ such that $x \in F(x)$.

We first note that the "(H1) condition" above is actually the same as "condition (C)" used in Theorem 1 of "condition (C)" [206]. Secondly, it was shown by Browder [32] that the nonexpansive mapping in a uniformly convex Banach X indeed satisfies condition (H1), as shown in the following.

Here, we recall again that, following Goebel and Reich [98], a normed vector space $(X, \|\cdot\|)$ is said to be uniformly convex if for every $\epsilon \in (0, 2]$, there is some $\delta > 0$ such that for any two vectors $x, y \in X$ with $\|x\| = \|y\| = 1$, the condition $\|x - y\| \geq \epsilon$ implies that $\|\frac{x+y}{2}\| \leq 1 - \delta$.

Indeed, by following the argument given by Petryshyn [206, pp. 320] (see also his proof of Theorem 2.2 and Corollary 2.1 in Ref. [206]), the mapping F is demiclosed, which means that the mapping $(I - F)$ is closed (at zero) under the product topology $(E, \sigma(E, E^*)) \times (E, \|\cdot\|)$, where I denotes the identity on E, $\sigma(E, E^*)$ is the weak topology, and $\|\cdot\|$ is the norm (or strong) topology) (which is the so-called Browder's demiclosedness principle), and thus F satisfies the following **condition (H1)**.

Condition (H1): If $\{x_n\}_{n \in \mathbb{N}}$ is any sequence in D such that for each x_n, there exists $y_n \in F(x_n)$ with $\lim_{n \to \infty}(x_n - y_n) = 0$, then we have $0 \in (I - F)(D)$, which means that there exists $x_0 \in D$ with $0 \in (I - F)(x_0)$.

For the convenience of the reader and making the text self-contained, here we provide the proof in detail for the demiclosedness principle of nonexpansive mappings in uniformly convex Banach spaces.

Lemma 7.2.3. *Let D be a nonempty bonded closed convex subset of a uniformly convex Banach space $(E, \|\cdot\|)$. Assume that $T : D \to E$ is a nonexpansive (single-valued) mapping with nonempty values. Then, the mapping $P := I - T$ defined by $P(x) := (x - T(x))$ for each $x \in D$ is demiclosed, and in particular, "condition (H1)" holds.*

Proof. It is indeed Theorem 3 of Browder [32], but here we give its proof with the notation of modulus of convexity for uniformly convex Banach spaces used by Goebel and Kirk [96] in two parts.

First, we claim that for given $\{u_n\}$ and $\{v_n\}$ in K and $z_n = \frac{1}{2}(u_n + v_n)$, if $\lim_{n \to \infty} \|u_n - Tu_n\| = 0$, $\lim_{n \to \infty} \|v_n - Tv_n\| = 0$, then we have $\lim_{n \to \infty} \|z_n - Tz_n\| = 0$. Indeed, if not, there exists $\epsilon > 0$, and we have $\|z_n - Tz_n\| \geq \epsilon > 0$. In view of this, we may suppose, by passing to a subsequence, that for some $r > 0$, $\lim_{n \to \infty} \|u_n - z_n\| = \lim_{n \to \infty} \|v_n - z_n\| = r$. Let d be the diameter of set D, i.e., $d = diam(D) > 0$. We choose some $t > 0$ such that $t < \frac{\epsilon}{d}$. Then, it is clear that $t < \frac{\epsilon}{\|u_n - z_n\|}$, and by the fact that $\lim_{n \to \infty} \|u_n - Tu_n\| = 0$ and thus for $n \in \mathbb{N}$ sufficiently large, we have

$$t < \frac{\epsilon}{\|u_n - Tu_n\| + \|u_n - z_n\|}.$$

In addition, we also have

$$\|u_n - Tz_n\| \leq \|u_n - Tu_n\| + \|Tu_n - Tz_n\|$$
$$\leq \|u_n - Tu_n\| + \|u_n - z_n\|.$$

By the fact that all the above inequalities hold if u_n is replaced by v_n, for $n \in \mathbb{N}$ large enough, by the equivalent formulation of uniform convexity with the notation δ for modulus of convexity for the space $(E, \|\cdot\|)$, we have

$$\|u_n - v_n\| \leq \left\|u_n - \frac{1}{2}(z_n + Tz_n)\right\| + \left\|v_n - \frac{1}{2}(z_n + Tz_n)\right\|$$
$$\leq [(\|u_n - Tu_n\| + \|u_n - z_n\|)$$
$$+ (\|v_n - Tv_n\| + \|v_n - z_n\|)](1 - \delta(t)).$$

Letting $n \to \infty$, we have the following inequality:

$$2r \leq (1 - \delta(t)).$$

This is impossible, and thus we must have $\lim_{n \to \infty} \|z_n - Tz_n\| = 0$.

Second, we claim that if $\inf\{\|x - Tx\| : x \in D\} = 0$, then T has a fixed point in K. First, we define R by

$$R := \{r > 0 : \inf\{\|x - Fx : x \in B(0; r) \cap K \neq \emptyset\} = 0\},$$

and let $r_0 = \inf R$. Since D is bounded, $r_0 < \infty$. Now, if $r_0 = 0$, then $0 \in K$ and $T(0) = 0$, which is a fixed point of T. So, we may assume that $r_0 > 0$. By the assumption that $\inf\{\|x - Tx\| : x \in D\} = 0$, it is clear that for each $n \in \mathbb{N}$, we can select $x_n \in B(0, r_0 + \frac{1}{n}) \cap K$ such that $\lim_{n \to \infty} \|x_n - Tx_n\| = 0$. Now, if $\{x_n\}$ strongly converges to x_0, then x_0 is a fixed point of T. So, we may assume that $\{x\}$ is not convergent. Then, there exists $\epsilon > 0$, and $\{x_n\}$ has a subsequence

$\{x_{n_k}\}$, such that $\|x_{n_k} - x_{n_{k+1}}\| \geq \epsilon$ for all $k = 1, 2, \ldots$. Now, for each $k \in \mathbb{N}$, let $m_k = \frac{1}{2}(x_{n_k} + x_{n_{k+1}})$. Then, let $t > 0$ be a positive number such that $t < \frac{\epsilon}{r_0}$, and also let $t \leq \frac{\epsilon}{\|x_k\|}$ for $k \in \mathbb{N}$ sufficiently large by using the notation δ for modulus of convexity for space E. We then have

$$\|m_k\| \leq \left(r_0 + \frac{1}{n_k}\right)(1 - \delta(t)).$$

Therefore, we have $\sup_{k \to \infty} \|m_k\| \leq r_0(1 - \delta(t)) < r_0$. By the claim proved in the first part above, it follows that $\lim_{k \to \infty} \|m_k - T m_k\| = 0$, which contradicts the assumption that r_0 is a positive number!

Now, by combining the claims proved above, it is time to show that the mapping $P := I - T$ is demiclosed. Indeed, let $\{u_n\}$ in K weakly converge to u, i.e., w-$\lim_{n \to \infty} u_n = u$, and $\lim_{n \to \infty} \|u_n - T u_n - w\| = 0$. Without loss of generality, we may assume that $w = 0$ (indeed by using the translation of $x \to x - w$ for each $x \in D$). For each $n \in \mathbb{N}$, let $K_n := \overline{conv}\{u_n, u_{n+1}, \ldots\}$. By the claim given in the second part above, there exists $y_n \in K_n$ such that $y_n = T y_n$. By the fact that the weakly limit y of subsequence $\{y_n\}$ must lie in $\cap_{n=1}^{\infty} K_n = \{u\}$, it implies that w-$\lim_{n \to} y_n = u$. Hence, u is in the weak closure of the set of fixed points denoted by $F(T)$ for T. As E is uniformly convex, by Theorem 3.3.17, E is also strictly convex and reflexive. So, the set $F(T)$ is closed and convex, and hence it is weakly closed. Thus, we have $u \in F(T)$.

Finally, as the mapping $T = (I - T)$ is demiclosed, the mapping T satisfies condition (H1). This completes the proof. $\square$

Remark 7.2.4. When a p-vector space E is with a p-norm, then "condition (H)" satisfies the "**(H1) condition**"; or in other words, the "**(H1) condition**" is guaranteed by the "demiclosedness principle" established by Browder [32].

Now, applying Theorem 6.2.3 (Theorem 6.1.9), we have the following result for non-self mappings in p-seminorm spaces for $p \in (0, 1]$.

Theorem 7.2.5. *Let U be a bounded open p-convex subset of a locally p-convex (or seminorm) space E, where $p \in (0, 1]$, and the zero element $\theta \in U$. Assume that $F : \overline{U} \to E$ is a continuous 1-set contractive single-valued mapping with nonempty values, satisfying condition (H) or (H1) above. In addition, for any $x \in \partial \overline{U}$, we have*

$\lambda x \neq F(x)$ *for any* $\lambda > 1$ *(i.e., the "Leray–Schauder boundary condition"), then F has at least one fixed point.*

Proof. By Theorem 6.2.3 (Theorem 6.1.9) with $C = E$, it follows that we have either (I) or (II) holding in the following:

(I) F has a fixed point $x_0 \in U$, i.e., $P_U(F(x_0) - x_0) = 0$.

(II) There exists $x_0 \in \partial(U)$ with $P_U(F(x_0) - x_0) = (P_U^{\frac{1}{p}}(F(x_0)) - 1)^p > 0$.

If F has no fixed point, then (II) holds and $x_0 \neq F(x_0)$. By the proof of Theorem 6.1.9 and the notation f used there, we have that $x_0 = f(F(x_0))$ and $F(x_0) \notin \overline{U}$. Thus, $P_U(F(x_0)) > 1$ and $x_0 = f(F(x_0)) = \dfrac{F(x_0)}{(P_U(F(x_0))^{\frac{1}{p}}}$, which means $F(x_0) = (P_U(F(x_0)))^{\frac{1}{p}} x_0$, where $(P_U(F(x_0)))^{\frac{1}{p}} > 1$, which contradicts the assumption. Thus, F must have a fixed point. The proof is complete. $\qquad\square$

By following the idea used and developed by Browder [32], Li [155], Li *et al.* [156], Goebel and Kirk [96], Petryshyn [205, 206], Tan and Yuan [258], Xu [278], Xu *et al.* [279], and the references therein, we have the following existence theorems for the principle of Leray–Schauder-type alternatives in locally p-convex spaces or p-seminorm spaces $(E, \| \cdot \|_p)$ for $p \in (0, 1]$.

Theorem 7.2.6. *Let U be a bounded open p-convex subset of a p-seminorm space $(E, \| \cdot \|_p)$, where $p \in (0, 1]$ and zero element $\theta \in U$. Assume that $F : \overline{U} \to E$ is a continuous 1-set contractive single-valued mapping with nonempty values, satisfying condition (H) or (H1) above. In addition, there exist $\alpha > 1$ and $\beta \geq 0$ such that for each $x \in \partial\overline{U}$, we have*

$$\|F(x) - x\|_p^{\alpha/p} \geq \|F(x)\|_p^{(\alpha+\beta)/p} \|x\|_p^{-\beta/p} - \|x\|_p^{\alpha/p}.$$

Then, F has at least one fixed point.

Proof. We prove the conclusion by showing that the Leray–Schauder boundary condition in Theorem 7.2.5 does not hold. If we assume that F has no fixed point, by the boundary condition of Theorem 7.2.5, there exist $x_0 \in \partial\overline{U}$ and $\lambda_0 > 1$ such that $F(x_0) = \lambda_0 x_0$.

Now, consider the function f defined by $f(t) := (t-1)^\alpha - t^{\alpha+\beta} + 1$ for $t \geq 1$. We observe that f is a strictly decreasing function for

$t \in [1, \infty)$, as the derivative of $f'(t) = \alpha(t-1)^{\alpha-1} - (\alpha+\beta)t^{\alpha+\beta-1} < 0$ by differentiation. Thus, we have $t^{\alpha+\beta} - 1 > (t-1)^{\alpha}$ for $t \in (1, \infty)$. By combining the boundary conditions, we have

$$\|F(x_0) - x_0\|_p^{\alpha/p} = \|\lambda_0 x_0 - x_0\|_p^{\alpha/p} = (\lambda_0 - 1)^{\alpha}\|x_0\|_p^{\alpha/p}$$
$$< (\lambda_0^{\alpha+\beta} - 1)\|x_0\|_p^{(\alpha+\beta)/p}\|x_0\|_p^{-\beta/p}$$
$$= \|F(x_0)\|_p^{(\alpha+\beta)/p}\|x_0\|_p^{-\beta/p} - \|x_0\|_p^{\alpha/p},$$

which contradicts the boundary condition given by Theorem 7.2.6. Thus, the conclusion follows and the proof is complete. $\square$

Theorem 7.2.7. *Let U be a bounded open p-convex subset of a p-seminorm space $(E, \|\cdot\|_p)$, where $p \in (0, 1]$ and the zero element $\theta \in U$. Assume that $F : \overline{U} \to E$ is a continuous 1-set contractive single-valued mapping with nonempty values, satisfying condition (H) or (H1) above. In addition, there exist $\alpha > 1$ and $\beta \geq 0$ such that for each $x \in \partial\overline{U}$, we have*

$$\|F(x) + x\|_p^{(\alpha+\beta)/p} \leq \|F(x)\|_p^{\alpha/p}\|x\|_p^{\beta/p} + \|x\|_p^{(\alpha+\beta)/p}.$$

Then, F has at least one fixed point.

Proof. We prove the conclusion by showing that the Leray–Schauder boundary condition in Theorem 7.2.5 does not hold. If we assume that F has no fixed point, by the boundary condition of Theorem 7.2.5, there exist $x_0 \in \partial\overline{U}$ and $\lambda_0 > 1$ such that $F(x_0) = \lambda_0 x_0$.

Now, consider the function f defined by $f(t) := (t+1)^{\alpha+\beta} - t^{\alpha} - 1$ for $t \geq 1$. We can then show that f is a strictly increasing function for $t \in [1, \infty)$, and thus we have $t^{\alpha} + 1 < (t+1)^{\alpha+\beta}$ for $t \in (1, \infty)$. By the boundary condition given in Theorem 7.2.7, we have that

$$\|F(x_0) + x_0\|_p^{(\alpha+\beta)/p} = (\lambda_0 + 1)^{\alpha+\beta}\|x_0\|_p^{(\alpha+\beta)/p}$$
$$> (\lambda_0^{\alpha} + 1)\|x_0\|_p^{(\alpha+\beta)/p}$$
$$= \|F(x_0)\|_p^{\alpha/p}\|x_0\|_p^{\beta/p} + \|x_0\|_p^{\alpha/p},$$

which contradicts the boundary condition given by Theorem 7.2.7. Thus, the conclusion follows and the proof is complete. $\square$

Theorem 7.2.8. *Let U be a bounded open p-convex subset of a p-seminorm space $(E, \| \cdot \|_p)$, where $p \in (0,1]$, and the zero element $\theta \in U$. Assume that $F : \overline{U} \to E$ is a continuous 1-set contractive single-valued mapping with nonempty values, satisfying condition (H) or (H1) above. In addition, there exist $\alpha > 1$ and $\beta \geq 0$ (or, alternatively, $\alpha > 1$ and $\beta \geq 0$) such that for each $x \in \partial \overline{U}$, we have*

$$\|F(x) - x\|_p^{\alpha/p} \|x\|_p^{\beta/p} \geq \|F(x)\|_p^{\alpha/p} \|F(x) + x\|_p^{\beta/p} - \|x\|_p^{(\alpha+\beta)/p}.$$

Then, F has at least one fixed point.

Proof. Similarly to above, we prove the conclusion by showing that the Leray–Schauder boundary condition in Theorem 7.2.5 does not hold. If we assume that F has no fixed point, by the boundary condition of Theorem 7.2.5, there exist $x_0 \in \partial \overline{U}$ and $\lambda_0 > 1$ such that $F(x_0) = \lambda_0 x_0$.

Now, consider the function f defined by $f(t) := (t - 1)^\alpha - t^\alpha (t-1)^\beta + 1$ for $t \geq 1$. We can then show that f is a strictly decreasing function for $t \in [1, \infty)$, and thus we have $(t-1)^\alpha < t^\alpha (t+1)^\beta - 1$ for $t \in (1, \infty)$.

By the boundary condition given in Theorem 7.2.8, we have

$$\begin{aligned}
\|F(x_0) - x_0\|_p^{\alpha/p} \|x_0\|_p^{\beta/p} &= (\lambda_0 - 1)^\alpha \|x_0\|_p^{(\alpha+\beta)/p} \\
&< (\lambda_0^\alpha (\lambda_0 + 1)^\beta - 1) \|x_0\|_p^{(\alpha+\beta)/p} \\
&= \|F(x_0)\|_p^{\alpha/p} \|F(x_0) + x_0\|_p^{\beta/p} - \|x_0\|_p^{(\alpha+\beta)/p},
\end{aligned}$$

which contradicts the boundary condition given by Theorem 7.2.8. Thus, the conclusion follows and the proof is complete. □

Theorem 7.2.9. *Let U be a bounded open p-convex subset of a p-seminorm space $(E, \| \cdot \|_p)$, where $p \in (0,1]$ and the zero element $\theta \in U$. Assume that $F : \overline{U} \to E$ is a continuous 1-set contractive single-valued mapping with nonempty values, satisfying condition (H) or (H1) above. In addition, there exist $\alpha > 1$ and $\beta \geq 0$, we have*

$$\|F(x) + x\|_p^{(\alpha+\beta)/p} \leq \|F(x) - x\|_p^{\alpha/p} \|x\|_p^{\beta/p} + \|F(x)\|_p^{\beta/p} \|x\|_p^{\alpha/p}.$$

Then, F has at least one fixed point.

Proof. Similarly to above, we prove the conclusion by showing that the Leray–Schauder boundary condition in Theorem 7.2.5 does not hold. If we assume that F has no fixed point, by the boundary condition of Theorem 7.2.5, there exist $x_0 \in \partial \overline{U}$ and $\lambda_0 > 1$ such that $F(x_0) = \lambda_0 x_0$.

Now, consider the function f defined by $f(t) := (t+1)^{\alpha+\beta} - (t-1)^{\alpha} - t^{\beta}$ for $t \geq 1$. We can then show that f is a strictly increasing function for $t \in [1, \infty)$, and thus we have $(t+1)^{\alpha+\beta} > (t-1)^{\alpha} + t^{\beta}$ for $t \in (1, \infty)$.

By the boundary condition given in Theorem 7.2.9, we have

$$
\begin{aligned}
\|F(x_0) + x_0\|_p^{(\alpha+\beta)/p} &= (\lambda_0 + 1)^{\alpha+\beta} \|x_0\|_p^{(\alpha+\beta)/p} \\
&> ((\lambda_0 - 1)^{\alpha} + \lambda_0^{\beta}) \|x_0\|_p^{(\alpha+\beta)/p} \\
&= \|\lambda_0 x_0 - x_0\|_p^{\alpha/p} \|x_0\|_p^{\beta/p} + \|\lambda_0 x_0\|_p^{\beta/p} \|x_0\|_p^{\alpha/p} \\
&= \|F(x_0) - x_0\|_p^{\beta/p} \|x_0\|_p^{\alpha/p} + \|F(x_0)\|_p^{\beta/p} \|x_9\|^{\alpha/p},
\end{aligned}
$$

which implies that

$$
\|F(x_0) + x_0\|_p^{(\alpha+\beta)/p} > \|F(x_0) - x_0\|_p^{\beta/p} \|x_0\|_p^{\alpha/p} + \|F(x_0)\|_p^{\beta/p} \|x_9\|^{\alpha/p}.
$$

This contradicts the boundary condition given by Theorem 7.2.9. Thus, the conclusion follows and the proof is complete. □

As an application of Theorems 7.2.5, by testing the Leray–Schauder boundary condition, we have the following conclusion and thus omit its detailed proof.

Corollary 7.2.10. *Let U be a bounded open p-convex subset of a p-seminorm space $(E, \|\cdot\|_p)$, where $p \in (0, 1]$ and the zero element $\theta \in U$. Assume that $F : \overline{U} \to E$ is a continuous 1-set contractive single-valued mapping with nonempty values, satisfying condition (H) or (H1) above. Then, F has at least one fixed point if one of the following conditions holds for $x \in \partial \overline{U}$:*

(i) $\|F(x)\|_p \leq \|x\|_p$,
(ii) $\|F(x)\|_p \leq \|F(x) - x\|_p$,
(iii) $\|F(x) + x\|_p \leq \|F(x)\|_p$,
(iv) $\|F(x) + x\|_p \leq \|x\|_p$,
(v) $\|F(x) + x\|_p \leq \|F(x) - x\|_p$,
(vi) $\|F(x)\|_p \cdot \|F(x) + x\|_p \leq \|x\|_p^2$,
(vii) $\|F(x)\|_p \cdot \|F(x) + x\|_p \leq \|F(x) - x\|_p \cdot \|x\|_p$.

Now, if the p-seminorm space E is a uniformly convex Banach space $(E, \| \cdot \|)$ (for p-norm space with $p = 1$), we have the following general fixed point theorems (and we discuss the corresponding results for nonexpansive set-valued mappings later in Chapter 8).

Theorem 7.2.11. *Let U be a bounded open convex subset of a uniformly convex Banach space $(E, \| \cdot \|)$ with zero $\theta \in U$. Assume that $F : \overline{U} \to E$ is a continuous semicontractive single-valued mapping with nonempty values. In addition, for any $x \in \partial \overline{U}$, we have $\lambda x \neq F(x)$ for any $\lambda > 1$ (i.e., the "Leray–Schauder boundary condition"). Then, F has at least one fixed point.*

Proof. By the assumption that F is a semicontractive and continuous single-valued mapping with nonempty values, it is nonexpansive. Moreover, by the assumption that E is a uniformly convex Banach, by Lemma 7.2.3, the mapping $(I - F)$ is closed (and so at zero, too) under the product topology $(E, \sigma(E, E^*)) \times (E, \| \cdot \|)$, where I denotes the identity on E, $\sigma(E, E^*)$ is the weak topology, $\| \cdot \|$ is the norm (or strong) topology, and F satisfies condition (H1).Thus, all the assumptions of Theorem 7.2.5 are satisfied with the (H1) condition by Lemma 7.2.3. Then, the conclusion follows by Theorem 7.2.5. This completes the proof. □

We conclude this section by noting that, according to Lemma 7.2.3, a single-valued nonexpansive mapping defined in a uniformly convex Banach space (see also Theorem 7.2.11) satisfies the (H1) condition. Actually, the nonexpansive set-valued mappings defined on a special class of Banach spaces with the so-called " Opial condition" [179] not only satisfy condition (H1) but also belong to the class of semiclosed 1-set contractive mappings, as shown in sections later in this chapter, as well as in Chapter 8.

7.3 Fixed Point Theorems for Semiclosed 1-Set Contractive Mappings

The goal of this section is to establish fixed point theorems for non-self semiclosed 1-set contractive mappings under general conditions in locally convex p-convex spaces for $p \in (0, 1]$, which would unify the corresponding results in the existing literature as special cases.

In order to do so, we first introduce the following definition, which is a set-valued generalization of the corresponding single-value semiclosed 1-set mappings. It was first discussed by Li [155], Li *et al.* [156], Xu [278], and Xu *et al.* [279] and related references therein.

Definition 7.3.1. Let D be a nonempty (bounded) closed subset of a given p-seminorm space $(E, \| \cdot \|_p)$, where $p \in (0, 1]$. We recall that a set-valued mapping $F : D \to 2^E$ is said to be a **semiclosed 1-set contractive mapping** if F is 1-set contraction, and $(I - F)$ is closed (at zero) under either the product topology $(D, \| \cdot \|_p) \times (E, \| \cdot \|_p)$ or the product topology $(D, \sigma(X, X^*)) \times (E, \| \cdot \|_p)$ (where $\sigma(X, X^*)$ denotes the weak topology of $(E, \| \cdot \|_p)$; and $(E, \| \cdot \|_p)$ denotes the p-norm (or say, strong) topology of E), which means that for a given net $\{x_n\}_{i \in I}$, for each $i \in I$, there exists $y_i \in F(x_i)$ with $\lim_{i \in I}(x_i - y_i) = 0$, and then $0 \in (I - F)(D)$, i.e., there exists $x_0 \in D$ such that $x_0 \in F(x_0)$.

By Definition 7.3.1 and Lemma 7.2.3, it follows that each nonexpansive (single-valued) mapping defined on a subset of uniformly convex Banach spaces (see also later sections in the following for nonexpansive set-valued mappings defined on a subset of Banach spaces satisfying the Opial condition) are semiclosed 1-set contractive mappings (see also the discussion given by Goebel [95], Goebel and Kirk [96], Petrusel *et al.* [208], Xu [273], Yangai [280], and related references therein). In particular, under the setting of metric spaces or Banach spaces with certain reasonable properties, we can see that semiclosed 1-set contractive mappings may satisfy condition (H1) above.

We know that by comparing nonexpansive mappings with single-valued case, based on a study in the literature about the approximation of fixed points for multi-valued mappings, a well-known counterexample was provided by Pietramala [210] (see also Muglia and Marino [172]), who proved in 1991 that the approximation Theorem 1 given by Browder [31] cannot be extended to the general multi-valued case even on a finite-dimensional space $\mathbb{R}^2$.

In addition, if a Banach space X satisfies the Opial property (see Opial [179]), that is, if x_n weakly converges to x, then we have that $\limsup \|x_n - x\| < \limsup \|x_n - y\|$ for all $x \in X$ and $y \neq x$), and then $I - f$ is demiclosed at θ (see also Lami Dozo [150], Yanagi [280], and the related references therein) provided $f : C :\to K(C)$ is nonexpansive (where $K(C)$ denotes the family of nonempty compact subsets of C). We know that all Hilbert spaces and L^p spaces $p \in (1, \infty)$ possess the Opial property. However, it seems that whether $I - f$ is demiclosed at zero θ depends on if f is a nonexpansive set-valued mapping defined on the space X, which is uniformly convex (e.g., $L[0, 1]$, $1 < p < \infty$, $\neq 2$) and $f : C \to K(C)$ is nonexpansive. Here, we remark that for a single-valued nonexpansive mapping f, the answer is "yes", which is indeed Lemma 7.2.3 above (namely the famous theorem of Browder [30]).

Here, we recall that a remarkable fixed point theorem for multivalued mappings established by Lim in [159] states the following: If C is a nonempty closed bounded convex subset of a uniformly convex Banach space X and $f : C \to K(C)$ is nonexpansive, then f has a fixed point.

Now, based on the concept for semiclosed 1-set contractive mappings, we give the existence result for their best approximation, fixed points, and related nonlinear alternatives under the framework of p-seminorm spaces for $p \in (0, 1]$.

Theorem 7.3.2 (Schauder fixed point theorem for semiclosed 1-set contractive mappings). *Let U be a nonempty bounded open subset of a (Hausdorff) locally p-convex space E and its zero element $\theta \in U$, and let $C \subset E$ be a closed p-convex subset of E such that zero element $\theta \in C$, with $p \in (0, 1]$. If $F : C \cap \overline{U} \to C \cap \overline{U}$ is a continuous semiclosed 1-set contractive single-valued mapping with nonempty values. Then, T has at least one fixed point in $C \cap \overline{U}$.*

Proof. As the mapping F is 1-set contractive, consider an increasing sequence $\{\lambda_n\}$ such that $0 < \lambda_n < 1$ and $\lim_{n \to \infty} \lambda_n = 1$, where $n \in \mathbb{N}$. Now, we define a mapping $F_n : C \to C$ by $F_n(x) := \lambda_n F(x)$ for each $x \in C$ and $n \in \mathbb{N}$. Then, it follows that F_n is a λ_n-set contractive mapping with $0 < \lambda_n < 1$. By Theorem 5.5.6 on the condensing mapping F_n in p-vector space with the p-seminorm P_U for each $n \in \mathbb{N}$, there exists $x_n \in C$ such that $x_n \in F_n(x_n) = \lambda_n F(x_n)$. Thus, we have $x_n = \lambda_n F(x_n)$. Let P_U be the Minkowski p-functional of U in E. It follows that P_U is continuous as $0 \in int(U) = U$. Note that for each $n \in \mathbb{N}$, $\lambda_n x_n \in \overline{U} \cap C$, which implies that $x_n = r(\lambda_n F(x_n)) = \lambda_n F(x_n)$, and thus $P_U(\lambda_n F(x_n)) \leq 1$ by Lemma 2.2.10.

Note that

$$P_U(F(x_n) - x_n) = P_U(F(x_n) - \lambda_n F(x_n)) = P_U\left(\frac{(1-\lambda_n)\lambda_n F(x_n)}{\lambda_n}\right)$$

$$\leq \left(\frac{1-\lambda_n}{\lambda_n}\right)^p P_U(\lambda_n F(x_n)) \leq \left(\frac{1-\lambda_n}{\lambda_n}\right)^p,$$

which implies that $\lim_{n\to\infty} P_U(F(x_n) - x_n) = 0$. Now, by the assumption that F is semiclosed, which means that $(I - F)$ is closed at zero, there exists a point $x_0 \in \overline{C}$ such that $0 \in (I - F)(\overline{C})$, and thus we have that $x_0 = F(x_0)$.

Indeed, without loss of generality, we assume that $\lim_{n\to\infty} x_n = x_0$, with $x_n = \lambda_n F(x_n)$, and $\lim_{n\to\infty} \lambda_n = 1$. This implies that $x_0 = \lim_{n\to\infty}(\lambda_n F(x_n))$, which means $F(x_0) := \lim_{n\to\infty} F(x_n) = x_0$, and thus $x_0 = F(x_0)$. We complete the proof. $\qquad\square$

Theorem 7.3.3 (Best approximation for semiclosed 1-set contractive mappings). *Let U be a bounded open p-convex subset of a locally p-convex space E, where $p \in (0,1]$, the zero element $\theta \in U$, and let C be a (bounded) closed p-convex subset of E also with zero element $\theta \in C$. Assume that $F : \overline{U} \cap C \to C$ is a continuous semiclosed 1-set contractive single-valued mapping with nonempty values, and for each $x \in \partial_C U$ with $F(x) \notin \overline{U}$, $(P_U^{\frac{1}{p}}(F(x)) - 1)^p \leq P_U(F(x) - x)$ for $0 < p \leq 1$ (which is trivial when $p = 1$). Then, we have that there exist $x_0 \in C \cap \overline{U}$ and $F(x_0)$ such that $P_U(F(x_0) - x_0) = d_p(F(x_0), \overline{U} \cap C) = d_p(F(x_0), \overline{I_U^p(x_0)} \cap C)$, where P_U is the Minkowski p-functional of U. More precisely, we have either (I) or (II) holding in the following:*

(I) *F has a fixed point $x_0 \in U \cap C$, i.e., $x_0 = F(x_0)$ (so that $0 = P_U(F(x_0) - x_0) = d_p(F(x_0), \overline{U} \cap C) = d_p(F(x_0), \overline{I_U^p(x_0)} \cap C)$).*

(II) *There exists $x_0 \in \partial_C(U)$ and $F(x_0) \notin \overline{U}$ with*

$$P_U(F(x_0) - x_0) = d_P(F(x_0), \overline{U} \cap C) = d_p(F(x_0), \overline{I_U^p(x_0)} \cap C)$$

$$= (P_U^{\frac{1}{p}}(F(x_0)) - 1)^p > 0.$$

Proof. Let $r : E \to U$ be a retraction mapping defined by $r(x) := \dfrac{x}{\max\{1, (P_U(x))^{\frac{1}{p}}\}}$ for each $x \in E$, where P_U is the Minkowski

p-functional of U. Since the space E's zero element $\theta \in U(= intU$, as U is open), it follows that r is continuous by Lemma 2.2.10. As the mapping F is 1-set contractive, consider an increasing sequence $\{\lambda_n\}$ such that $0 < \lambda_n < 1$ and $\lim_{n\to\infty} \lambda_n = 1$, where $n \in \mathbb{N}$. Now, we define a mapping $F_n : C \cap \overline{U} \to C$ by $F_n(x) := \lambda_n F \circ r(x)$ for each $x \in C \cap \overline{U}$ and $n \in \mathbb{N}$. Then, it follows that F_n is a λ_n-set contractive mapping with $0 < \lambda_n < 1$ for each $n \in \mathbb{N}$. As C and $\overline{U}$ are p-convex, we have $r(C) \subset C$ and $r(\overline{U}) \subset \overline{U}$, so $r(C \cap \overline{U}) \subset C \cap \overline{U}$. Thus, F_n is a self-mapping defined on $C \cap \overline{U}$. By Theorem 5.5.6 for condensing mapping F_n, for each $n \in \mathbb{N}$, there exists $z_n \in C \cap \overline{U}$ such that $z_n \in F_n(z_n) = \lambda_n F \circ r(z_n)$. Let $x_n = r(z_n)$. Then, we have $x_n \in C \cap \overline{U}$ with $x_n = r(\lambda_n F(x_n))$ such that either (1) or (2) holds in the following for each $n \in \mathbb{N}$:

(1) $\lambda_n F(x_n) \in C \cap \overline{U}$;
(2) $\lambda_n F(x_n) \in C \backslash \overline{U}$.

Now, we prove the conclusion by considering the following two cases:

- Case (I) For each $n \in N$, $\lambda_n F(x_n) \in C \cap \overline{U}$.
- Case (II) There exists a positive integer n such that $\lambda_n F(x_n) \in C \backslash \overline{U}$.

First, according to case (I), for each $n \in \mathbb{N}$, $\lambda_n F(x_n) \in \overline{U} \cap C$, which implies that $x_n = r(\lambda_n F(x_n)) = \lambda_n F(x_n)$, and thus $P_U(\lambda_n F(x_n)) \leq 1$ by Lemma 2.2.10. Note that

$$P_U(F(x_n) - x_n) = P_U(F(x_n) - \lambda_n F(x_n)) = P_U\left(\frac{(1 - \lambda_n)\lambda_n F(x_n)}{\lambda_n}\right)$$

$$\leq \left(\frac{1 - \lambda_n}{\lambda_n}\right)^p P_U(\lambda_n F(x_n)) \leq \left(\frac{1 - \lambda_n}{\lambda_n}\right)^p,$$

which implies that $\lim_{n\to\infty} P_U(F(x_n) - x_n) = 0$. Now, by the fact that F is semiclosed, it implies that there exists a point $x_0 \in \overline{U}$ (i.e., the consequence $\{x_n\}_{n\in\mathbb{N}}$ has a convergent subsequence with the limit x_0) such that $x_0 = F(x_0)$. Indeed, without loss of generality, we assume that $\lim_{n\to\infty} x_n = x_0$, with $x_n = \lambda_n F(x_n)$, and $\lim_{n\to\infty} \lambda_n = 1$, and as $x_0 = \lim_{n\to\infty}(\lambda_n F(x_n))$, it implies that $F(x_0) = \lim_{n\to\infty} F(x_n) = x_0$. Thus, there exists $F(x_0) = x_0$, and we have $0 = d_p(x_0, F(x_0)) = d(F(x_0), \overline{U} \cap C) = d_p(F(x_0), \overline{I_{\overline{U}}^p(x_0)} \cap C))$, as indeed $x_0 = F(x_0) \in \overline{U} \cap C \subset \overline{I_{\overline{U}}^p(x_0)} \cap C$.

Second, according to case (II), there exists a positive integer n such that $\lambda_n F(x_n) \in C \backslash \overline{U}$. Then, we have that $P_U(\lambda_n F(x_n)) > 1$, and also $P_U(F(x_n)) > 1$, as $\lambda_n < 1$. As $x_n = r(\lambda_n F(x_n)) = \frac{\lambda_n F(x_n)}{(P_U(\lambda_n F(x_n)))^{\frac{1}{p}}}$, which implies that $P_U(x_n) = 1$, $x_n \in \partial_C(U)$. Note that

$$P_U(F(x_n) - x_n) = P_U \left(\frac{(P_U(F(x_n))^{\frac{1}{p}} - 1)F(x_n)}{P_U(F(x_n))^{\frac{1}{p}}} \right)$$
$$= \left(P_U^{\frac{1}{p}}(F(x_n)) - 1 \right)^p .$$

By the assumption, we have $(P_U^{\frac{1}{p}}(F(x_n)) - 1)^p \leq P_U(F(x_n) - x)$ for $x \in C \cap \partial \overline{U}$, and it follows that

$$P_U(F(x_n)) - 1 \leq P_U(F(x_n)) - \sup\{P_U(z) : z \in C \cap \overline{U}\}$$
$$\leq \inf\{P_U(F(x_n) - z) : z \in C \cap \overline{U}\}$$
$$= d_p(F(x_n), C \cap \overline{U}).$$

Thus, we have the best approximation $P_U(F(x_n) - x_n) = d_P(F(x_n), \overline{U} \cap C) = (P_U^{\frac{1}{p}}(F(x_n)) - 1)^p > 0$.

Now, we want to show that $P_U(F(x_n) - x_n) = d_P(F(x_n), \overline{U} \cap C) = d_p(F(x_n), I_{\overline{U}}^p(x_0) \cap C) > 0$.

By the fact that $(\overline{U} \cap C) \subset I_{\overline{U}}^p(x_n) \cap C$, let $z \in I_{\overline{U}}^p(x_n) \cap C \backslash (\overline{U} \cap C)$, and we first claim that $P_U(F(x_n) - x_n) \leq P_U(F(x_n) - z)$. If not, we have $P_U(F(x_n) - x_n) > P_U(F(x_n) - z)$. As $z \in I_{\overline{U}}^p(x_n) \cap C \backslash (\overline{U} \cap C)$, there exists $y \in \overline{U}$ and a nonnegative number c (actually $c \geq 1$, as will be soon shown) with $z = x_n + c(y - x_n)$. Since $z \in C$, but $z \notin \overline{U} \cap C$, it implies that $z \notin \overline{U}$. By the fact that $x_n \in \overline{U}$ and $y \in \overline{U}$, we must have the constant $c \geq 1$; otherwise, it implies that $z(= (1-c)x_n + cy) \in \overline{U}$, which is impossible by our assumption, i.e., $z \notin \overline{U}$. Thus, we have that $c \geq 1$, which implies that $y = \frac{1}{c}z + (1 - \frac{1}{c})x_n \in C$ (as both $x_n \in C$ and $z \in C$). On the other hand, as $z \in I_{\overline{U}}^p(x_n) \cap C \backslash (\overline{U} \cap C)$ and $c \geq 1$ with $(\frac{1}{c})^p + (1 - \frac{1}{c})^p = 1$, combining with our assumption that for each $x \in \partial_C \overline{U}$ and $F(x) \notin \overline{U}$, $P_U^{\frac{1}{p}}(F(x)) - 1 \leq P_U^{\frac{1}{p}}(F(x) - x)$

for $0 < p \leq 1$, it then follows that

$$
\begin{aligned}
P_U(F(x_n) - y) &= P_U \left[\frac{1}{c}(F(x_n) - z) + \left(1 - \frac{1}{c}\right)(F(x_n) - x_n) \right] \\
&\leq \left[\left(\frac{1}{c}\right)^p P_U(F(x_n) - z) + \left(1 - \frac{1}{c}\right)^p P_U(F(x_n) - x_n) \right] \\
&< P_U(F(x_n) - x_n),
\end{aligned}
$$

which contradicts that $P_U(F(x_n) - x_n) = d_P(F(x_n), \overline{U} \cap C)$. As shown above, we know that $y \in \overline{U} \cap C$, and we should have $P_U(F(x_n) - x_n) \leq P_U(F(x_n) - y)$! This helps us to complete the claim $P_U(F(x_n) - x_n) \leq P_U(F(x_n) - z)$ for any $z \in I_{\overline{U}}^p(x_n) \cap C \backslash (\overline{U} \cap C)$, which means that the following best approximation of Fan type (see [84, 85]) holds:

$$
0 < d_P(F(x_n), \overline{U} \cap C) = P_U(F(x_n) - x_n) = d_p(F(x_n), I_{\overline{U}}^p(x_n) \cap C).
$$

Now, by the continuity of P_U, it follows that the following best approximation of Fan type is also true:

$$
\begin{aligned}
0 < P_U(F(x_n) - x_n) &= d_P(F(x_n), \overline{U} \cap C) = d_p(F(x_n), I_{\overline{U}}^p(x_n) \cap C) \\
&= d_p(F(x_n), \overline{I_{\overline{U}}^p(x_n)} \cap C),
\end{aligned}
$$

and we have that

$$
P_U(F(x_0) - x_0) = d_P(F(x_0), \overline{U} \cap C) = d_p(F(x_0), I_{\overline{U}}^p(x_0) \cap C > 0.
$$

The proof is complete. $\qquad \square$

Let $p = 1$ in Theorem 7.3.3. We then have the following best approximation in locally convex spaces.

Theorem 7.3.4 (Best approximation in locally convex spaces). *Let U be a bounded open convex subset of a locally convex space E (i.e., $p = 1$) with zero element $\theta \in \text{int}U = U$ (the interior $\text{int}U = U$ as U is open), and let C be a closed p-convex subset of E also with zero element $\theta \in C$. Assume that $F : \overline{U} \cap C \to C$ is a continuous semiclosed 1-set contractive single-valued mapping with nonempty values. Then, there exist $x_0 \in \overline{U \cap X}$ such that $P_U(F(x_0) - x_0) = d_P(F(x_0), \overline{U} \cap C) = d_p(F(x_0), \overline{I_{\overline{U}}(x_0)} \cap C)$, where P_U is the Minkowski p-functional of U.*

More precisely, we have either (I) or (II) holding in the following:

(I) *F has a fixed point* $x_0 \in \overline{U} \cap C$, *i.e.*, $x_0 = F(x_0)$ *(so that* $P_U(F(x_0) - x_0) = d_P(F(x_0), \overline{U} \cap C) = d_p(F(x_0), \overline{I_{\overline{U}}(x_0)} \cap C)) = 0)$.

(II) *There exists* $x_0 \in \partial_C(U)$ *and* $F(x_0) \notin \overline{U}$ *with*

$$P_U(F(x_0) - x_0) = d_p(F(x_0), I_{\overline{U}}(x_0) \cap C)$$
$$= d_p(F(x_0), \overline{I_{\overline{U}}(x_0)} \cap C) > 0.$$

Proof. By applying Theorem 7.3.3 with $p = 1$, the conclusion follows. This completes the proof. □

Now, as an application of Theorem 7.3.3, we have the following general principle for Birkhoff–Kellogg problems in p-seminorm spaces, where $p \in (0, 1]$.

Theorem 7.3.5 (Principle of the Birkhoff–Kellogg alternative). *Let U be a bounded open p-convex subset of a locally p-convex space E, where $p \in (0, 1]$ with zero element $\theta \in intU = (U)$ (the interior $intU$, as U is open), and let C be a closed p-convex subset of E also with zero $\theta \in C$. Assume that $F : \overline{U} \cap C \to C$ is a continuous semiclosed 1-set contractive single-valued mapping with nonempty values. Then, F has at least one of the following two properties:*

(I) *F has a fixed point $x_0 \in \overline{U} \cap C$ such that $x_0 = F(x_0)$.*
(II) *There exist $x_0 \in \partial_C(U)$ and $F(x_0) \notin \overline{U}$, and $\lambda = \dfrac{1}{(P_U(F(x_0)))^{\frac{1}{p}}} \in$ $(0, 1)$ such that $x_0 = \lambda F(x_0)$. In addition, if for each $x \in \partial_C U$, $P_U^{\frac{1}{p}}(F(x)) - 1 \le P_U^{\frac{1}{p}}(F(x) - x)$ for $0 < p \le 1$ (which is trivial when $p = 1$), then the best approximation between x_0 and $F(x_0)$ is given by*

$$P_U(F(x_0) - x_0) = d_p(F(x_0), \overline{I_{\overline{U}}^p(x_0)} \cap C) = (P_U^{\frac{1}{p}}(F(x_0)) - 1)^p > 0.$$

Proof. If (I) is not the case, then (II) is proved by Remark 6.2.4 and by following the proof in Theorem 7.3.3 (actually Theorem 6.1.9) for case (ii): $F(x_0) \in C \backslash \overline{U}$, with $y_0 = f(F(x_0))$, where f is the restriction of the continuous mapping r restriction to the subset U in E. Indeed, as $y_0 \notin \overline{U}$, it follows that $P_U(y_0) > 1$, and $x_0 = f(y_0) =$

$F(x_0) \dfrac{1}{(P_U(F(x_0)))^{\frac{1}{p}}}$. Now, let $\lambda = \dfrac{1}{(P_U(F(x_0)))^{\frac{1}{p}}}$, we have $\lambda < 1$ and $x_0 = \lambda F(x_0)$. Finally, the additionally assumption in (II) allows us to obtain the best approximation between x_0 and $F(x_0)$ by following the proof of Theorem 7.3.3 (see Theorem 6.1.9), as $P_U(F(x_0) - x_0) = d_P(F(x_0), \overline{U} \cap C) = d_p(F(x_0), \overline{I_{\overline{U}}^p}(x_0) \cap C) > 0$. This completes the proof. □

As another application of Theorem 7.3.3 for non-self mappings, we have the following general principle of Birkhoff–Kellogg alternative in locally convex spaces.

Theorem 7.3.6 (Principle of the Birkhoff–Kellogg alternative in LCS). *Let U be a bounded open p-convex subset of a locally convex space E with the zero element $\theta \in U$, and let C be a closed convex subset of E also with zero element $\theta \in C$. Assume that $F : \overline{U} \cap C \to C$ is a continuous semiclosed 1-set contractive single-valued mapping with nonempty values. Then, it has at least one of the following two properties:*

(I) *F has a fixed point $x_0 \in U \cap C$ such that $x_0 = F(x_0)$.*
(II) *There exists $x_0 \in \partial_C(U)$ and $F(x_0) \notin \overline{U}$ and $\lambda \in (0,1)$ such that $x_0 = \lambda F(x_0)$, and the best approximation between $\{x_0\}$ and $F(x_0)$ is given by $P_U(F(x_0) - x_0) = d_P(F(x_0), \overline{U} \cap C) = d_p(F(x_0), \overline{I_{\overline{U}}^p}(x_0) \cap C) > 0$.*

We note that by the proof of Theorems 7.3.3, for case (II) of Theorem 7.3.3, the assumption "each $x \in \partial_C U$ with $y \in F(x)$, $P_U^{\frac{1}{p}}(y) - 1 \le P_U^{\frac{1}{p}}(y - x)$" is only used to guarantee the best approximation "$P_U(y_0 - x_0) = d_P(y_0, \overline{U} \cap C) = d_p(y_0, \overline{I_{\overline{U}}^p}(x_0) \cap C) > 0$", and thus we have the following Leray–Schauder alternative in p-vector spaces, which, of course, includes the corresponding results in LCSs as special cases.

Theorem 7.3.7 (Leray–Schauder nonlinear alternative). *Let C a closed p-convex subset of P-seminorm space E, where $p \in (0,1]$ and the zero element $\theta \in C$. Assume that $F : C \to C$ is a continuous semiclosed 1-set contractive single-valued mapping with nonempty values. Let $\varepsilon(F) := \{x \in C : x \in \lambda F(x), \text{ for some } 0 < \lambda < 1\}$. Then, either F has a fixed point in C or the set $\varepsilon(F)$ is unbounded.*

Proof. By assuming that case (I) is not true, i.e., F has no fixed point, we claim that the set $\varepsilon(F)$ is unbounded. Otherwise, assume the set $\varepsilon(F)$ is bounded. Assuming that P is the continuous p-seminorm for E, then there exists $r > 0$ such that the set $B(0,r) := \{x \in E : P(x) < r\}$, which contains the set $\varepsilon(F)$, i.e., $\varepsilon(F) \subset B(0,r)$, which means for any $x \in \varepsilon(F)$, $P(x) < r$. Then, $B(0.r)$ is an open p-convex subset of E and the zero $0 \in B(0,r)$ by Lemma 2.2.10 and Remark 2.2.12. Now, let $U := B(0,r)$ in Theorem 7.3.5. It follows that for the mapping $F : B(0,r) \cap C \to C$ satisfies all general conditions of Theorem 7.3.5, and we have that for any $x_0 \in \partial_C B(0,r)$, no $\lambda \in (0,1)$ such that $x_0 = \lambda y_0$, where $y_0 \in F(x_0)$. Indeed, for any $x \in \varepsilon(F)$, it follows that $P(x) < r$ as $\varepsilon(F) \subset B(0,r)$, but for any $x_0 \in \partial_C B(0,r)$, we have $P(x_0) = r$, and thus conclusion (II) of Theorem 7.3.5 does not hold. By Theorem 7.3.5 again, F must have a fixed point, but this contradicts our assumption that F is fixed point free. This completes the proof. $\square$

Now, assume a given p-vector space E equipped with the P-seminorm (by assuming it is continuous at zero) for $0 < p \le 1$. Then, we know that $P : E \to \mathbb{R}^+$, $P^{-1}(0) = 0$, $P(\lambda x) = |\lambda|^p P(x)$ for any $x \in E$ and $\lambda \in \mathbb{R}$. Then, we have the following useful result for fixed points due to Rothe and Altman in p-vector spaces, which plays important roles in optimization problem, variational inequality, and complementarity problems.

Corollary 7.3.8. *Let U be a bounded open p-convex subset of a locally p-convex space E and zero element $\theta \in U$, and let C be a closed p-convex subset of E with $U \subset C$, where $p \in (0,1]$. Assume that $F : \overline{U} \to C$ is a continuous semiclosed 1-set contractive single-valued mapping with nonempty values. We consider if one of the following conditions is satisfied for P_U (which is the Minkowski p-functional of U):*

(1) *(Rothe-type condition):* $P_U(F(x)) \le P_U(x)$ *for any* $x \in \partial U$.
(2) *(Petryshyn-type condition):* $P_U(F(x)) \le P_U(F(x) - x)$ *for any* $x \in \partial U$.
(3) *(Altman-type condition):* $|P_U(F(x))|^{\frac{2}{p}} \le [P_U(F(x)) - x)]^{\frac{2}{p}} + [P_U(x)]^{\frac{2}{p}}$ *for any* $x \in \partial U$.

Then, F has at least one fixed point.

Proof. By conditions (1), (2), and (3), it follows that the conclusion of (II) in Theorem 7.3.5 "there exist $x_0 \in \partial_C(U)$ and $\lambda \in (0,1)$ such that $x_0 \neq F(x_0)$" does not hold, and thus by the alternative of Theorem 7.3.5, F has a fixed point. This completes the proof. $\square$

By the fact that when $p = 1$ in p-vector space being a LCS, we have the following classical Fan's best approximation (see [84]). It is a powerful tool for the study of optimization, mathematical programming, game theory, mathematical economics, and others related topics in applied mathematics.

Corollary 7.3.9 (Fan's best approximation). *Let U be a bounded open convex subset of a LCS E with the zero element $\theta \in U$, and let C be a closed convex subset of E also with zero element $\theta \in C$. Assume that $F : \overline{U} \cap C \to C$ is a continuous semiclosed 1-set contractive single-valued mapping with nonempty values. Then, there exists $x_0 \in \overline{U} \cap X$ such that $P_U(F(x_0) - x_0) = d_P(F(x_0, \overline{U} \cap C) = d_p(F(x_0), \overline{I_{\overline{U}}(x_0)} \cap C)$, where P_U is the Minkowski functional of U in E.*

More precisely, we have either (I) or (II) holding in the following, where $W_{\overline{U}}(x_0)$ is either the inward set $I_{\overline{U}}(x_0)$ or the outward set $O_{\overline{U}}(x_0)$:

(I) *F has a fixed point $x_0 \in \overline{U} \cap C$, i.e., $x_0 = F(x_0)$.*
(II) *There exists $x_0 \in \partial_C(U)$ with $F(x_0) \notin \overline{U}$ such that*

$$P_U(F(x_0) - x_0) = d_p(F(x_0), \overline{I_{\overline{U}}(x_0)} \cap C) = P_U(F(x_0)) - 1 > 0.$$

Proof. When $p = 1$, then it automatically satisfies the inequality $P_U^{\frac{1}{p}}(F(x)) - 1 \leq P_U^{\frac{1}{p}}(F(x) - x)$ for each $x \in \overline{U} \cap C$. Indeed, we have that for $x_0 \in \partial_C(U)$, $P_U(F(x_0) - x_0) = d_P(F(x_0), \overline{U} \cap C) = d_p(F(x_0), \overline{I_{\overline{U}}(x_0)} \cap C) = P_U(F(x_0)) - 1$. The conclusions are given by Theorem 7.3.3 (or Theorem 7.3.4). The proof is complete. $\square$

Before we end this section, we would like to point out similar results on Rothe and Leray–Schauder alternatives have been developed by Isac [118], Park [185], Potter [213], Shahzad [238, 240], and Xiao and Zhu [270], which are useful tools of nonlinear analysis in topological vector spaces. As mentioned above, when $p = 1$ and F as a continuous mapping, then we can also obtain the version of Leray–Schauder in LCSs.

7.4 Principle of Alternative for Semiclosed 1-Set Contractive Mappings

The goal of this section is to establish the alternative principle and related fixed point theorems for non-self semiclosed 1-set contractive mappings associated with various boundary conditions in locally p-convex spaces, where $p \in (0, 1]$.

First, using the results in Section 7.3 above, we establish the following general existence of solutions for Birkhoff–Kellogg problems and the principle of Leray–Schauder alternatives for semiclosed 1-set contractive mappings for locally p-convex spaces, where $p \in (0, 1]$.

Theorem 7.4.1 (Birkhoff–Kellogg alternative). *Let U be a bounded open p-convex subset of a locally p-convex space E, where $p \in (0, 1]$ with the zero element $\theta \in U$, and let C be a closed p-convex subset of E also with zero element $\theta \in C$. Assume that $F : \overline{U} \cap C \to C$ is a continuous semiclosed 1-set contractive single-valued mapping with nonempty values and for each $x \in \partial_C(U)$ with*

$$(P_U^{\frac{1}{p}}(F(x)) - 1)^p \leq P_U^{\frac{1}{p}}(F(x) - x) \ \text{for} \ 0 < p \leq 1 \ \text{(which is trivial}$$

when $p = 1$), where P_U is the Minkowski p-functional of U. Then, we have that either (I) or (II) holding in the following:

(I) *There exists $x_0 \in \overline{U} \cap C$ such that $x_0 = F(x_0)$.*
(II) *There exists $x_0 \in \partial_C(U)$ with $F(x_0) \notin \overline{U}$ and $\lambda > 1$ such that $\lambda x_0 = F(x_0)$, i.e., $F(x_0) \in \{\lambda x_0 : \lambda > 1\}$.*

Proof. By following the argument and notations used in Theorem 7.3.3, we have that either:

(1) F has a fixed point $x_0 \in U \cap C$ or
(2) there exists $x_0 \in \partial_C(U)$ with $x_0 = f(F(x_0))$ such that

$$P_U(F(x_0) - x_0) = d_p(F(x_0), \overline{I_{\overline{U}}^p(x_0)} \cap C)$$

$$= (P_U^{\frac{1}{p}}(F(x_0)) - 1)^p > 0,$$

where $\partial_C(U)$ denotes the boundary of U relative to C in E and f is the restriction of the continuous retraction r with respect to the set U in E.

If F has no fixed point, then (2) holds and $x_0 \neq F(x_0)$. As given by the proof of Theorem 7.3.3, we have that $F(x_0) \notin \overline{U}$, thus $P_U(F(x_0)) > 1$ and $x_0 = f(y_0) = \dfrac{F(x_0)}{P_U^{\frac{1}{p}}(F(x_0))}$, which means

$F(x_0) = P_U^{\frac{1}{p}}(F(x_0))x_0$. Let $\lambda = P_U^{\frac{1}{p}}(F(x_0))$. Then, $\lambda > 1$, and we have $\lambda x_0 = F(x_0)$. This completes the proof. $\qquad\square$

Theorem 7.4.2 (Birkhoff–Kellogg alternative in LCS). *Let U be a bounded open convex subset of a LCS E with the zero element $\theta \in U$, and let C be a closed convex subset of E also with zero element $\theta \in C$. Assume that $F : \overline{U} \cap C \to C$ is a continuous semiclosed 1-set contractive single-valued mapping with nonempty values. Then, we have either (I) or (II) holding the following:*

(I) *There exists $x_0 \in \overline{U} \cap C$ such that $x_0 = F(x_0)$.*
(II) *There exists $x_0 \in \partial_C(U)$ with $F(x_0) \notin \overline{U}$ and $\lambda > 1$ such that $\lambda x_0 = F(x_0)$, i.e., $F(x_0) \in \{\lambda x_0 : \lambda > 1\}$.*

Proof. When $p = 1$, it automatically satisfies the inequality $P_U^{\frac{1}{p}}(F(x)) - 1 \le P_U^{\frac{1}{p}}(F(x) - x)$ for all $x \in \overline{U} \cap C$. Indeed, we have that for $x_0 \in \partial_C(U)$, $(P_U(F(x_0) - x_0))^p = d_p(F(x_0), \overline{I_{\overline{U}}(x_0)} \cap C) = P_U(F(x_0)) - 1$. The conclusions are given by Theorems 7.3.4 and 7.3.5. The proof is complete. $\qquad\square$

Indeed, we have the following fixed points for non-self mappings in locally p-convex spaces for $0 < p \le 1$ under different boundary conditions.

Theorem 7.4.3 (Fixed points of non-self mappings). *Let U be a bounded open p-convex subset of a locally p-convex space E, where $p \in (0, 1]$ with the zero element $\theta \in U$, and let C be a closed p-convex subset of E also with zero element $\theta \in C$. Assume that $F : \overline{U} \cap C \to C$ is a continuous semiclosed 1-set contractive single-valued mapping with nonempty values. In addition, for each $x \in \partial_C(U)$, $(P_U^{\frac{1}{p}}(F(x)) - 1)^p \le P_U^{\frac{1}{p}}(F(x) - x)$ for $0 < p \le 1$ (which is trivial when $p = 1$), where P_U is the Minkowski p-functional of U. We consider if F satisfies any one of the following conditions for any $x \in \partial_C(U) \backslash F(x)$.*

(i) *$P_U(F(x) - z) < P_U(F(x) - x)$ for some $z \in \overline{I_{\overline{U}}(x)} \cap C$;*
(ii) *there exists λ with $|\lambda| < 1$ such that $\lambda x + (1 - \lambda)F(x) \in \overline{I_{\overline{U}}(x)} \cap C$;*

(iii) $F(x) \in \overline{I_{\overline{U}}(x)} \cap C$;

(iv) $F(x) \in \{\lambda x : \lambda > 1\} = \emptyset$;

(v) $F(\partial U) \subset \overline{U} \cap C$;

(vi) $P_U(F(x) - x) \neq ((P_U(F(x)))^{\frac{1}{p}} - 1)^p$.

Then, F must has a fixed point.

Proof. By following the argument and symbols used in the proof of Theorem 7.3.3 (see actually Theorem 6.1.9), we have that either:

(1) F has a fixed point $x_0 \in \overline{U} \cap C$ or

(2) there exists $x_0 \in \partial_C(U)$ with $x_0 = f(F(x_0))$ such that

$$P_U(F(x_0) - x_0) = d_p(F(x_0), \overline{I_{\overline{U}}^p(x_0)} \cap C) = (P_{\overline{U}}^{\frac{1}{p}}(F(x_0)) - 1)^p > 0,$$

where $\partial_C(U)$ denotes the boundary of U relative to C in E and f is the restriction of the continuous retraction r with respect to the set U in E.

First, suppose that F satisfies condition (i). If F has no fixed point, then (2) holds and $x_0 \neq F(x_0)$. Then, by condition (i), it follows that $P_U(F(x_0) - z) < P_U(F(x_0) - x_0)$ for some $z \in \overline{I_{\overline{U}}(x)} \cap C$, which contradicts he best approximation equations given by (2) above, and thus F must have a fixed point.

Second, suppose that F satisfies condition (ii). If F has no fixed point, then (2) holds and $x_0 \neq F(x_0)$. Then, by condition (ii), there exists $|\lambda| < 1$ such that $\lambda x_0 + (1 - \lambda)F(x_0) \in \overline{I_{\overline{U}}(x)} \cap C$. It follows that

$$P_U(F(x_0) - x_0) \leq P_U(F(x_0) - (\lambda x_0 + (1 - \lambda F(x_0))))$$
$$= P_U(\lambda(F(x_0) - x_0))$$
$$= |\lambda|^p P_U(F(x_0) - x_0) < P_U(F(x_0) - x_0),$$

which is impossible, and thus F must have a fixed point in $\overline{U} \cap C$.

Third, suppose that F satisfies condition (iii), i.e., $F(x) \in \overline{I_{\overline{U}}(x)} \cap C$; then by (2), we have that $P_U(F(x_0) - x_0) = 0$, and thus $x_0 = F(x_0)$, which means F has a fixed point.

Fourth, suppose that F satisfies condition (iv). If F has no fixed point, then (2) holds and $x_0 \neq F(x_0)$. As given by the proof of Theorem 7.3.3, we have that $F(x_0) \notin \overline{U}$, and thus $P_U(F(x_0)) > 1$ and $x_0 = f(F(x_0)) = \dfrac{F(x_0)}{P_U^{\frac{1}{p}}(F(x_0))}$, which means $F(x_0) = P_U^{\frac{1}{p}} F(x_0) x_0$, where $P_U^{\frac{1}{p}} F(x_0) > 1$. This contradicts assumption (iv), and thus F must have a fixed point in $\overline{U} \cap C$.

Fifth, suppose that F satisfies condition (v). Then, $x_0 \neq F(x_0)$. As $x_0 \in \partial_C U$, now by condition (v), we have that $F(\partial U) \subset \overline{U} \cap C$, and it follows that for any we have $F(x_0) \in \overline{U} \cap C$. Thus, $F(x) \notin \overline{U} \setminus \cap C$, which implies that $0 < P_U(F(x_0) - x_0) = d_P(F(x_0), \overline{U} \cap C) = 0$. This is impossible, and thus F must have a fixed point. Here, as pointed out in Remark 6.2.4, based on condition (v), applying $F(\partial U) \subset \overline{U} \cap C$ is enough to know that the mapping F has a fixed point by, making the following general hypothesis unnecessary: "for each $x \in \partial_C(U)$, $(P_U^{\frac{1}{p}}(F(x)) - 1)^p \leq P_U^{\frac{1}{p}}(F(x) - x)$ for $0 < p \leq 1$".

Finally, suppose that F satisfies condition (vi). If F has no fixed point, then (2) holds and $x_0 \neq F(x_0)$. Then, condition (v) implies that $P_U(F(x_0) - x_0) \neq ((P_U(F(x))^{\frac{1}{p}} - 1)^p$, but our proof in Theorem 5.2 shows that $P_U(F((x_0) - x_0) = ((P_U(F(x_0)))^{\frac{1}{p}} - 1)^p$, which is impossible, and thus F must have a fixed point. Then. the proof is complete. $\square$

Now, by taking the set C in Theorem 7.4.1 as the whole locally p-convex space E itself, we have the following general results for non-self upper semicontinuous mappings, which include results of fixed point types given by Rothe, Petryshyn, Altman, and Leray–Schauder as special cases.

Taking $p = 1$ and $C = E$ in Theorem 7.4.3, we have the following fixed points for non-self continuous mappings associated with inward or outward sets for LCSs which are locally p-convex spaces for $p = 1$.

Theorem 7.4.4 (Fixed points of non-self mappings with boundary conditions). *Let U be a bounded open convex subset of the LCS E with the zero element $\theta \in U$, and assume that $F : \overline{U} \to E$ is a continuous semiclosed 1-set contractive single-valued mapping* *with nonempty values. We consider if* F *satisfies any one of the following conditions for any* $x \in \partial(U) \backslash F(x)$:

 (i) $P_U(F(x) - z) < P_U(F(x) - x)$ *for some* $z \in \overline{I_{\overline{U}}(x)}$;
 (ii) *there exists* λ *with* $|\lambda| < 1$ *such that* $\lambda x + (1 - \lambda)F(x) \in \overline{I_{\overline{U}}(x)}$;
 (iii) $F(x) \in \overline{I_{\overline{U}}(x)}$;
 (iv) $F(x) \in \{\lambda x : \lambda > 1\} = \emptyset$;
 (v) $F(\partial(U) \subset \overline{U}$;
 (vi) $P_U(F(x) - x) \neq P_U(F(x)) - 1$.

Then, F *must has a fixed point.*

In what follows, based on the best approximation theorem in p-seminorm space, we also give some fixed point theorems for non-self continuous mappings with various boundary conditions, which are related to the study of the existence of solutions for PDEs and differential equations with boundary problems (see, Browder [32], Petryshyn [205, 206], and Reich [222]), which play key roles in non-linear analysis for p-seminorm space, as shown in the following.

First, as discussed in Remark 6.2.4, the proof of Theorem 7.4.3 with only the strong boundary condition "$F(\partial(U)) \subset \overline{U} \cap C$", we can show that F has a fixed point. Thus, we have the following fixed point theorem of Rothe type in locally p-convex spaces.

Theorem 7.4.5 (Rothe type). *Let* U *be a bounded open p-convex subset of a locally p-convex space* E, *where* $p \in (0, 1]$ *with the zero element* $\theta \in U$. *Assume that* $F : \overline{U} \to E$ *is a continuous semiclosed 1-set contractive single-valued mapping with nonempty values, such that* $F(\partial(U)) \subset \overline{U}$. *Then,* F *must have a fixed point.*

Now, as applications of Theorem 7.4.5, we give the following Leray–Schauder alternative in p-vector spaces for non-self single-valued mappings associated with boundary conditions, which often appear in applications (see Isac [118] and references therein for the study of complementary problems and related topics in optimization).

By using the same argument used in the proof of Theorem 7.1.6, we have the following result.

Theorem 7.4.6 (Leray–Schauder alternative in locally p-convex spaces). *Let* E *be a locally p-convex space* E, *where*

$p \in (0, 1]$ and $B \subset E$ is a bounded closed p-convex such that zero element $\theta \in intB$. Let $F : [0, 1] \times B \to E$ be a continuous semiclosed 1-set contractive single-valued mapping with nonempty values such that the set $F([0, 1] \times B)$ being relatively compact in E. We consider if the following assumptions are satisfied:

(1) $x \neq F(t, x)$ for all $x \notin \partial B$ and $t \in [0, 1]$.
(2) $F(\{0\} \times \partial B) \subset B$.

Then, there is an element $x^* \in B$ such that $x^* = F(1, x^*)$.

Proof. The conclusion is proved by following the same argument used in Theorem 7.1.6. The proof is complete. □

As a special case of Theorem 7.4.6, we have the following principle for the implicit form of Leray–Schauder-type alternative in locally p-convex spaces for $0 < p \leq 1$.

Corollary 7.4.7 (Implicit Leray–Schauder alternative). *Let E be a locally p-convex space E, where $p \in (0, 1]$ and $B \subset E$ isa bounded closed p-convex such that the zero element $\theta \in intB$. Let $F : [0, 1] \times B \to E$ be a continuous semiclosed 1-set contractive single-valued mapping with nonempty values, and the set $F([0, 1] \times B)$ is relatively compact in E. If the following assumptions are satisfied:*

(1) $F(\{0\} \times \partial B) \subset B$,
(2) $x \notin F(0, x)$ for all $x \in \partial B$,

Then at least one of the following properties is satisfied:

(i) There exists $x^* \in B$ such that $x^* = F(1, x^*)$.
(ii) There exists $(\lambda^*, x^*) \in (0, 1) \times \partial B$ such that $x^* = F(\lambda^*, x^*)$.

Proof. The result is an immediate consequence of Theorem 7.4.6. This completes the proof. □

Before ending this section, we give a brief summary as follows.

First, we note that similar results on Rothe- and Leray–Schauder-type alternatives have been developed by Furi and Pera [92], Granas and Dugundji [102], Górniewicz [100], Górniewicz *et al.* [101], Isac [118], Li *et al.* [156], Liu [160], Park [185], Potter [213], Shahzad [238, 240], Xu [278], Xu *et al.* [279], and related references therein. These are used as tools of nonlinear analysis in the Banach space setting and applications to the boundary value problems for ordinary differential equations in noncompact problems, a general class of mappings for nonlinear alternative of Leray–Schauder type in normal topological spaces. Some Birkhoff–Kellogg-type theorems for general class mappings in topological vector spaces are also established by Agarwal *et al.* [1], Agarwal and O'Regan [2,3], Park [186], and the references therein. In particular, recently, O'Regan [180] uses the Leray–Schauder-type coincidence theory to establish some new results of the Birkhoff–Kellogg problem type [23] and the Furi–Pera type [92] for a general class of mappings in locally p-convex spaces, where $p \in (0, 1]$.

Secondly, we would like to share that as applications of the best approximation result for 1-set contractive mappings, we can establish the fixed point theorems and general principles of Leray–Schauder alternative [154] for non-self mappings, which could play an important role in the development of nonlinear analysis under the framework of p-vector spaces, where $p \in (0, 1]$.

7.5 Fixed Point Theorems for Non-Self Semiclosed 1-Set Contractive and Nonexpansive Mappings

The goal of this section is to establish fixed point theorems for nonself semiclosed 1-set contractive, in particular for nonexpansive mappings.

Actually, in this section, based on the best approximation Theorem 7.3.3 established for the 1-set contractive mappings in Section 7.3 above, we first show how we can use it as a tool for developing fixed point theorems for semiclosed 1-set contractive non-self upper semicontinuous mappings in p-seminorm spaces for $p \in (0, 1]$, and nonexpansive mappings in LCSs which satisfy the Opial condition or have strictly convex structures.

By following Definition 7.2.1 above, we first observe that if f is a continuous demicompact mapping, then $(I - f)$ is closed, where I is the identity mapping on X. It is also clear from definitions that every demicompact map is hemicompact in seminorm spaces, but the converse is not true in general (e.g., see the example by Tan and Yuan [258, p. 380]). It is evident that if f is demicompact, then $I - f$ is demiclosed. It is know that for each condensing mapping f, when D or $f(D)$ is bounded, then f is hemicompact; and also

f is demicompact in metric spaces by Lemmas 2.1 and 2.2 of Tan and Yuan [258], respectively. In addition, it is known that every non-expansive map is a 1-set contractive mapping; and also if f is a hemicompact 1-set contractive mapping, then f is a 1-set contractive mapping satisfying the following "**condition (H1)**" (which is the same as that introduced in Chapter 5 under the framework of p-seminorm spaces, where $p \in (0, 1]$).

"**Condition (H1)**": Let D be a nonempty bounded closed subset of a space E, and assume that $F : D \to 2^E$ a set-valued mapping. If $\{x_n\}_{n \in \mathbb{N}}$ is any sequence in D such that for each x_n, there exists $y_n \in F(x_n)$ with $\lim_{n \to \infty}(x_n - y_n) = 0$, then there exists a point $x \in D$ such that $x \in F(x)$.

We first note that "condition (H1)" above is actually "condition (C)" used in Theorem 1 of Petryshyn [206]. Indeed, by following Goebel and Kirk [97] (see also Xu [273] and the reference therein), Browder [32] (see also p. 103 of Browder [35]) proved that if K is a closed and convex subset of a uniformly convex Banach space X and if $T : K \to X$ is nonexpansive, then the mapping $f := I - T$ is demiclosed on X. This result, known as Browder's demiclosedness principle (Browder's proof in [30] inspired by the technique of Göhde in [99]), is one of the fundamental results in the theory of nonexpansive mappings, which satisfies "condition (H1)".

For the convenience of our discussion, we recall the following result which is indeed Lemma 7.2.3 (also called Browder's demiclosedness principle established by Browder [32]), which states that a nonexpansive mapping in a uniformly convex Banach X enjoys condition (H1), as shown in the following.

Lemma 7.5.1. *Let D be a nonempty bonded closed convex subset of a uniformly convex Banach space E. Assume that $F : D \to E$ is a nonexpansive single-valued mapping with nonempty values. Then the mapping $P := I - F$ defined by $U(x) := (x - F(x))$ for each $x \in D$ is demiclosed, and in particular, "condition (H1)" holds.*

Proof. It is indeed Lemma 7.2.3. The proof is complete. $\square$

On the other hand, by following the notion called the "Opial condition" given by Opial [179], we recall its definition as follows. A Banach space X is said to satisfy the Opial condition if

$\liminf_{n\to\infty} \|w_n - w\| < \liminf_{n\to\infty} \|w_n - p\|$ whenever (w_n) is a sequence in X weakly convergent to w and $p \neq w$.

We all know that the Opial condition plays an important role in fixed point theory, e.g., see Lami Dozo [150], Goebel and Kirk [97], Xu [273], and the references therein. Actually, the following result tells us that there exists a class of nonexpansive set-valued mappings (which include singe-valued mappings as a special class) in Banach spaces with the Opial condition (see Lami Dozo [150] satisfying the "(H1) condition". In addition, we recall Lemma 7.2.12 above and restate it as the following Lemma 7.5.2.

Lemma 7.5.2. *Let C is a nonempty convex weakly compact subset of a Banach space X which satisfies the Opial condition. Let $T : C \to K(X)$ be a nonexpansive set-valued mapping with nonempty compact-values. Then, the graph of $(I - T)$ is closed $(X, \sigma(X, X^*) \times (X, \|\cdot\|))$, and thus T satisfies the "(H1) condition", where I denotes the identity on X, $\sigma(X, X^*)$ is the weak topology, and $\|\cdot\|$ is the norm (or strong) topology.*

Proof. By following the proof of Theorem 3.1 given by Lami Dozo [150], we can prove that the mapping T is demiclosed, and thus T satisfies "condition (H1)".

We now prove that the graph of $U = I - T$ (denoted by $G(U)$) is closed in X, $\sigma(X, X^*) \times (X, \|\cdot\|)$, where I denotes the identity on Z, $\sigma(X, X^*)$ is the weak topology, and $\|\cdot\|$ is the norm (or strong) topology. As the domain of the mapping $U = I - T$ is weakly compact, we must prove that the graph of U (denoted by $G(U)$) is only sequentially closed. Let $(x_n, y_n) \in G(U)$ be such that x_n weakly converges to x and y_n strongly converges to y. We must see that $x \in C$ and $y \in U(x) = x - T(x)$. That $x \in C$ is clear. As $y_n \in x_n - T(x_n)$, we can write $y_n = x_n - v_n$, where $v_n \in T(x_n)$. Then, by the definition of Hausdorff metrics and as T is nonexpansive, we can find $v'_n \in T(x)$ such that

$$\|v_n - v'_n\| \leq d_H(T(x_n), T(x)) \leq \|x_n - x\|.$$

Now, by putting all the above together and passing to limits on n, it implies that

$$\liminf \|x_n - x\| \geq \liminf \|v_n - v'_n\| = \liminf \|x_n - y_n - v'_n\|.$$

As x_n weakly converges to x, the Opial condition implies that $y + v = x$, so $y = x - v \in x - T(x) = U(x)$, and thus we prove that $G(U)$ is closed in X. Thus, T satisfies the "(H1) condition", and the proof is complete. $\square$

By Lemma 7.5.2 above (see also Theorem 3.1 of Lami Dozo [150]), we have the following statement, which is an another version of Lemma 7.5.2 using the term of "distance convergence" for self-mappings.

Lemma 7.5.3. *Let C be a nonempty convex weakly compact subset of a Banach space (X, d) which satisfies the Opial condition. Let $T : C \to K(C)$ be a nonexpansive set-valued mapping with nonempty compact values (with fixed points existing). Let $(y_n)_{n \in \mathbb{N}}$ be a bounded sequence, such that $\lim_{n \to \infty} d(y_n, T(y_n)) = 0$, and assume that y is the weak cluster points of $(y_n)_{n \in \mathbb{N}}$ (i.e., $y_n \overset{*}{\to} y$, which means y_n weakly converges to y). Then, y is indeed a fixed point of the mapping T.*

Proof. By Lemma 7.5.2, the graph of $U = I - T$ (denoted by $G(U)$) is closed in X, $\sigma(X, X^*) \times (X, \| \cdot \|)$, where I denotes the identity on Z, $\sigma(X, X^*)$ is the weak topology, and $\| \cdot \|$ is the norm (or strong) topology, i.e., $G(U)$ is closed in X. By the assumption that y is the weak cluster point of $(y_n)_{n \in \mathbb{N}}$, i.e., $y_n \overset{*}{\to} y$, $\lim_{n \to \infty} d(y_n, T(y_n)) = 0$. Now, by the fact that $G(U)$ is closed in X, $\sigma(X, X^*) \times (X, \| \cdot \|)$ and $\lim_{n \to \infty} d(y_n, T(y_n)) = 0$, it follows that

$$\lim_{n \to \infty} d(U(y_n), 0) = \lim_{n \to \infty} d(y_n - T(y_n), 0) = \lim_{n \to \infty} d(y_n, T(y_n)) = 0,$$

which implies that $0 \in U(y)$. This means $y \in T(y)$, and thus y is a fixed point of T. This completes the proof. $\square$

Definition 7.5.4 (∗-Nonexpansive mappings). Let C be a subset of a Banach space $(X, \| \cdot \|)$ and $K(C)$ be the family of all nonempty compact subsets of C (see Husain and Tarafdar [112] or Husain and Latif [111]). A mapping $W : C \to K(C)$ is said to be ∗-nonexpansive if for all $x, y \in C$ and $x^W \in W(x)$ such that $\|x - x^W\| = d(x, W(x))$, there exists $y^W \in W(y)$ with $\|y - y^W\| = d(y, W(y))$ such that $\|x^W - y^W\| \leq \|x - y\|$.

We note that the class of ∗-nonexpansive set-valued mappings is proved to hold the demiclosedness principle in reflexive Banach spaces satisfying the Opial condition by Muglia and Marino (i.e., see Lemma 3.4 in [172]. Thus, the demiclosedness principle also holds in reflexive Banach spaces with duality mapping that is weakly sequentially continuous since these satisfy the Opial condition.

On the other hand, as pointed by Muglia and Marino [172], however, the class of *-nonexpansive multi-valued mappings may neither be continuous (see Example 1.1 of Hussain and Khan [113] for nonexpansive mappings with respect to the definition obtained by using the Hausdorff metric (see also Xu [272]). However, by Theorem 3 of López-Acdeo and Xu [161], it is proved that a multi-valued mapping $W : C \to K(C)$ is *-nonexpansive if and only if the metric projection $P_W(x); = \{u_x \in W(x) : \|x - u_x\| = \inf_{y \in W(x)} \|x - y\|\}$ is nonexpansive.

We now have the following result which is the demiclosedness principle for multi-valued *-nonexpansive mapping given by Lemma 3.4 of Muglia and Marino [172] for *-nonexpansive multi-valued mappings.

Lemma 7.5.5. *Let X be a reflexive space satisfying the Opial condition, and let $W : X \to K(X)$ be a *-nonexpansive multi-valued mapping with fixed points (existing) (denoted by $Fix(W)$). Let $(y_n)_{n \in \mathbb{N}}$ be a bounded sequence such that $\lim_{n \to \infty} d(y_n, W(y_n)) \to 0$. Then, the weak cluster points of $(y_n)_{n \in \mathbb{N}}$ belong to $Fix(W)$.*

Proof. Following the proof for Lemma 3.4 given by Muglia and Marino [172], since X is reflexive and $(y_n)_{n \in \mathbb{N}}$ is bounded, we may assume that $(y_{n_k})_{k \in \mathbb{N}} \subset (y_n)_{n \in \mathbb{N}}$ weakly converges to z (also denoted by $y_n \overset{*}{\rightharpoonup} z$). As $W(z)$ is compact, it is closed, and there exists $(z_{n_{k_j}})_{j \in \mathbb{N}} \subset (z_{n_k})_k \subset W(z)$ such that

$$\|y_{n_k} - z_{n_k}\| = d(y_{n_k}, W(z)).$$

Still by the compactness of $W(z)$, there exists a subsequence $(z_{n_{k_j}})_j \subset (z_{n_k})_k \subset W(z)$ that strongly converges to $\hat{z} \in W(z)$.

Now, by the definition of *-nonexpansivity, for any $j \in \mathbb{N}$ and $y_{n_{k_j}}$, there exists $u_{y(j)} \in W(y_{n_{k_j}})$ with $\|y_{n_{k_j}} - u_{(j)}\| = d(y_{n_{k_j}}, W(y_{n_{k_j}}))$ and $u_{z(j)} \in W(z)$ with $\|u_{z(j)} - z\| = d(z, W(z))$ such that

$$\|u_{y(j)} - u_{z(j)}\| \leq \|y_{n_{k_j}} - z\|.$$

We now prove that $z = \hat{z}$, and so will be $z \in W(z)$, which means that z is a fixed point of W. If not, i.e., assuming that $z \neq \hat{z}$, as X satisfies the Opial condition, we have $\limsup_{j \to \infty} \|y_{n_{k_j}} - z\| < \limsup_{j \to \infty} \|y_{n_{k_j}} - \hat{z}\|$. Then, we would arrive at a contradiction. Indeed, since $d(y_n, W(y_n)) \to 0$ and using the Opial inequality, we have that

$$\limsup_{j \to \infty} \|y_{n_{k_j}} - \hat{z}\| \leq \limsup_{j \to \infty} [\|y_{n_{k_j}} - z_{n_{k_j}}\| + \|z_{n_{k_j}} - \hat{z}\|]$$

$$= \limsup_{j \to \infty} d(y_{n_{k_j}}, W(z)) \leq \limsup_{j \to \infty} \|y_{n_{k_j}} - u_{z(j)}\|$$

$$\leq \limsup_{j \to \infty} [\|y_{n_{k_j}} - u_{y(j)}\| + \|u_{y(j)} - u_{z(j)}\|]$$

$$\leq \limsup_{j \to \infty} [d(y_{n_{k_j}}, W(y_{n_{k_j}})) + \|u_{y(j)} - u_{z(j)}\|]$$

$$\leq \limsup_{j \to \infty} \|y_{n_{k_j}} - z\| < \limsup_{j \to \infty} \|y_{n_{k_j}} - \hat{z}\|,$$

which is impossible. Therefore, we must have $z \in W(z)$. This completes the proof. □

Remark 7.5.6. We would like to point out that indeed Xu [275], Lópezo-Acdeo and Xu [161] established fixed point theorems for *-nonexpansive in Banach spaces either satisfying the Opial condition or being uniformly convex. Thus, the assumption on the existence of fixed points for the mappings F or T in Lemma 7.2.3 or Lemma 7.5.2 makes sense under the framework of either being uniformly convex or satisfying the Opial condition.

Let E denote a Hausdorff LCS, $\mathfrak{F}$ denote the family of continuous seminorms generating the topology of E, and $K(E)$ denote the family of nonempty compact subsets of E. For each $p \in \mathfrak{F}$ and $A, B \in K(E)$, by following Ko and Tsai [143], we can define the metric $\delta(A, B)$ between the subsets A and B by

$$\delta(A, B) := \sup\{p(a - b) : a \in A, b \in B\};$$

and the Hausdorff metric $d_{Hp}(A, B)$ by

$$d_{Hp}(A, B) := \max\{\sup_{a \in A} \inf_{b \in B} p(a - b), \sup_{b \in B} \inf_{a \in A} p(a - b)\}.$$

It is clear that when p is only a seminorm in E (thus E is a normed space), d_{Hp} is a traditional Hausdorff metric on $K(E)$.

Secondly, by the definition of the Hausdorff metric d_{Hp} for each $p \in \mathfrak{F}$, we know that if $A, B \in C(E)$, then each $a \in A$, and there exists $b \in B$ such that $p(a - b) \le d_{Hp}(A, B)$ (e.g., by the definition of the Hausdorff metric d_{Hp} and the compactness of subsets $A, B \in K(E)$ for the same idea used by Nadler [173]).

Definition 7.5.7 (Contraction mappings in LCS). Let E denote a Hausdorff locally convex topological vector space, and let $\mathfrak{F}$ denote the family of continuous seminorms generating the topology of E. Let M he a nonempty subset of E. A mapping $T : M \to K(E)$ is said to be a $\mathfrak{F}$ contractive multi-valued mapping if for each $p \in \mathfrak{F}$, there exists a constant $k_p \in (0, 1)$ such that $d_{Hp}(T(x), T(y)) \le k_p p(x - y)$. When $k_p = 1$ for all $p \in \mathfrak{F}$, T is said to be a nonexpansive set-valued mapping if for any $x, y \in M$, we have $d_{Hp}(T(x), T(y))) \le p(x - y)$.

Referring to Chen and Singh [57], we recall the following definition for the Opial condition in LCSs (denoted by P-Opial condition, where the letter P is in the upper case, and not the lower case $p \in (0, 1]$ used throughout this book).

Definition 7.5.8 (P-Opial condition in LCS). The LCS E is said to satisfy the P-Opial condition if for each $x \in E$ and every net (x_α) converging weakly to x, then for each $P \in \mathfrak{F}$, we have $\liminf P(x_\alpha - y) > \liminf P(x_\alpha - x)$ for any $y \ne x$.

Now, we have the following demiclosedness principle for LCSs which satisfy the P-Opial condition due to Chen and Singh [57]. for nonexpansive set-valued mappings in (Hausdorff) local convex spaces E, which is indeed Theorem 1 of Chen and Singh [57].

Lemma 7.5.9. *Let M be a nonempty convex and weakly compact subset of E. Let $T : M \to K(E)$ be a nonexpansive set-valued mapping with nonempty compact values. If E satisfies the P-Opial condition, then the graph $(I - T)$ is closed in $E_w \times E$, where E_w is E with its weak topology and I is the identity mapping.*

Proof. By following the proof of Theorem 1 given by Chen and Singh [57], for any given $P \in \mathfrak{F}$, and let (x_α, y_α) be a net in the graph of $I - T$, denoted by $G(I - T)$ such that x_α weakly converges to x, i.e., $x_\alpha \overset{*}{\rightharpoonup} x$; and y_α (strongly) converges to y, i.e., $y_\alpha \to y$.

It is easy to see that $x \in M$, so we much show that $y \in (I - T)(x)$. Now, for every α, there exists a $v_\alpha \in T(x_\alpha)$ such that $y_\alpha = x_\alpha - v_\alpha$. Then by the definition of the Hausdorff metric d_{Hp}, we can find a $v'_\alpha \in T(x)$ such that

$$P(v_\alpha - v'_\alpha) \leq d_{Hp}(T(x_\alpha), T(x)) \leq P(x_\alpha - x),$$

as T is nonexpansive. Thus, we have

$$\lim P(x_\alpha - x) \geq \lim P(v_\alpha - v'_\alpha) = \lim P(x_\alpha - y_\alpha - v'_\alpha).$$

By the compactness of $T(x)$, this is a convergent subnet in $T(x)$, still denoted by (v'_α) such that $v'_\alpha \to v$ for some $v \in T(x)$. Therefore, we have that

$$\lim P(x_\alpha - x) \geq \lim P(x_\alpha - y - v).$$

Since E satisfies the Opial condition, we obtain $y + v = x$, which means that $y = x - v \in (I - T)(x)$, i.e., the graph $G(I - T)$ of the mapping $(I - G)$ is closed in $E_w \times E$. This completes the proof. $\square$

If the p-normed space E is a uniformly convex Banach space $(E, \|\cdot\|)$, where $p = 1$, then we have the following general existence result, which can be applied to general nonexpansive (single-valued or set-valued) mappings, too.

Theorem 7.5.10. *Let U be a bounded open convex subset of a uniformly convex Banach space $(E, \|\cdot\|)$ (a complete p-normed space with $p = 1$) with zero element $\theta \in U$. Assume that $F : \overline{U} \to E$ is a continuous semi-contractive single-valued mapping with nonempty values. In addition, for any $x \in \partial \overline{U}$, $\lambda x \neq F(x)$ for any $\lambda > 1$ (i.e., the "Leray–Schauder boundary condition"). Then, F has at least one fixed point in $\overline{U}$.*

Proof. By Lemma 7.5.1, F is a semiclosed 1-set contractive mapping. Moreover, by the assumption that E is a uniformly convex Banach space, the mapping $(I - F)$ is closed at zero by Lemma 7.2.3, and thus F is semiclosed at zero element (see Browder [32] or Goebel and Kirk [96]). Thus, all the assumptions of Theorem 7.4.2 are satisfied. The conclusion follows by Theorem 7.4.2. The proof is complete. $\square$

By assuming that the Banach space $(E, \| \cdot \|)$ in Theorem 7.5.10 satisfies the Opial condition instead of being uniformly convex, the conclusion of Theorem 7.5.10 still holds, as given by the following Theorem 7.5.11 for nonexpansive set-valued mappings; however, we list here for the purpose of comparison with nonexpansive set-valued (instead of single-valued) mappings in Banach spaces.

Theorem 7.5.11. *Let C be a nonempty convex weakly compact subset of a local convex space E which satisfies the P-Opial condition and the zero element $\theta \in intC$. Let $T : C \to K(E)$ be a nonexpansive set-valued mapping with nonempty compact convex values. In addition, for any $x \in \partial\overline{C}$, we have $\lambda x \neq F(x)$ for any $\lambda > 1$ (i.e., the "Leray–Schauder boundary condition"). Then, F has at least one fixed point in $\overline{U}$.*

Proof. As T is nonexpansive, it is 1-set contractive. By Lemma 7.5.2, condition (H1) of Theorem 6.2.6 is satisfied. Now, by applying Theorem 6.2.6 in LCS, the conclusion follows. Thus, the proof is complete. □

By using Lemma and Remark 7.5.6, we have the following result in local convex spaces for *-nonexpansive single-valued mappings.

Theorem 7.5.12. *Let C be a nonempty (bonded) convex closed subset of a local convex space E which is either uniformly convex or satisfies the P-Opial condition. Let $T : C \to E$ be a continuous *-nonexpansive single-valued mapping with nonempty values. In addition, for any $x \in \partial\overline{C}$, we have $\lambda x \neq F(x)$ for any $\lambda > 1$ (i.e., the "Leray–Schauder boundary condition"). Then, F has at least one fixed point in $\overline{U}$.*

Proof. As T is *-nonexpansive, by the demiclosedness principle given by Xu [272] for uniformly convex spaces (see also Lemma 7.2.3) and Lópezo-Acdeo and Xu in [161] for Banach spaces satisfying the Opial condition and by the demiclosedness principle for *-nonexpansive mappings given by Lemma 7.5.5, it follows that T satisfies the (H1) condition of Theorem 7.2.5. Then, all the conditions of Theorem 7.2.5 are satisfied, and thus the conclusion follows by Theorem 7.2.5. The proof is complete. □

By considering p-seminorm space $(E, \| \cdot \|)$ with a seminorm for $p = 1$ and by the fact that each continuous semi-closed 1-set contractive single-valued mapping satisfies condition (H1), we have the following result.

Corollary 7.5.13. *Let U be a bounded open convex subset of a norm space $(E, \| \cdot \|)$. Assume that $F : \overline{U} \to E$ is a continuous semi-closed 1-set contractive single-valued mapping with nonempty values. Then, F has at least one fixed point in $\overline{U}$ if there exist $\alpha > 1$ and $\beta \geq 0$ such that any one of the following conditions is satisfied:*

(i) *for each $x \in \partial \overline{U}$, $\|F(x) - x\|^{\alpha} \geq \|F(x)\|^{(\alpha+\beta)}\|x\|^{-\beta} - \|x\|^{\alpha}$;*

(ii) *for each $x \in \partial \overline{U}$, $\|F(x) + x\|^{(\alpha+\beta)} \leq \|F(x)\|^{\alpha}\|x\|^{\beta} + \|x\|^{(\alpha+\beta)}$;*

(iii) *for each $x \in \partial \overline{U}$, $\|F(x) - x\|^{\alpha}\|x\|^{\beta} \geq \|F(x)\|^{\alpha}\|y + x\|^{\beta} - \|x\|^{(\alpha+\beta)}$;*

(iv) *for each $x \in \partial \overline{U}$, $\|F(x) + x\|^{(\alpha+\beta)} \leq \|F(x) - x\|^{\alpha}\|x\|^{\beta} + \|F(x)\|^{\beta}\|x\|^{\alpha}$.*

Remark 7.5.14. As discussed in Lemma 7.5.1 and the proof of Theorem 7.5.10, when a p-vector space is a uniformly convex Banach space, the semi-contractive or nonexpansive mappings automatically satisfy the conditions (see (H1)) required by Theorem 10.1, that is, the mappings are indeed semiclosed. Moreover, the results from Theorem 7.4.1 to Corollary 7.5.13 improve or unify the corresponding results given by Browder [32], Li [155], Li *et al.* [156], Goebel and Kirk [96], Petryshyn [205, 206], Reich [222], Tan and Yuan [258], Xu [272], Xu [278], Xu *et al.* [279], Yuan [286–291], and results from the reference therein by extending the non-self mappings to the classes of semiclosed 1-set contractive set-valued mappings in p-seminorm spaces with $p \in (0.1]$ (including the norm space or Banach space when $p = 1$ for p-seminorm spaces).

To conclude this chapter, we would like to share with readers that the main goal of this part is to develop new results and tools in a natural way for nonlinear analysis of three kinds of mappings, namely (1) condensing; (2) 1-set contractive; and (3) semiclosed mappings under the general framework of locally p-convex spaces, where $p \in (0, 1]$ for (single-valued) continuous mappings, instead of set-valued mappings (so that without the strong condition being "closed p-convex values")!

We do also expect that these new results would become very useful tools for the development of nonlinear functional analysis under the general framework of p-vector spaces, which include the topological vector spaces as a special class, and also its related applications for nonlinear problems on optimization, nonlinear programming, variational inequality, complementarity, game theory, mathematical economics, and so on.

As mentioned at the beginning of this chapter, we do expect that nonlinear results and principles of the best approximation theorem established in this book would play a very important role in nonlinear analysis under the general framework of p-vector spaces for $p \in (0, 1]$, as shown by those results given in Chapters 6 and 7 for both condensing and 1-set contractive mapping. More general, new results in nonlinear analysis for semiclosed 1-set contractive mappings for the development of fixed point theorems for non-self mappings, principles of nonlinear alternative, Rothe-type, Leray–Schauder alternative, and related topics, which not only include corresponding results in the existing literature as special cases but also expected to be important tools for the study of its nonlinear analysis.

Finally, we note that the work presented in this book focuses on the development of nonlinear analysis for single-valued (instead of set-valued) mappings for locally p-convex spaces. This is essentially an important continuation of the work presented recently by Yuan [286]. There, the author aims to establish new results on fixed points and principles of nonlinear alternative for nonlinear mappings, primarily on set-valued mappings developed in locally p-convex spaces for $p \in (0, 1]$.

Though some new results for set-valued mappings in locally p-convex spaces have been developed, e.g., see Gholizadeh *et al.* [94], Park [189–200], Qiu and Rolewicz [221], Xiao and Zhu [270], Yuan [286–291], and the references wherein, we still like to emphasize that results obtained for set-valued mappings for p-vector spaces may face some challenging in dealing with true nonlinear problems. One example is that the assumption used for "set-valued mappings with closed p-convex values" seems too strong, as it always means that the zero element is a trivial fixed point of set-valued mappings, as given by Lemma 2.2.8 in Chapter 2 (see also some discussion by Yuan [286, pp. 40–41]) for $p \in (0, 1]$.

Chapter 8

Nonlinear Analysis of Set-Valued Mappings in Locally p-Convex Spaces

The goal of this chapter is to establish the general fixed point theorem and nonlinear alternatives, which are main tools in nonlinear functional analysis for quasi upper semicontinuous (in short, QUSC) set-valued mappings in locally p-convex spaces, where $p \in (0,1]$. These results are new, and they also provide an affirmative answer to the Schauder conjecture in QUSC set-valued mappings defined on s-convex subsets in Hausdorff locally p-convex spaces, where $s, p \in (0,1]$, which would be fundamental for nonlinear functional analysis in mathematics. This chapter consists of six sections, which are briefly introduced as follows.

Section 8.1 considers the study of fixed point theorems for QUSC condensing set-valued mappings in locally p-convex spaces. Section 8.2 discusses the fixed point theorems and best approximation for QUSC 1-set contractive mappings. In Section 8.3, we study the fixed point theorems and best approximation for QUSC semi-closed 1-set contractive mappings. Section 8.4 discusses the principle of the Birkhoff–Kellogg alternative [23] for QUSC semi-closed 1-set contractive mappings and related fixed points for non-self mappings under various boundary conditions. Then, Section 8.5 focuses on the fixed point theorems of non-self semi-closed 1-set contractive set-valued mappings. Finally, the goal of Section 8.6 is to explore the corresponding fixed point theorems for non-self set-valued nonexpansive mappings under the general locally complete convex spaces by using the Caristi fixed point theorem [39] as a fundamental and powerful tool.

We would like to mention that by comparing with the topological degree approach or other related methods used or developed by Cauty [42, 43], Nhu [175], and others, the analysis method developed in this chapter actually provides an accessible way for the study of nonlinear analysis for p-convex vector spaces, where $p \in (0, 1]$. The results given in this part are new and may be easily understandable for general readers in the mathematical community. In addition, the general fixed point theorems established for QUSC set-valued mappings in locally p-convex spaces for $p \in (0, 1]$ or in topological vector spaces are important for the study of functional analysis. This is illustrated by the works of Agarwal *et al.* [1], Ben-El-Mechaiekh and Mechaiekh [17], Ben-El-Mechaiekh and Saidi [18], Browder [33], Cellina [45], Chang [47], Chang *et al.* [49], Ennassik *et al.* [78], Fan [81, 82], Górniewicz [100], Granas and Dugundji [102], Guo [103], Guo *et al.* [104], Nhu [175], Park [189], Reich [222], Smart [250], Tychonoff [260], Weber [264, 265], Xiao and Lu [268], Xiao and Zhu [270], Xu [273], Yuan [283–285], Zeidler [297], and related references therein. We would also like to point out that the results given in this chapter are new and are a continuation of related work discussed in Chapter 7; see also the recent works of Yuan [286–291] on locally p-convex spaces, for $p \in (0, 1]$.

8.1   Fixed Point Theorems for QUSC Condensing Set-Valued Mappings in Locally p-Convex Spaces

The goal of this section is to establish fixed point theorems for QUSC set-valued mappings in locally p-convex for $p \in (0, 1]$. The new results in this section also provide an affirmative answer to the Schauder conjecture for QUSC set-valued mappings defined on s-convex subsets in locally p-convex spaces, which would be fundamental for nonlinear functional analysis in mathematics, where $s, p \in (0, 1]$.

In this section, as an application of graph approximation for QUSC mappings, which is Lemma 5.3.4 in Chapter 5, we first establish general fixed point theorems for general (compact) QUSC set-valued mappings in locally p-convex spaces, which allow us not only to answer the Schauder conjecture [237] in the affirmative under the general framework of locally p-convex spaces but also unify or improve the corresponding results in the existing literature for nonlinear analysis, where $p \in (0,1]$.

Here, we recall and gather the necessary definitions, notions, and known facts needed in this section. For the convenience of our discussion, we first recall the following result.

Lemma 8.1.1. *If $U \in \mathfrak{V}_0$ is p-convex for some $0 < p \leq 1$, then $\alpha(C_p(A)) = \alpha(A)$ for every $A \subset E$.*

Proof. It is followed by Lemma 5.5.5 in Chapter 5. The proof is complete. □

Now, based on the definition for the measure of noncompactness given by Definition 5.5.3 (originally introduced by Machrafi and Oubbi [162]), we have the following general extended version of the Schauder, Darbo, and Sadovskii type fixed point theorems in the context of locally p-convex vector spaces for condensing mappings.

Theorem 8.1.2. *Let $C \subset E$ be a complete s-convex subset of a locally p-convex space E, with $s, p \in (0,1]$. If $T : C \to 2^C$ is a QUSC and (α) condensing set-valued mappings with nonempty p-convex values and with a closed graph, then T has a fixed point in C.*

Proof. Let $\mathfrak{B}$ be a sufficient collection of p-convex zero neighborhoods in E with respect to which T is condensing and for any given $U \in \mathfrak{B}$. We choose some $x_0 \in C$, and let $\mathfrak{F}$ be the family of all closed p-convex subsets A of C with $x_0 \in A$ and $T(A) \subset A$. Note that $\mathfrak{F}$ is not empty since $C \in \mathfrak{F}$. Let $A_0 = \cap_{A \in \mathfrak{F}} A$. Then, A_0 is a nonempty closed p-convex subset of C, such that $T(A_0) \subset A_0$. We shall show that A_0 is compact. Let $A_1 = \overline{C_p(T(A_0) \cup \{x_0\})}$. Since $T(A_0) \subset A_0$ and A_0 is closed and p-convex, $A_1 \subset A_0$. Hence, $T(A_1) \subset T(A_0) \subset A_1$. It follows that $A_1 \in \mathfrak{F}$ and therefore $A_1 = A_0$. Now, by Lemma 8.1.1 (and the common facts listed on Proposition 1 of Machrafi and Oubbi [162] and their Theorems 1 and 2 in Ref. [162]), we get $\alpha_U(T(A_0)) = \alpha_U(A)$. Our assumption on T shows that $\alpha_U(A_0) = 0$ since T is condensing. As U is arbitrary from the family $\mathfrak{B}$, A_0 is p-convex and compact (see Proposition 4 of Machrafi and Oubbi [162]). Now, the conclusion follows by Theorem 5.4.3 above. The proof is complete. □

As an application of Theorem 8.1.2, we have the following general result which answers the Schauder conjecture for QUSC set-valued mappings defined on s-convex subsets in locally convex spaces (LCSs), where $p \in (0, 1]$.

Theorem 8.1.3 (Schauder fixed point theorem). *Let K be a nonempty closed p-convex subset of a locally p convex space, where $p \in (0, 1]$, then any QUSC set-valued, (α) condensing mapping $T : K \to 2^K$ with nonempty p-convex values, and with a closed graph has at least one fixed point.*

Proof. Let $s = p$ in Theorem 8.1.2. Then, the conclusion follows. Thus, we complete the proof.   □

As a special case of Theorem 8.1.3, we have the following result.

Theorem 8.1.4. *Let K be a closed p-convex subset of a Hausdorff locally p-convex space X, where $p \in (0, 1]$. If $T : K \to 2^K$ is an upper continuous condensing set-valued mapping with nonempty closed p-convex values, then T has a fixed point in K.*

Proof. By the fact each upper semicontinuous (USC) set-valued mapping is QUSC and, secondly, each USC with closed value has a closed graph, the conclusion follows by Theorem 8.1.3. This completes the proof.   □

As applications of Theorem 8.1.4, we have the following fixed point theorems for condensing mappings in locally p-convex spaces for $p \in (0, 1]$.

Corollary 8.1.5 (Darbo-type fixed point theorem). *Let C be a complete p-convex subset of a Hausdorff locally p-convex space E with $p \in (0, 1]$. If $T : C \to 2^C$ is a (k)-set contraction (where $k \in (0, 1)$) with closed and p-convex values, then T has a fixed point.*

Corollary 8.1.6 (Sadovskii-type fixed point theorem). *Let $(E, \|\cdot\|)$ be a complete p-normed space and C be a bounded, closed, and p-convex subset of E, where $p \in (0, 1]$. Then, every QUSC and condensing mapping $T : C \to 2^C$ with closed and p-convex values has a fixed point.*

Proof. In Theorem 8.1.3, let $\mathfrak{B} := \{B_p(0,1)\}$, where $B_p(0,1)$ stands for the closed unit ball of E, and by the fact that it is clear that $\alpha(A) = (\alpha_{\mathfrak{B}}(A))^p$ for each $A \subset E$, then, T satisfies all the conditions of Theorem 8.1.3. This completes the proof. □

Corollary 8.1.7 (Darbo type). *Let $(E, \|\cdot\|)$ be a complete p-normed space and C be a bounded, closed, and p-convex subset of E, where $p \in (0,1]$. Then, each mapping $T : C \to C$ which is continuous and set-contractive has a fixed point.*

Before concluding this section, we note that Theorems 8.1.2 and 8.1.3 improve Theorem 5 of Machrafi and Oubbi [162] for general condensing mappings, which are general USC mappings with closed p-convex values and also unify the corresponding results in the existing literature, e.g., see Alghamdi *et al.* [5], Górniewicz [100], Górniewicz *et al.* [101], Nussbaum [176], Silva *et al.* [249], Xiao and Lu [268], Xiao and Zhu [270], and the references therein.

Secondly, as an application of the KKM principle for abstract convex spaces with graph approximation Lemma 5.3.4 for QUSC set-valued mappings in locally p-convex spaces, we establish the general fixed point theorems for QUSC set-valued mappings, which allow us to answer the Schauder conjecture [237] in the affirmative way under the framework of locally p-convex spaces for $p \in (0,1]$.

Finally, we also remark that by comparing with the topological method or other related arguments used by Askoura *et al.* [8], Cauty [42,43], Dobrowolski [67], Nhu [175], and Reich [222], the fixed points theorems given in this section improve or unify the corresponding ones given by Alghamdi *et al.* [5], Darbo [63], Liu [160], Machrafi and Oubbi [162], Sadovskii [235], Silva *et al.* [249], Xiao and Lu [268], and Yuan [285, 286], as well as those from the references therein, by applying this analysis method.

8.2 Fixed Point and Best Approximation for QUSC 1-Set Contractive Mappings

The goal of this section is first to establish a general best approximation result for 1-set upper semicontinuous and hemicompact (see its definition in the following) non-self set-valued mappings, which in turn is used as a tool to derive the general principle for the existence of solutions for Birkhoff–Kellogg problems (formulated by Birkhoff and Kellogg [23] in 1922) and fixed points for non-self 1-set contractive set-valued mappings.

Since Birkhoff and Kellogg developed their theorem, studies on Birkhoff–Kellogg problems have received considerable attention from scholars. For example, one of the fundamental results in nonlinear functional analysis, called the Leray–Schauder alternative, which was proposed by Leray and Schauder [154] in 1934, was established via the topological degree. Thereafter, certain other types of Leray–Schauder alternatives were proved using different techniques other than the topological degree; see the works of Granas and Dugundji [102] and Furi and Pera [92] in the Banach space setting and applications to the boundary value problems for ordinary differential equations, a general class of mappings for nonlinear alternative of Leray–Schauder type in normal topological spaces, and also Birkhoff–Kellogg-type theorems for general class of mappings in topological vector spaces by Agarwal *et al.* [1], Agarwal and O'Regan [2, 3], Park [186–200], and the references therein. In particular, recently, O'Regan [180] used the Leray–Schauder-type coincidence theory to establish some Birkhoff–Kellogg problems and Furi–Pera-type results for a general class of set-valued mappings, too.

In this section, a best approximation result for 1-set contractive mappings in p-seminorm spaces was first established, which is then used for the general principle for solutions of Birkhoff–Kellogg problems and related nonlinear alternatives. Then, they allows us to give general existence results for the Leray–Schauder-type and related fixed point theorems of non-self mappings in p-seminorm spaces for $p \in (0, 1]$. The new results given in this part not only include the corresponding results in the existing literature as special cases but are also expected to be useful tools for the study of nonlinear problems arising from theory to practice for 1-set contractive mappings.

We also note that the general nonlinear alternative related to the Leray–Schauder alternative under the framework of p-seminorm spaces for $p \in (0, 1]$ given in this section would be a useful tool for the study of nonlinear problems. In addition, we also note that the corresponding results in the existing literature for Birkhoff–Kellogg problems and the Leray–Schauder alternative have been studied comprehensively by Granas and Dugundji [102], Isac [118], Park [187–189], Carbone and Conti [38], Chang and Yen [55], Chang *et al.* [53, 54], Kim *et al.* [129], Shahzad [238, 240], Singh [248]; in particular, many general forms were recently obtained by O'Regan [181] and the references therein.

By following the original idea by Tan and Yuan [258] for hemicompact mappings in metric spaces, we recall the following definition for a mapping being hemicompact in p-seminorm spaces for $p \in (0,1]$, which is indeed the "**(H) condition**" used in Theorem 6.1.5 in Chapter 6, which will be used to establish existence results for best approximation of 1-set contractive set-valued mappings in p-seminorm vector spaces for $p \in (0,1]$.

Definition 8.2.1 (Hemicompact mappings in p-seminorm spaces). Let E be a p-vector space with p-seminorm for $1 < p \leq 1$. For a given bounded (closed) subset D in E, a mapping $F : D \to 2^E$ is said to be **hemicompact** if each sequence $\{x_n\}_{n \in N}$ in D has a convergent subsequence with limit x_0 such that $x_0 \in F(x_0)$, whenever $\lim_{n \to \infty} d_{P_U}(x_n, F(x_n)) = 0$ for each $U \in \mathfrak{U}$, where $d_{P_U}(x, C) := \inf\{P_U(x - y) : y \in C\}$ is the distance of a single point x with the subset C in E based on P_U. P_U is the Minkowski p-functional in E for $U \in \mathfrak{U}$, which is the base of the family consisting of all subsets of θ-neighborhoods in E.

We would like to point out that Definition 8.2.1 is indeed an extension of Definition 6.1.3 given for a "hemicompact mapping" defined from a metric space to p-vector spaces with the p-seminorm in general, where $p \in (0,1]$. By the monotonicity of Minkowski p-functionals, i.e., the larger the θ-neighborhoods, the smaller the values of the Minkowski p-functionals (see Balachandran [13, p. 178]), indeed, the "**Hemicompact mapping F**" given by Definition 6.1.3 describes the convergence for the distance between x_n and $F(x_n)$ by using the language of seminorms in terms of Minkowski p-functionals for each θ-neighborhood in $\mathfrak{U}$ (the base), which is the family consisting of its θ-neighborhoods in p-vector space E.

Now, we have the following Schauder fixed point theorem for 1-set contractive mappings in locally p-convex spaces for $p \in (0,1]$.

Theorem 8.2.2 (Schauder fixed point theorem for 1-set contractive mappings). *Let U be a nonempty bounded open p-convex subset of a (Hausdorff) locally p-convex space E and its zero element $\theta \in U$, and let $C \subset E$ be a closed p-convex subset of E such that $\theta \in C$, with $p \in (0,1]$. If $F : C \cap \overline{U} \to 2^{C \cap \overline{U}}$ is a QUSC 1-set* *contractive set-valued mapping with nonempty p-convex values, with a closed graph, and satisfying the following conditions (H) or (H1):*

(H) condition: *The sequence $\{x_n\}_{n\in\mathbb{N}}$ in $\overline{U}$ has a convergent subsequence with limit $x_0 \in \overline{U}$ such that $x_0 \in F(x_0)$, whenever $\lim_{n\to\infty} d_{P_U}(x_n, F(x_n)) = 0$, where, $d_{P_U}(x_n, F(x_n)) := \inf\{P_U(x_n - z) : z \in F(x_n)\}$, with P_U being the Minkowski p-functional for any $U \in \mathfrak{U}$, which is the family of all nonempty open p-convex subsets containing the zero in E;*

(H1) condition: *There exists x_0 in $\overline{U}$ with $x_0 \in F(x_0)$ if there exists $\{x_n\}_{n\in\mathbb{N}}$ in $\overline{U}$ such that $\lim_{n\to\infty} d_{P_U}(x_n, F(x_n)) = 0$, where P_U is the Minkowski p-functional for any $U \in \mathfrak{U}$, which is the family of all nonempty open p-convex subsets containing the zero in E.*

Then, F has at least one fixed point in $C \cap \overline{U}$.

Proof. Let $\mathfrak{U}$ be the family of all nonempty open p-convex subsets containing the zero in E, and let U be any element in $\mathfrak{U}$. As the mapping T is 1-set contractive, consider an increasing sequence $\{\lambda_n\}$ such that $0 < \lambda_n < 1$ and $\lim_{n\to\infty} \lambda_n = 1$, where $n \in \mathbb{N}$. Now, we define a mapping $F_n : C \to 2^C$ by $F_n(x) := \lambda_n F(x)$ for each $x \in C$ and $n \in \mathbb{N}$. Then, it follows that F_n is a λ_n-set contractive mapping with $0 < \lambda_n < 1$, and also QUSC has p-convex values, and its graph is also closed. Now, by Theorem 8.1.2 on the condensing mapping F_n in locally p-convex spaces with the p-seminorm P_U (which is the Minkowski p-functional for $U \in \mathfrak{U}$), for each $n \in \mathbb{N}$, there exists $x_n \in C$ such that $x_n \in F_n(x_n) = \lambda_n F(x_n)$. Thus, there exists $y_n \in F(x_n)$ such that $x_n = \lambda_n y_n$. As P_U is the Minkowski p-functional of U in E, it follows that P_U is continuous as $0 \in int(U) = U$. Note that for each $n \in \mathbb{N}$, $\lambda_n x_n \in \overline{U} \cap C$, which implies that $x_n = r(\lambda_n y_n) = \lambda_n y_n$, and thus $P_U(\lambda_n y_n) \leq 1$ by Lemma 2.2.10. Note that

$$P_U(y_n - x_n) = P_U(y_n - x_n) = P_U(y_n - \lambda_n y_n)$$
$$= P_U\left(\frac{(1-\lambda_n)\lambda_n y_n}{\lambda_n}\right)$$
$$\leq \left(\frac{1-\lambda_n}{\lambda_n}\right)^p P_U(\lambda_n y_n)$$
$$\leq \left(\frac{1-\lambda_n}{\lambda_n}\right)^p,$$

which implies that $\lim_{n\to\infty} P_U(y_n - x_n) = 0$ for all $U \in \mathfrak{U}$.

Now,

(i) If F satisfies the (H) condition, it implies that the consequence $\{x_n\}_{n\in\mathbb{N}}$ has a convergent subsequence which converges to x_0 such that $x_0 \in F(x_0)$. Without loss of generality, we assume that $\lim_{n\to\infty} x_n = x_0$, where $y_n \in F(x_n)$ is with $x_n = \lambda_n y_n$, and $\lim_{n\to\infty} \lambda_n = 1$. It implies that $x_0 = \lim_{n\to\infty}(\lambda_n y_n)$, which means $y_0 := \lim_{n\to\infty} y_n = x_0$. There exists $y_0(= x_0) \in F(x_0)$.

(ii) If F satisfies the (H1) condition, then by that condition, it follows that there exists x_0 in $\overline{U}$ such that $x_0 \in F(x_0)$, which is a fixed point of F. We complete the proof. $\qquad\square$

Theorem 8.2.3 (Best approximation for 1-set contractive mappings). *Let U be a bounded open p-convex subset of a locally p-convex space E, where $p \in (0,1]$ and the zero element $\theta \in U$, and let C be a (bounded) closed convex subset of E also with zero element $\theta \in C$. Assume that $F : \overline{U} \cap C \to 2^C$ is a QUSC 1-set contractive set-valued mapping with nonempty p-convex values and with a closed graph, and for each $x \in \partial_C U$ with $y \in F(x) \cap (C \setminus \overline{U}))$,*

$$(P_U^{\frac{1}{p}}(y) - 1)^p \le P_U(y-x) \text{ for } 0 < p \le 1 \text{ (which is trivial when } p = 1).$$

In addition, we consider if F satisfies the following condition (H) or (H1):

(H) condition: *The sequence $\{x_n\}_{n\in\mathbb{N}}$ in $\overline{U}$ has a convergent subsequence with limit $x_0 \in \overline{U}$ such that $x_0 \in F(x_0)$, whenever $\lim_{n\to\infty} d_{P_U}(x_n, F(x_n)) = 0$, where $d_{P_U}(x_n, F(x_n)) := \inf\{P_U(x_n - z) : z \in F(x_n)\}$, with P_U being the Minkowski p-functional for any $U \in \mathfrak{U}$, which is the family of all nonempty open p-convex subsets containing the zero in E.*

(H1) condition: *There exists x_0 in $\overline{U}$ with $x_0 \in F(x_0)$ if there exists $\{x_n\}_{n\in\mathbb{N}}$ in $\overline{U}$ such that $\lim_{n\to\infty} d_{P_U}(x_n, F(x_n)) = 0$, where P_U is the Minkowski p-functional for any $U \in \mathfrak{U}$, which is the family of all nonempty open p-convex subsets containing the zero in E.*

Then, we have that there exist $x_0 \in C \cap \overline{U}$ and $y_0 \in F(x_0)$ such that

$$P_U(y_0 - x_0) = d_P(y_0, \overline{U} \cap C) = d_p(y_0, \overline{I_U^p(x_0)} \cap C),$$

where P_U is the Minkowski p-functional of U. More precisely, we have either (I) or (II) holding in the following:

(I) *F has a fixed point $x_0 \in \overline{U} \cap C$, i.e., $0 = P_U(y_0 - x_0) = d_P(y_0, \overline{U} \cap C) = d_p(y_0; \overline{I_U^p(x_0)} \cap C)$.*

(II) *There exists $x_0 \in \partial_C(U)$ and $y_0 \in F(x_0)\backslash\overline{U}$ with*

$$P_U(y_0 - x_0) = d_P(y_0, \overline{U} \cap C) = d_p(y_0, \overline{I_{\overline{U}}^p(x_0)} \cap C)$$

$$= (P_{\overline{U}}^{\frac{1}{p}}(y_0) - 1)^p > 0.$$

Proof. As E is a p-convex space and U is a bounded open p-convex subset of E, it suffices to prove that there exists a sequence $(x_n)_{n \in \mathbb{N}}$ in $\overline{U}$ and $y_n \in F(x_n)$ such that $\lim_{n \to \infty} P_U(y_n - x_n) = 0$, and the conclusion follows by applying the (H) condition.

Let $r : E \to U$ be a retraction mapping defined by $r(x) := \dfrac{x}{\max\{1, (P_U(x))^{\frac{1}{p}}\}}$ for each $x \in E$, where P_U is the Minkowski p-functional of U. Since the space E's zero element $\theta \in U(= intU$, as U is open), it follows that r is continuous by Lemma 2.2.10. As the mapping F is 1-set contractive, consider an increasing sequence $\{\lambda_n\}$ such that $0 < \lambda_n < 1$ and $\lim_{n \to \infty} \lambda_n = 1$, where $n \in \mathbb{N}$. Now, for each $n \in \mathbb{N}$, we define a mapping $F_n : C \cap \overline{U} \to 2^C$ by $F_n(x) := \lambda_n F \circ r(x)$ for each $x \in C \cap \overline{U}$. By the fact that C and $\overline{U}$ are p-convex, it follows that $r(C) \subset C$ and $r(\overline{U}) \subset \overline{U}$, and thus $r(C \cap \overline{U}) \subset C \cap \overline{U}$. Therefore, F_n is a mapping from $\overline{U} \cap C$ to itself. For each $n \in \mathbb{N}$, by the fact that F_n is a λ_n-set contractive mapping with $0 < \lambda_n < 1$ and is also QUSC, with nonempty p-convex and its graph is also closed. Then, it follows by Theorem 8.1.2 for the condensing mapping that there exists $z_n \in C \cap \overline{U}$ such that $z_n \in F_n(z_n) = \lambda_n F \circ r(z_n)$. As $r(C \cap \overline{U}) \subset C \cap \overline{U}$, let $x_n = r(z_n)$. Then, we have that $x_n \in C \cap \overline{U}$, and there exists $y_n \in F(x_n)$ with $x_n = r(\lambda_n y_n)$ such that the following (1) or (2) holds for each $n \in \mathbb{N}$:

(1) $\lambda_n y_n \in C \cap \overline{U}$;
(2) $\lambda_n y_n \in C \backslash \overline{U}$.

Now, we prove the conclusion by considering the following two cases under conditions (H) and (H1):

- Case (I) For each $n \in N$, $\lambda_n y_n \in C \cap \overline{U}$.
- Case (II) There exists a positive integer n such that $\lambda_n y_n \in C \backslash \overline{U}$.

First, according to case (I), for each $n \in \mathbb{N}$, $\lambda_n y_n \in \overline{U} \cap C$, which implies that $x_n = r(\lambda_n y_n) = \lambda_n y_n$, and thus $P_U(\lambda_n y_n) \leq 1$ by

Lemma 2.2. Note that

$$P_U(y_n - x_n) = P_U(y_n - x_n) = P_U(y_n - \lambda_n y_n) = P_U\left(\frac{(1 - \lambda_n)\lambda_n y_n}{\lambda_n}\right)$$

$$\leq \left(\frac{1 - \lambda_n}{\lambda_n}\right)^p P_U(\lambda_n y_n) \leq \left(\frac{1 - \lambda_n}{\lambda_n}\right)^p,$$

which implies that $\lim_{n \to \infty} P_U(y_n - x_n) = 0$. Now, for any $V \in \mathbb{U}$, without loss of generality, let $U_0 = V \cap U$. Then, we have the following conclusion:

$$P_{U_0}(y_n - x_n) = P_{U_0}(y_n - x_n) = P_{U_0}(y_n - \lambda_n y_n)$$

$$= P_{U_0}\left(\frac{(1 - \lambda_n)\lambda_n y_n}{\lambda_n}\right)$$

$$\leq \left(\frac{1 - \lambda_n}{\lambda_n}\right)^p P_{U_0}(\lambda_n y_n) \leq \left(\frac{1 - \lambda_n}{\lambda_n}\right)^p,$$

which implies that $\lim_{n \to \infty} P_{U_0}(y_n - x_n) = 0$, where P_{U_0} is the Minkowski p-functional of U_0 in E.

Now, if F satisfies the (H) condition, it follows that the consequence $\{x_n\}_{n \in \mathbb{N}}$ has a convergent subsequence which converges to x_0 such that $x_0 \in F(x_0)$. Without loss of generality, we assume that $\lim_{n \to \infty} x_n = x_0$, where $y_n \in F(x_n)$ is with $x_n = \lambda_n y_n$, and $\lim_{n \to \infty} \lambda_n = 1$, and as $x_0 = \lim_{n \to \infty}(\lambda_n y_n)$, it implies that $y_0 = \lim_{n \to \infty} y_n = x_0$. Thus, there exists $y_0(= x_0) \in F(x_0)$, and we have $0 = d_p(x_0, F(x_0)) = d(y_0, \overline{U} \cap C) = d_p(y_0, \overline{I_{\overline{U}}^p}(x_0) \cap C))$, as indeed $x_0 = y_0 \in F(x_0) \in \overline{U} \cap C \subset \overline{I_{\overline{U}}^p(x_0)} \cap C$.

If F satisfies the (H1) condition, it follows that there exists $x_0 \in \overline{U} \cap C$ with $x_0 \in F(x_0)$. Then, we have $0 = P_U(y_0 - x_0) = d_P(y_0, \overline{U} \cap C) = d_p(y_0, \overline{I_{\overline{U}}^p}(x_0) \cap C)$.

Second, according to case (II), there exists a positive integer n such that $\lambda_n y_n \in C \backslash \overline{U}$. Then, we have that $P_U(\lambda_n y_n) > 1$, and also $P_U(y_n) > 1$, as $\lambda_n < 1$. As $x_n = r(\lambda_n y_n) = \frac{\lambda_n y_n}{(P_U(\lambda_n y_n))^{\frac{1}{p}}}$, which implies that $P_U(x_n) = 1$, and thus $x_n \in \partial_C(U)$. Note that $P_U(y_n - x_n) = P_U(\frac{(P_U(y_n)^{\frac{1}{p}} - 1) y_n}{P_U(y_n)^{\frac{1}{p}}}) = (P_U^{\frac{1}{p}}(y_n) - 1)^p$. By the assumption, we have

$(P_U^{\frac{1}{p}}(y_n) - 1)^p \leq P_U(y_n - x)$ for $x \in C \cap \partial \overline{U}$. It follows that

$$(P_U(y_n) - 1)^p \leq P_U(y_n) - \sup\{P_U(z) : z \in C \cap \overline{U}\}$$
$$\leq \inf\{P_U(y_n - z) : z \in C \cap \overline{U}\} = d_p(y_n, C \cap \overline{U}).$$

Thus, we have the following best approximation:

$$P_U(y_n - x_n) = d_P(y_n, \overline{U} \cap C) = (P_U^{\frac{1}{p}}(y_n) - 1)^p > 0.$$

Now, we want to show that $P_U(y_n - x_n) = d_P(y_n, \overline{U} \cap C) = d_p(y_n, \overline{I_{\overline{U}}^p(x_0)} \cap C) > 0$.

By the fact that $(\overline{U} \cap C) \subset I_{\overline{U}}^p(x_n) \cap C$, let $z \in I_{\overline{U}}^p(x_n) \cap C \backslash (\overline{U} \cap C)$. We first claim that $P_U(y_n - x_n) \leq P_U(y_n - z)$. If not, we have $P_U(y_n - x_n) > P_U(y_n - z)$. As $z \in I_{\overline{U}}^p(x_n) \cap C \backslash (\overline{U} \cap C)$, there exists $y \in \overline{U}$ and a non-negative number c (actually $c \geq 1$ as will be soon shown) with $z = x_n + c(y - x_n)$. Since $z \in C$, but $z \notin \overline{U} \cap C$, it implies that $z \notin \overline{U}$. By the fact that $x_n \in \overline{U}$ and $y \in \overline{U}$, we must have the constant $c \geq 1$; otherwise, it implies that $z(= (1-c)x_n + cy) \in \overline{U}$, which is impossible by our assumption, i.e., $z \notin \overline{U}$. Thus, we have that $c \geq 1$, which implies that $y = \frac{1}{c}z + (1 - \frac{1}{c})x_n \in C$ (as both $x_n \in C$ and $z \in C$). On the other hand, as $z \in I_{\overline{U}}^p(x_n) \cap C \backslash (\overline{U} \cap C)$, and $c \geq 1$ with $(\frac{1}{c})^p + (1 - \frac{1}{c})^p = 1$, combining with our assumption that for each $x \in \partial_C \overline{U}$ and $y \in F(x_n) \backslash \overline{U}$, $(P_U^{\frac{1}{p}}(y) - 1)^p \leq P_U^{\frac{1}{p}}(y - x)$ for $0 < p \leq 1$, it then follows that

$$P_U(y_n - y) = P_U \left[\frac{1}{c}(y_n - z) + \left(1 - \frac{1}{c}\right)(y_n - x_n) \right]$$
$$\leq \left[\left(\frac{1}{c}\right)^p P_U(y_n - z) + \left(1 - \frac{1}{c}\right)^p P_U(y_n - x_n) \right]$$
$$< P_U(y_n - x_n),$$

which contradicts that $P_U(y_n - x_n) = d_P(y_n, \overline{U} \cap C)$ as shown above. We know that $y \in \overline{U} \cap C$, and we should have $P_U(y_n - x_n) \leq P_U(y_n - y)$! This helps us complete the claim $P_U(y_n - x_n) \leq P_U(y_n - z)$ for any $z \in I_{\overline{U}}^p(x_n) \cap C \backslash (\overline{U} \cap C)$, which means that the following best approximation of Fan type (see [84, 85]) holds:

$$0 < d_P(y_n, \overline{U} \cap C) = P_U(y_n - x_n) = d_p(y_n, I_{\overline{U}}^p(x_n) \cap C).$$

Now, by the continuity of P_U, it follows that the following best approximation of Fan type is also true:

$$0 < P_U(y_n - x_n) = d_P(y_n, \overline{U} \cap C) = d_p(y_n, I_{\overline{U}}^p(x_n) \cap C)$$

$$= d_p(y_n, \overline{I_{\overline{U}}^p(x_n)} \cap C).$$

The proof is complete.   □

Remark 8.2.4. Based on the Proof of Theorem 8.2.3, we have the following general conclusions:

First, for the condition "$x \in \partial_C U$ with $y \in F(x)$, $P_U^{\frac{1}{p}}(y) - 1 \le P_U^{\frac{1}{p}}(y - x)$ for $0 < p \le 1$", indeed we only need that "$x \in \partial_C U$ with $y \in F(x) \cap (C \backslash \overline{U})$, $(P_U^{\frac{1}{p}}(y) - 1)^p \le P_U^{\frac{1}{p}}(y - x)$ for $0 < p \le 1$".

Second, Theorem 8.2.3 also improves the corresponding best approximation for 1-set contractive mappings given by Li *et al.* [156], Liu [160], Xu [278], Xu *et al.* [279], and the results from the references therein.

Third, when $p = 1$, we have a similar best approximation result for the mapping F in the LCSs with the following outward set boundary condition (see Theorem 3 of Park [185] and related discussion in the references therein).

For the p-vector space with $p = 1$ being a topological vector space E, we have the following best approximation for the outward set $O_{\overline{U}}(x_0)$ based on the point $\{x_0\}$ with respect to the convex subset U in E.

Theorem 8.2.5 (Best approximation for outward sets). *Let U be a bounded open convex subset of an LCS E with zero element $\theta \in int U = U$ (the interior $int U = U$, as U is open), and let C be a closed p-convex subset of E also with zero element $\theta \in C$. Assume that $F : \overline{U} \cap C \to 2^C$ is a QUSC 1-set contractive set-valued mapping with nonempty p-convex values, with a closed graph, and satisfying condition (H) or (H1) above. Then, there exist $x_0 \in \overline{U} \cap X$ and $y_0 \in F(x_0)$ such that $P_U(y_0 - x_0) = d_P(y_0, \overline{U} \cap C) = d_p(y_0, \overline{O_{\overline{U}}(x_0)} \cap C)$, where P_U is the Minkowski p-functional of U. More precisely, we have either (I) or (II) holding in the following:*

(I) *F has a fixed point $x_0 \in \overline{U} \cap C$, i.e., $P_U(y_0 - x_0) = P_U(y_0 - x_0) = d_P(y_0, \overline{U} \cap C) = d_p(y_0, \overline{O_{\overline{U}}(x_0) \cap C})) = 0$.*

(II) *There exists $x_0 \in \partial_C(U)$ and $y_0 \in F(x_0) \backslash \overline{U}$ with*

$$P_U(y_0 - x_0) = d_P(y_0, \overline{U} \cap C) = d_p(y_0, O_{\overline{U}}(x_0) \cap C)$$

$$= d_p(y_0, \overline{O_{\overline{U}}(x_0)} \cap C) > 0.$$

Proof. We define a new mapping $F_1 : \overline{U} \cap C \to 2^C$ by $F_1(x) := \{2x\} - F(x)$ for each $x \in \overline{U} \cap C$. Then, F_1 is also compact and upper semicontinuous mapping with nonempty closed convex values, and F_1 satisfies all the hypotheses of Theorem 8.2.3 with $p = 1$. It follows by Theorem 8.2.3 that there exists $x_0 \in \overline{U} \cap X$ and $y_1 \in F_1(x_0)$ such that $P_U(y_1 - x_0) = d_p(y_1, \overline{U} \cap C) = d_p(y_1, \overline{I_{\overline{U}}(x_0)} \cap C)$. More precisely, we have either (I) or (II) holding in the following:

(I) F_1 has a fixed point $x_0 \in U \cap C$ (so $0 = P_U(y_1 - x_0) = P_U(y_1 - x_0) = d_P(y_1, \overline{U} \cap C) = d_p(y_1, \overline{I_{\overline{U}}(x_0)} \cap C))$.

(II) There exists $x_0 \in \partial_C(U)$ and $y_1 \in F_1(x_0) \backslash \overline{U}$ with

$$P_U(y_1 - x_0) = d_P(y_1, \overline{U} \cap C) = d_p(y_1, \overline{O_{\overline{U}}(x_0)} \cap C) > 0.$$

Now, for any $x \in O_{\overline{U}}(x_0)$, there exists $r < 0, u \in \overline{U}$ such that $x = x_0 + r(u - x_0)$. Let $x_1 = 2x_0 - x$. Then, $x_1 = 2x_0 - x_0 - r(u - x_0) = x_0 + (-r)(u - x_0) \in I_{\overline{U}}(x_0)$. Let $y_1 = 2x_0 - y_0$, for some $y_0 \in F(x_0)$. As we have $P_U(y_1 - x_0) = d_P(y_1, \overline{U} \cap C) = d_p(y_1, \overline{I_{\overline{U}}(x_0)} \cap C)$, it follows that $P_U(y_1 - x_0) \leq P_U(y_1 - x_1)$, which implies that

$$P_U(x_0 - y_0) = P_U(y_1 - x_0) \leq P_U(y_1 - x_1)$$

$$= P_U(2x_0 - y_0 - (2x_0 - x)) = P_U(y_0 - x)$$

for all $x \in O_{\overline{U}}(x_0)$. Thus, we have $P_U(y_0 - x_0) = d_P(y_0, \overline{U} \cap C) = d_p(y_0, O_{\overline{U}}(x_0) \cap C)$, and by the continuity of P_U, it follows that

$$P_U(y_0 - x_0) = d_P(y_0, \overline{U} \cap C) = d_p(y_0, \overline{O_{\overline{U}}(x_0)} \cap C) = (P_U(y_0) - 1) > 0.$$

This completes the proof. □

Now, by the application of Theorems 8.2.3 and 8.2.5, Remark 8.2.4, and the argument used in Theorem 8.2.3, we have the following general principle for the existence of solutions for Birkhoff–Kellogg problems in p-seminorm spaces, where $(0 < p \leq 1)$.

Theorem 8.2.6 (Principle of the Birkhoff–Kellogg alternative).

Let U be a bounded open p-convex subset of a locally p-convex space E, where $p \in (0,1]$ with zero element $\theta \in intU = U$, and let C be a closed p-convex subset of E also with zero element $\theta \in C$. Assume that $F : \overline{U} \cap C \to 2^C$ is a QUSC 1-set contractive set-valued mapping with nonempty p-convex values, with a closed graph, and satisfying the condition (H) or (H1) above. Then, F has at least one of the following two properties:

(I) *F has a fixed point $x_0 \in \overline{U} \cap C$ such that $x_0 \in F(x_0)$.*

(II) *There exist $x_0 \in \partial_C(U)$, $y_0 \in F(x_0) \backslash \overline{U}$, and $\lambda = \dfrac{1}{(P_U(y_0))^{\frac{1}{p}}} \in$*

$(0,1)$ such that $x_0 = \lambda y_0 \in \lambda F(x_0)$. In addition, if for each $x \in \partial_C U$, $(P_U^{\frac{1}{p}}(y) - 1)^p \le P_U^{\frac{1}{p}}(y - x)$ for $0 < p \le 1$ (which is trivial when $p = 1$), then the best approximation between x_0 and y_0 is given by

$$P_U(y_0 - x_0) = d_P(y_0, \overline{U} \cap C) = d_p(y_0, \overline{I_U^p(x_0)} \cap C)$$

$$= (P_U^{\frac{1}{p}}(y_0) - 1)^p > 0.$$

Proof. If (I) is not the case, then (II) is proved by Remark 8.2.4 and by following the proof of Theorem 8.2.3 for case (ii): $y_0 \in C \backslash \overline{U}$, with $y_0 := f(x_0) \in F(x_0)$. Indeed, as $y_0 \notin \overline{U}$, it follows that $P_U(y_0) > 1$, and $x_0 = f(y_0) = y_0 \dfrac{1}{(P_U(y_0))^{\frac{1}{p}}}$. Now, let $\lambda = \dfrac{1}{(P_U(y_0))^{\frac{1}{p}}}$, and we have $\lambda < 1$ and $x_0 = \lambda y_0$ with $y_0 \in F(x_0)$. Finally, the additional assumption in (II) allows us to obtain the best approximation between x_0 and y_0 by following the proof of Theorem 8.2.3, as $P_U(y_0 - x_0) = d_P(y_0, \overline{U} \cap C) = d_p(y_0, \overline{I_U^p(x_0)} \cap C) > 0$. This completes the proof. $\square$

As an application of Theorem 8.2.3 for the non-self set-valued mappings discussed in Theorem 8.2.5 with the outward set condition, we have the following general principle of Birkhoff–Kellogg alternative in LCSs.

Theorem 8.2.7 (Principle of Birkhoff–Kellogg alternative in locally p-convex spaces).

Let U be a bounded open p-convex subset of a locally p-convex space E with the zero element $\theta \in U$, and let C be a closed p-convex subset of E also with zero element $\theta \in C$, where $p \in (0,1]$. Assume that $F : \overline{U} \cap C \to 2^C$ is a QUSC 1-set contractive set-valued mapping with nonempty p-convex values,. with *a closed graph, and satisfying the condition (H) or (H1) above Then, it has at least one of the following two properties:*

(I) *F has a fixed point $x_0 \in \overline{U} \cap C$ such that $x_0 \in F(x_0)$.*

(II) *There exists $x_0 \in \partial_C(U)$ and $y_0 \in F(x_0) \backslash \overline{U}$ and $\lambda \in (0,1)$ such that $x_0 = \lambda y_0$, and the best approximation between x_0 and y_0 is given by $P_U(y_0 - x_0) = d_P(y_0, \overline{U} \cap C) = d_p(y_0, \overline{I_{\overline{U}}^p(x_0)} \cap C) > 0$.*

On the other hand, we note that for case (II) of Theorem 8.2.3, the assumption "each $x \in \partial_C U$ with $y \in F(x)$, $(P_U^{\frac{1}{p}}(y) - 1)^p \leq P_U^{\frac{1}{p}}(y-x)$" is only used to guarantee the best approximation "$P_U(y_0 - x_0) = d_P(y_0, \overline{U} \cap C) = d_p(y_0, \overline{I_{\overline{U}}^p(x_0)} \cap C) > 0$". Thus, we have the following Leray–Schauder alternative in p-vector spaces, which, of course, includes the corresponding results in LCSs as special cases.

Theorem 8.2.8 (Leray–Schauder nonlinear alternative). *Let C be a closed p-convex subset of p-seminorm space E and the zero element $\theta \in C$, where $p \in (0,1]$. Assume that $F : C \to 2^C$ is a QUSC 1-set contractive set-valued mapping with nonempty p-convex values, with a closed graph, and satisfying the condition (H) or (H1) above. Let $\varepsilon(F) := \{x \in C : x \in \lambda F(x), \text{ for some } 0 < \lambda < 1\}$. Then, either F has a fixed point in C or the set $\varepsilon(F)$ is unbounded.*

Proof. We prove the conclusion by assuming that F has no fixed point. Then, we claim that the set $\varepsilon(F)$ is unbounded. Otherwise, assume that the set $\varepsilon(F)$ is bounded. Also, assuming that P is the continuous p-seminorm for E, then there exists $r > 0$ such that the set $B(0,r) := \{x \in E : P(x) < r\}$, which contains the set $\varepsilon(F)$, i.e., $\varepsilon(F) \subset B(0,r)$, which means for any $x \in \varepsilon(F)$, $P(x) < r$. Then, $B(0.r)$ is an open p-convex subset of E and the zero $0 \in B(0,r)$ by Lemma 2.2.10 and Remark 2.2.12. Now, let $U := B(0,r)$ in Theorem 8.2.7. It follows that the mapping $F : B(0,r) \cap C \to 2^C$ satisfies all the general conditions of Theorem 8.2.7, and we have that for any $x_0 \in \partial_C B(0,r)$, no $\lambda \in (0,1)$ such that $x_0 = \lambda y_0$, where $y_0 \in F(x_0)$. Indeed, for any $x \in \varepsilon(F)$, it follows that $P(x) < r$, as $\varepsilon(F) \subset B(0,r)$, but for any $x_0 \in \partial_C B(0,r)$, we have $P(x_0) = r$. Thus, conclusion (II) of Theorem 8.2.7 does not hold. By Theorem 8.2.7 again, F must have a fixed point, but this contradicts our assumption that F is fixed point free. This completes the proof.   □

Now, assume a given p-vector space E equipped with the P-seminorm (by assuming it is Then, continuous at zero) for $0 < p \leq 1$.

we know that $P : E \to \mathbb{R}^+$, $P^{-1}(0) = 0$, $P(\lambda x) = |\lambda|^p P(x)$ for any $x \in E$ and $\lambda \in \mathbb{R}$. Then, we have the following useful result for fixed points due to Rothe and Altman in locally p-convex spaces, which are crucial to the study of optimization problems, variational inequality, and complementarity problems (see Isac [118], Yuan [284], and the references therein for more details).

Corollary 8.2.9. *Let U be a bounded open p-convex subset of a locally p-convex space E and zero element $\theta \in U$, and let C be a closed p-convex subset of E with $U \subset C$, where $p \in (0, 1]$. Assume that $F : \overline{U} \to 2^C$ is a QUSC 1-set contractive set-valued mapping with nonempty p-convex values, with a closed graph, and satisfying the condition (H) or (H1) above. We consider if one of the following is satisfied:*

(1) *(Rothe type)*: $P_U(y) \leq P_U(x)$ *for $y \in F(x)$, where $x \in \partial U$.*
(2) *(Petryshyn type)*: $P_U(y) \leq P_U(y - x)$ *for $y \in F(x)$, where $x \in \partial U$.*
(3) *(Altman type)*: $|P_U(y)|^{\frac{2}{p}} \leq [P_U(y) - x)]^{\frac{2}{p}} + [P_U(x)]^{\frac{2}{p}}$ *for $y \in F(x)$,*

where $x \in \partial U$.

Then, F has at least one fixed point in $\overline{U}$.

Proof. By conditions (1), (2), and (3), it follows that the conclusion of (II) in Theorem 8.2.7, "there exist $x_0 \in \partial_C(U)$ and $\lambda \in (0, 1)$ such that $x_0 \notin \lambda F(x_0)$", does not hold, and thus by the alternative of Theorem 8.2.7, F has a fixed point. This completes the proof. $\square$

When $p = 1$, each locally p-convex space is an LCS, and thus we have the following classical Fan's best approximation in LCSs (see Ref. [84]). It is a powerful tool for the study of optimization, mathematical programming, game theory, mathematical economics, and others related topics in applied mathematics.

Corollary 8.2.10 (Fan's best approximation). *Let U be a bounded open convex subset of an LCS E with the zero element $\theta \in U$, and let C be a closed convex subset of E also with zero element $\theta \in C$. Assume that $F : \overline{U} \cap C \to 2^C$ is a QUSC 1-set contractive set-valued mapping with nonempty closed convex values, with a closed graph, and satisfying condition (H) or (H1) above. Assuming P_U as the Minkowski p-functional of U in E, then there exist*

$x_0 \in \overline{U} \cap X$ *and* $y_0 \in T(x_0)$ *such that* $P_U(y_0 - x_0) = d_P(y_0, \overline{U} \cap C) = d_p(y_0, \overline{I_{\overline{U}}(x_0)} \cap C)$. *More precisely, we have either* (I) *or* (II) *holding in the following, where* $W_{\overline{U}}(x_0)$ *is either the inward set* $I_{\overline{U}}(x_0)$ *or the outward set* $O_{\overline{U}}(x_0)$:

(I) *F has a fixed point* $x_0 \in \overline{U} \cap C$, $0 = P_U(y_0 - x_0) = P_U(y_0 - x_0) = d_P(y_0, \overline{U} \cap C) = d_p(y_0, \overline{W_{\overline{U}}(x_0)} \cap C))$.

(II) *There exists* $x_0 \in \partial_C(U)$ *and* $y_0 \in F(x_0) \backslash \overline{U}$ *with*

$$P_U(y_0 - x_0) = d_P(y_0, \overline{U} \cap C)$$
$$= d_p(y_0, \overline{W_{\overline{U}}(x_0)} \cap C) = P_U(y_0) - 1 > 0.$$

Proof. As $p = 1$, each locally p-convex space is an LCS. Then, it automatically satisfies the inequality $P_U^{\frac{1}{p}}(y) - 1 \leq P_U^{\frac{1}{p}}(y - x)$, and indeed we have that for $x_0 \in \partial_C(U)$, with $y_0 \in F(x_0)$, $P_U(y_0 - x_0) = d_P(y_0, \overline{U} \cap C) = d_p(y_0, \overline{W_{\overline{U}}(x_0)} \cap C) = P_U(y_0) - 1$. The conclusions are given by Theorem 8.2.3 (or Theorem 8.2.5). The proof is complete. $\square$

Before concluding this section, we would like to point out that similar results on Rothe and Leray–Schauder alternatives have been developed by Isac [118], Park [185], Potter [213], Shahzad [238,240], Xiao and Zhu [270], and related references therein. These are used as tools of nonlinear analysis in locally p-convex spaces. As mentioned above, when $p = 1$ and taking F as a continuous mapping, we then obtain the traditional version of the Leray–Schauder alternative in LCSs.

8.3 Fixed Point and Best Approximation for QUSC Semiclosed 1-Set Contractive Mappings

The goal of this section is to establish fixed point theorems and best approximation for QUSC semi-closed 1-set contractive mappings in locally p-convex spaces for $p \in (0, 1]$. In this section, based on the best approximation Theorem 8.2.3 for classes of semiclosed 1-set contractive mappings developed in Section 8.2 above, we show how it can be used as a useful tool to establish fixed point theorems for non-self upper semicontinuous mappings in locally p-convex spaces for $p \in (0, 1]$, including norm spaces and uniformly convex Banach spaces as special classes.

We also note that each nonexpansive mapping is a 1-set contractive mapping; and also if f is a hemicompact 1-set contractive mapping, then f is a 1-set contractive mapping satisfying the following condition:

(H1) condition: Let D be a nonempty bounded subset of a space E, and assume that $F : \overline{D} \to 2^E$ is a set-valued mapping. If $\{x_n\}_{n \in \mathbb{N}}$ is any sequence in D such that for each x_n, there exists $y_n \in F(x_n)$ with $\lim_{n \to \infty}(x_n - y_n) = 0$, then there exists a point $x \in \overline{D}$ such that $x \in F(x)$.

As mentioned in Chapter 7, we know that the "(H1) condition" above is the same as "**condition (C)**" used in Theorem 1 of Petryshyn [206]. In addition, it was shown by Browder [32] that nonexpansive mappings in a uniformly convex Banach space X exhibits condition (H1), as shown in the following (which is Lemma 7.5.1 in Chapter 7).

Lemma 8.3.1. *Let D be a nonempty bounded closed convex subset of a uniformly convex Banach space E. Assume that $F : D \to E$ is a nonexpansive single-valued mapping. Then, the mapping $U := I - F$ defined by $U(x) := (x - F(x))$ for each $x \in D$ is demiclosed, and in particular, the "(H1) condition" holds.*

Proof. It is indeed Lemma 7.5.1 in Chapter 7. But here, we point out that by following the argument used for the proof of Theorem 2.2 and Corollary 2.1 given by Petryshyn [206, p. 329], the mapping F is demiclosed (which is actually the demiclosedness principle established by Browder [32] in 1968), which says that by the assumption of (H1) condition, if $\{x_n\}_{n \in \mathbb{N}}$ is any sequence in D such that for each x_n, there exists $y_n \in F(x_n)$ with $\lim_{n \to \infty}(x_n - y_n) = 0$, then we have $0 \in (I - F)(D)$, which means that there exists $x_0 \in D$ with $0 \in (I - F)(x_0)$. This implies that $x_0 \in F(x_0)$. The proof is complete.

$\square$

Remark 8.3.2. When a p-vector space E is with a p-norm, then "condition (H)" satisfies "condition (H1)". In fact, condition (H1) is mainly supported by the so-called demiclosedness principle established by Browder [32].

Secondly, Lemma 7.5.1 in Chapter 7 shows that s single-valued nonexpansive mapping defined in a uniformly convex Banach space satisfied the (H1) condition. Actually, the nonexpansive set-valued mappings defined on a special class of Banach spaces with the so-called "Opial condition" not only satisfy condition (H1) but also belong to the class of semiclosed 1-set contractive mappings, as shown in Chapter 7.

Now, based on the concept of the semiclosed 1-set contractive mappings, we give the existence results for their best approximation, fixed points, and related nonlinear alternative under the framework of p-seminorm spaces for $p \in (0, 1]$.

Theorem 8.3.3 (Schauder fixed point theorem for semiclosed 1-set contractive mappings). *Let U be a nonempty bounded open p-subset of a (Hausdorff) locally p-convex space E and its zero element $\theta \in U$, and let $C \subset E$ be a closed p-convex subset of E such that zero element $\theta \in C$, where $p \in (0, 1]$. If $F : C \cap \overline{U} \to 2^{C \cap \overline{U}}$ is a QUSC semiclosed 1-set contractive set-valued mappings with nonempty closed p-convex values, with a closed graph, then T has at least one fixed point in $C \cap \overline{U}$.*

Proof. As the mapping T is 1-set contractive, consider an increasing sequence $\{\lambda_n\}$ such that $0 < \lambda_n < 1$ and $\lim_{n \to \infty} \lambda_n = 1$, where $n \in \mathbb{N}$. Now, we define a mapping $F_n : C \to 2^C$ by $F_n(x) := \lambda_n F(x)$ for each $x \in C$ and $n \in \mathbb{N}$. Then, it follows that F_n is a λ_n-set contractive mapping with $0 < \lambda_n < 1$, QUSC with nonempty p-convex, and its graph is closed. Now, by Theorem 8.1.2 on the condensing mapping F_n in p-vector space with the p-seminorm P_U for each $n \in \mathbb{N}$, there exists $x_n \in C$ such that $x_n \in F_n(x_n) = \lambda_n F(x_n)$. Thus, there exists $y_n \in F(x_n)$ such that $x_n = \lambda_n y_n$. Let P_U be the Minkowski p-functional of U in E. It follows that P_U is continuous as $0 \in int(U) = U$. Note that for each $n \in \mathbb{N}$, $\lambda_n x_n \in \overline{U} \cap C$, which implies that $x_n = r(\lambda_n y_n) = \lambda_n y_n$, and thus $P_U(\lambda_n y_n) \leq 1$ by Lemma 2.2.10. Note that

$$P_U(y_n - x_n) = P_U(y_n - x_n) = P_U(y_n - \lambda_n y_n)$$

$$= P_U\left(\frac{(1 - \lambda_n)\lambda_n y_n}{\lambda_n}\right) \leq \left(\frac{1 - \lambda_n}{\lambda_n}\right)^p P_U(\lambda_n y_n)$$

$$\leq \left(\frac{1 - \lambda_n}{\lambda_n}\right)^p,$$

which implies that $\lim_{n\to\infty} P_U(y_n - x_n) = 0$. Now, by the assumption that F is semiclosed, which means that $(I - F)$ is closed at zero, there exists a point $x_0 \in \overline{C}$ such that $0 \in (I - F)(\overline{C})$, and thus we have $x_0 \in F(x_0)$.

Indeed, without loss of generality, we assume that $\lim_{n\to\infty} x_n = x_0$, where $y_n \in F(x_n)$ is with $x_n = \lambda_n y_n$, and $\lim_{n\to\infty} \lambda_n = 1$, implying that $x_0 = \lim_{n\to\infty}(\lambda_n y_n)$, which means $y_0 := \lim_{n\to\infty} y_n = x_0$. There exists $y_0(= x_0) \in F(x_0)$. We complete the proof. $\qquad\square$

Theorem 8.3.4 (Best approximation for semiclosed 1-set contractive mappings). *Let U be a bounded open p-convex subset of a locally p-convex space E, where $p \in (0,1]$, the zero element $\theta \in U$, and let C be a (bounded) closed p-convex subset of E also with zero element $\theta \in C$. Assume that $F : \overline{U} \cap C \to 2^C$ is a QUSC semiclosed 1-set contractive set-valued mapping with nonempty closed p-convex values, with a closed graph, and for each $x \in \partial_C U$ with $y \in F(x) \cap (C \backslash \overline{U})), (P_U^{\frac{1}{p}}(y) - 1)^p \le P_U(y - x)$ for $0 < p \le 1$ (which is trivial when $p = 1$). Then, we have that there exist $x_0 \in C \cap \overline{U}$ and $y_0 \in F(x_0)$ such that $P_U(y_0 - x_0) = d_P(y_0, \overline{U} \cap C) = d_p(y_0, \overline{I_{\overline{U}}^p(x_0)} \cap C)$, where P_U is the Minkowski p-functional of U. More precisely, we have either (I) or (II) holding in the following:*

(I) *F has a fixed point $x_0 \in \overline{U} \cap C$, i.e., $0 = P_U(y_0 - x_0) = d_P(y_0, \overline{U} \cap C) = d_p(y_0, \overline{I_{\overline{U}}^p(x_0)} \cap C)$.*

(II) *There exists $x_0 \in \partial_C(U)$ and $y_0 \in F(x_0) \backslash \overline{U}$ with*

$$P_U(y_0 - x_0) = d_P(y_0, \overline{U} \cap C) = d_p(y_0, \overline{I_{\overline{U}}^p(x_0)} \cap C)$$

$$= (P_U^{\frac{1}{p}}(y_0) - 1)^p > 0.$$

Proof. Let $r : E \to U$ be a retraction mapping defined by $r(x) := \dfrac{x}{\max\{1, (P_U(x))^{\frac{1}{p}}\}}$ for each $x \in E$, where P_U is the Minkowski p-functional of U. Since the space E's zero $0 \in U(= intU$, as U is open), it follows that r is continuous by Lemma 2.2.10. As the mapping F is 1-set contractive, consider an increasing sequence $\{\lambda_n\}$ such that $0 < \lambda_n < 1$ and $\lim_{n\to\infty} \lambda_n = 1$, where $n \in \mathbb{N}$. Now, we define a mapping $F_n : C \cap \overline{U} \to 2^C$ by $F_n(x) := \lambda_n F \circ r(x)$ for each

$x \in C \cap \overline{U}$ and $n \in \mathbb{N}$. Then, it follows that F_n is a λ_n-set contractive mapping with $0 < \lambda_n < 1$ for each $n \in \mathbb{N}$. As C and $\overline{U}$ are p-convex, we have $r(C) \subset C$ and $r(\overline{U}) \subset \overline{U}$, so $r(C \cap \overline{U}) \subset C \cap \overline{U}$. Thus, F_n is a self-mapping defined on $C \cap \overline{U}$, and we can also show that F_n satisfies all the conditions of Theorem 8.1.2. By Theorem 8.1.2 for condensing mapping F_n, for each $n \in \mathbb{N}$, there exists $z_n \in C \cap \overline{U}$ such that $z_n \in F_n(z_n) = \lambda_n F \circ r(z_n)$. Let $x_n = r(z_n)$. Then, we have $x_n \in C \cap \overline{U}$, and there exists $y_n \in F(x_n)$ with $x_n = r(\lambda_n y_n)$ such that in the following (1) or (2) holds for each $n \in \mathbb{N}$:

(1) $\lambda_n y_n \in C \cap \overline{U}$;
(2) $\lambda_n y_n \in C \backslash \overline{U}$.

Now, we prove the conclusion by considering the following two cases:

- Case (I) For each $n \in N$, $\lambda_n y_n \in C \cap \overline{U}$.
- Case (II) There exists a positive integer n such that $\lambda_n y_n \in C \backslash \overline{U}$.

First, according to case (I), for each $n \in \mathbb{N}$, $\lambda_n y_n \in \overline{U} \cap C$, which implies that $x_n = r(\lambda_n y_n) = \lambda_n y_n$, and thus $P_U(\lambda_n y_n) \leq 1$ by Lemma 2.2. Note that

$$P_U(y_n - x_n) = P_U(y_n - x_n) = P_U(y_n - \lambda_n y_n) = P_U\left(\frac{(1-\lambda_n)\lambda_n y_n}{\lambda_n}\right)$$

$$\leq \left(\frac{1-\lambda_n}{\lambda_n}\right)^p P_U(\lambda_n y_n) \leq \left(\frac{1-\lambda_n}{\lambda_n}\right)^p,$$

which implies that $\lim_{n\to\infty} P_U(y_n - x_n) = 0$. Now, by the fact that F is semiclosed, it implies that there exists a point $x_0 \in \overline{U}$ (i.e., the consequence $\{x_n\}_{n\in\mathbb{N}}$ has a convergent subsequence with the limit x_0) such that $x_0 \in F(x_0)$. Indeed, without loss of generality, we assume that $\lim_{n\to\infty} x_n = x_0$, where $y_n \in F(x_n)$ is with $x_n = \lambda_n y_n$, and $\lim_{n\to\infty} \lambda_n = 1$, and as $x_0 = \lim_{n\to\infty}(\lambda_n y_n)$, it implies that $y_0 = \lim_{n\to\infty} y_n = x_0$. Thus, there exists $y_0 (= x_0) \in F(x_0)$, and we have $0 = d_p(x_0, F(x_0)) = d(y_0, \overline{U} \cap C) = d_p(y_0, I_{\overline{U}}^p(x_0) \cap C))$, as indeed $x_0 = y_0 \in F(x_0) \in \overline{U} \cap C \subset I_{\overline{U}}^p(x_0) \cap C)$.

Second, according to case (II), there exists a positive integer n such that $\lambda_n y_n \in C \backslash \overline{U}$. Then, we have that $P_U(\lambda_n y_n) > 1$, and also $P_U(y_n) > 1$, as $\lambda_n < 1$. As $x_n = r(\lambda_n y_n) = \dfrac{\lambda_n y_n}{(P_U(\lambda_n y_n))^{\frac{1}{p}}}$, it implies that $P_U(x_n) = 1$, and thus $x_n \in \partial_C(U)$. Note that

$$P_U(y_n - x_n) = P_U \left(\frac{(P_U(y_n)^{\frac{1}{p}} - 1)y_n}{P_U(y_n)^{\frac{1}{p}}} \right) = (P_U^{\frac{1}{p}}(y_n) - 1)^p.$$

By the assumption, we have $(P_U^{\frac{1}{p}}(y_n) - 1)^p \leq P_U(y_n - x)$ for $x \in C \cap \partial \overline{U}$. It follows that

$$P_U(y_n) - 1 \leq P_U(y_n) - \sup\{P_U(z) : z \in C \cap \overline{U}\}$$
$$\leq \inf\{P_U(y_n - z) : z \in C \cap \overline{U}\} = d_p(y_n, C \cap \overline{U}).$$

Thus, we have the best approximation: $P_U(y_n - x_n) = d_P(y_n, \overline{U} \cap C) = (P_U^{\frac{1}{p}}(y_n) - 1)^p > 0$.

Now, we want to show that $P_U(y_n - x_n) = d_P(y_n, \overline{U} \cap C) = d_p(y_n, \overline{I_{\overline{U}}^p(x_0)} \cap C) > 0$.

By the fact that $(\overline{U} \cap C) \subset I_{\overline{U}}^p(x_n) \cap C$, let $z \in I_{\overline{U}}^p(x_n) \cap C \backslash (\overline{U} \cap C)$. We first claim that $P_U(y_n - x_n) \leq P_U(y_n - z)$. If not, we have $P_U(y_n - x_n) > P_U(y_n - z)$. As $z \in I_{\overline{U}}^p(x_n) \cap C \backslash (\overline{U} \cap C)$, there exists $y \in \overline{U}$ and a non-negative number c (actually $c \geq 1$ as will be soon shown) with $z = x_n + c(y - x_n)$. Since $z \in C$, but $z \notin \overline{U} \cap C$, it implies that $z \notin \overline{U}$. By the fact that $x_n \in \overline{U}$ and $y \in \overline{U}$, we must have the constant $c \geq 1$; otherwise, it implies that $z(= (1 - c)x_n + cy) \in \overline{U}$, which is impossible by our assumption, i.e., $z \notin \overline{U}$. Thus, we have that $c \geq 1$, which implies that $y = \frac{1}{c}z + (1 - \frac{1}{c})x_n \in C$ (as both $x_n \in C$ and $z \in C$). On the other hand, as $z \in I_{\overline{U}}^p(x_n) \cap C \backslash (\overline{U} \cap C)$, and $c \geq 1$ with $(\frac{1}{c})^p + (1 - \frac{1}{c})^p = 1$, combining with our assumption that for each $x \in \partial_C \overline{U}$ and $y \in F(x_n) \backslash \overline{U}$, $(P_U^{\frac{1}{p}}(y) - 1)^p \leq P_U^{\frac{1}{p}}(y - x)$ for $0 < p \leq 1$, it then follows that

$$P_U(y_n - y) = P_U \left[\frac{1}{c}(y_n - z) + \left(1 - \frac{1}{c}\right)(y_n - x_n) \right]$$
$$\leq \left[\left(\frac{1}{c}\right)^p P_U(y_n - z) + \left(1 - \frac{1}{c}\right)^p P_U(y_n - x_n) \right]$$
$$< P_U(y_n - x_n),$$

which contradicts that $P_U(y_n - x_n) = d_P(y_n, \overline{U} \cap C)$. As shown above, we know that $y \in \overline{U} \cap C$, and we should have $P_U(y_n - x_n) \leq P_U$

$(y_n - y)$! This helps us complete the claim $P_U(y_n - x_n) \leq P_U(y_n - z)$ for any $z \in I_{\overline{U}}^p(x_n) \cap C\backslash(\overline{U} \cap C)$, which means that the following best approximation of Fan type (see [84,85]) holds:

$$0 < d_P(y_n, \overline{U} \cap C) = P_U(y_n - x_n) = d_p(y_n, I_{\overline{U}}^p(x_n) \cap C).$$

Now, by the continuity of P_U, it follows that the following best approximation of Fan type is also true:

$$0 < P_U(y_n - x_n) = d_P(y_n, \overline{U} \cap C) = d_p(y_n, I_{\overline{U}}^p(x_n) \cap C)$$

$$= d_p(y_n, \overline{I_{\overline{U}}^p(x_n)} \cap C).$$

The proof is complete. □

For a p-vector space when $p = 1$, which is a (Hausdorff) topological vector space E, we have the following best approximation for the outward set $\overline{O_{\overline{U}}(x_0)}$ based on the point $\{x_0\}$ with respect to the convex subset U in E.

Theorem 8.3.5 (Best approximation for outward sets). *Let U be a bounded open convex subset of an LCS E with zero element $\theta \in intU = U$ (the interior $intU = U$, as U is open), and let C be a closed convex subset of E also with zero element $\theta \in C$. Assume that $F : \overline{U} \cap C \to 2^C$ is a QUSC semiclosed 1-set contractive set-valued mapping with nonempty closed p-convex values, with a closed graph. Then, there exist $x_0 \in \overline{U} \cap X$ and $y_0 \in F(x_0)$ such that $P_U(y_0 - x_0) = d_P(y_0, \overline{U} \cap C) = d_p(y_0, \overline{O_{\overline{U}}(x_0)} \cap C)$, where P_U is the Minkowski functional of U. More precisely, we have either (I) or (II) holding in the following:*

(I) *F has a fixed point $x_0 \in U \cap C$, i.e., $P_U(y_0 - x_0) = P_U(y_0 - x_0) = d_P(y_0, \overline{U} \cap C) = d_p(y_0, \overline{O_{\overline{U}}(x_0)} \cap C)) = 0$.*
(II) *There exists $x_0 \in \partial_C(U)$ and $y_0 \in F(x_0)\backslash\overline{U}$ with*

$$P_U(y_0 - x_0) = d_P(y_0, \overline{U} \cap C) = d_p(y_0, O_{\overline{U}}(x_0) \cap C)$$

$$= d_p(y_0, \overline{O_{\overline{U}}(x_0)} \cap C) > 0.$$

Proof. We define a new mapping $F_1 : \overline{U} \cap C \to 2^C$ by $F_1(x) := \{2x\} - F(x)$ for each $x \in \overline{U} \cap C$. Then, F_1 is also a compact and upper semicontinuous mapping with nonempty closed convex values, and F_1 satisfies all the hypotheses of Theorem 8.2.3 wit $p = 1$. It follows by

Theorem 8.2.3 that there exist $x_0 \in \overline{U} \cap X$ and $y_1 \in F_1(x_0)$ such that $P_U(y_1 - x_0) = d_P(y_1, \overline{U} \cap C) = d_p(y_1, I_{\overline{U}}(x_0) \cap C)$. More precisely, we have either (I) or (II) holding in the following:

(I) F_1 has a fixed point $x_0 \in U \cap C$ (so $0 = P_U(y_1 - x_0) = P_U(y_1 - x_0) = d_P(y_1, \overline{U} \cap C) = d_p(y_1, \overline{I_{\overline{U}}(x_0)} \cap C)$).

(II) There exists $x_0 \in \partial_C(U)$ and $y_1 \in F_1(x_0) \backslash \overline{U}$ with

$$P_U(y_1 - x_0) = d_P(y_1, \overline{U} \cap C) = d_p(y_1, \overline{O_{\overline{U}}(x_0)} \cap C) > 0.$$

Now, for any $x \in O_{\overline{U}}(x_0)$, there exist $r < 0, u \in \overline{U}$ such that $x = x_0 + r(u - x_0)$. Let $x_1 = 2x_0 - x$. Then $x_1 = 2x_0 - x_0 - r(u - x_0) = x_0 + (-r)(u - x_0) \in I_{\overline{U}}(x_0)$. Let $y_1 = 2x_0 - y_0$, for some $y_0 \in F(x_0)$. As we have $P_U(y_1 - x_0) = d_P(y_1, \overline{U} \cap C) = d_p(y_1, \overline{I_{\overline{U}}(x_0)} \cap C)$, it follows that $P_U(y_1 - x_0) \leq P_U(y_1 - x_1)$, which implies that

$$P_U(x_0 - y_0) = P_U(y_1 - x_0) \leq P_U(y_1 - x_1)$$
$$= P_U(2x_0 - y_0 - (2x_0 - x)) = P_U(y_0 - x)$$

for all $x \in O_{\overline{U}}(x_0)$. Thus, we have $P_U(y_0 - x_0) = d_P(y_0, \overline{U} \cap C) = d_p(y_0, O_{\overline{U}}(x_0) \cap C)$, and by the continuity of P_U, it follows that

$$P_U(y_0 - x_0) = d_P(y_0, \overline{U} \cap C) = d_p(y_0, \overline{O_{\overline{U}}(x_0)} \cap C) = (P_U(y_0) - 1) > 0.$$

This completes the proof.   □

Now, by the application of Theorems 8.3.4 and 8.3.5, we have the following general principle for the existence of solutions to Birkhoff–Kellogg problems in p-seminorm spaces, where $(0 < p \leq 1)$.

Theorem 8.3.6 (Principle of the Birkhoff–Kellogg alternative). *Let U be a bounded open p-convex subset of a locally p-convex space E, where $p \in (0, 1]$ with zero element $\theta \in intU = (U)$ (the interior $intU$, as U is open), and let C be a closed p-convex subset of E also with zero element $\theta \in C$. Assume that $F : \overline{U} \cap C \to 2^C$ is a QUSC semiclosed 1-set contractive set-valued mapping with nonempty closed p-convex values, with a closed graph. Then, F has at least one of the following two properties:*

(I) *F has a fixed point $x_0 \in \overline{U} \cap C$ such that $x_0 \in F(x_0)$.*

(II) *There exist $x_0 \in \partial_C(U)$, $y_0 \in F(x_0)\backslash \overline{U}$, and $\lambda = \dfrac{1}{(P_U(y_0))^{\frac{1}{p}}} \in$ $(0,1)$ such that $x_0 = \lambda y_0 \in \lambda F(x_0)$. In addition, if for each $x \in \partial_C U$, $(P_U^{\frac{1}{p}}(y) - 1)^p \le P_U^{\frac{1}{p}}(y-x)$ for $0 < p \le 1$ (this is trivial when $p = 1$), then the best approximation between x_0 and y_0 is given by*

$$P_U(y_0 - x_0) = d_P(y_0, \overline{U} \cap C) = d_p(y_0, \overline{I_{\overline{U}}^p(x_0)} \cap C)$$

$$= (P_U^{\frac{1}{p}}(y_0) - 1)^p > 0.$$

Proof. If (I) is not the case, then (II) is proved by Remark 5.2 and by following the proof in Theorem 8.3.4 for case (ii): $y_0 \in C\backslash \overline{U}$, with $y_0 := f(x_0) \in F(x_0)$. Indeed, as $y_0 \notin \overline{U}$, it follows that $P_U(y_0) > 1$, and $x_0 = f(y_0) = y_0 \dfrac{1}{(P_U(y_0))^{\frac{1}{p}}}$. Now, let $\lambda = \dfrac{1}{(P_U(y_0))^{\frac{1}{p}}}$, and we have $\lambda < 1$ and $x_0 = \lambda y_0$ with $y_0 \in F(x_0)$. Finally, the additional assumption in (II) allows us to obtain the best approximation between x_0 and y_0 by following the proof of Theorem 8.3.4, as $P_U(y_0 - x_0) = d_P(y_0, \overline{U} \cap C) = d_p(y_0, \overline{I_{\overline{U}}^p(x_0)} \cap C) > 0$. This completes the proof. $\qquad\square$

As an application of Theorem 8.3.4 for set-valued mappings, we have the following result, which is a general principle for the Birkhoff–Kellogg alternative in locally p-convex spaces.

Theorem 8.3.7 (Principle of the Birkhoff–Kellogg alternative in LCS). *Let U be a bounded open p-convex subset of a locally p-convex space E ($0 < p \le 1$) with the zero element $\theta \in U$, and let C be a closed convex subset of E also with zero element $\theta \in C$. Assume that $F : \overline{U} \cap C \to 2^C$ is a semiclosed 1-set contractive and QUSC mapping with nonempty closed p-convex values, with a closed graph. Then, it has at least one of the following two properties:*

(I) *F has a fixed point $x_0 \in \overline{U} \cap C$ such that $x_0 \in F(x_0)$.*
(II) *There exists $x_0 \in \partial_C(U)$, $y_0 \in F(x_0)\backslash \overline{U}$, and $\lambda \in (0,1)$ such that $x_0 = \lambda y_0$, and the best approximation between x_0 and y_0 is given by $P_U(y_0 - x_0) = d_P(y_0, \overline{U} \cap C) = d_p(y_0, \overline{I_{\overline{U}}^p(x_0)} \cap C) > 0$.*

As an application of Theorem 8.3.4 for set-valued mappings and by noting case (II) of Theorem 8.3.4, the assumption "each $x \in \partial_C U$ with $y \in F(x)$, $(P_U^{\frac{1}{p}}(y) - 1)^p \le P_U^{\frac{1}{p}}(y - x)$" is only used to guarantee the best approximation "$P_U(y_0 - x_0) = d_P(y_0, \overline{U} \cap C) = d_p(y_0, \overline{I_{\overline{U}}^p(x_0)} \cap C) > 0$". Thus, we have the following Leray–Schauder alternative in p-vector spaces, which, of course, includes the corresponding results in LCSs as special cases.

Theorem 8.3.8 (Leray–Schauder nonlinear alternative). *Let C be a closed p-convex subset of p-seminorm space E, where $p \in (0,1]$ and the zero element $\theta \in C$. Assume that $F : C \to 2^C$ is QUSC semiclosed 1-set contractive set-valued mappings with nonempty closed p-convex values, with a closed graph. Let $\varepsilon(F) := \{x \in C : x \in \lambda F(x), \text{ for some } 0 < \lambda < 1\}$. Then, either F has a fixed point in C or the set $\varepsilon(F)$ is unbounded.*

Proof. By assuming that case (I) is not true, i.e., F has no fixed point, we claim that the set $\varepsilon(F)$ is unbounded. Otherwise, assume the set $\varepsilon(F)$ is bounded and that P is the continuous p-seminorm for E. Then, there exists $r > 0$ such that the set $B(0,r) := \{x \in E : P(x) < r\}$, which contains the set $\varepsilon(F)$, i.e., $\varepsilon(F) \subset B(0,r)$, which means for any $x \in \varepsilon(F)$, $P(x) < r$. Then, $B(0.r)$ is an open p-convex subset of E and the zero $0 \in B(0,r)$ by Lemma 2.2.10 and Remark 2.2.12. Now, let $U := B(0,r)$ in Theorem 8.3.6. It follows that the mapping $F : B(0,r) \cap C \to 2^C$ satisfies all the general conditions of Theorem 8.3.6, and we have that any $x_0 \in \partial_C B(0,r)$, no any $\lambda \in (0,1)$ such that $x_0 = \lambda y_0$, where $y_0 \in F(x_0)$. Indeed, for any $x \in \varepsilon(F)$, it follows that $P(x) < r$, as $\varepsilon(F) \subset B(0,r)$, but for any $x_0 \in \partial_C B(0,r)$, we have $P(x_0) = r$. Thus, conclusion (II) of Theorem 8.3.6 does not hold. By Theorem 8.3.6 again, F must have a fixed point, but this contradicts our assumption that F is fixed point free. This completes the proof. $\square$

Now, assume a given p-vector space E equipped with the P-seminorm (by assuming it is continuous at zero) for $0 < p \leq 1$. Then, we know that $P : E \to \mathbb{R}^+$, $P^{-1}(0) = 0$, $P(\lambda x) = |\lambda|^p P(x)$ for any $x \in E$ and $\lambda \in \mathbb{R}$. Then, we have the following useful result for fixed points due to Rothe and Altman in p-vector spaces, which plays an important role in optimization problems, variational inequality, and complementarity problems.

Corollary 8.3.9. *Let U be a bounded open p-convex subset of a locally p-convex space E and zero element $\theta \in U$, and let C be a closed p-convex subset of E with $U \subset C$, where $p \in (0,1]$. Assume* that $F : \overline{U} \to 2^C$ is a QUSC semiclosed 1-set contractive set-valued mapping with nonempty closed p-convex values, with a closed graph. Then, one of the following conditions is satisfied:

(1) *Rothe-type condition:* $P_U(y) \le P_U(x)$ for $y \in F(x)$, where $x \in \partial U$.
(2) *Petryshyn-type condition:* $P_U(y) \le P_U(y-x)$ for $y \in F(x)$, where $x \in \partial U$.
(3) *Altman-type condition:* $|P_U(y)|^{\frac{2}{p}} \le [P_U(y) - x)]^{\frac{2}{p}} + [P_U(x)]^{\frac{2}{p}}$ for $y \in F(x)$, where $x \in \partial U$.

Then, F has at least one fixed point.

Proof. By conditions (1), (2), and (3), it follows that the conclusion of (II) in Theorem 8.3.6, "there exist $x_0 \in \partial_C(U)$ and $\lambda \in (0,1)$ such that $x_0 \notin \lambda F(x_0)$", does not hold. Thus, by the alternative of Theorem 8.3.6, F has a fixed point. This completes the proof. □

By the fact that when $p = 1$, each p-vector space is a topological vector space, we have the following classical Fan's best approximation (see [84]). It is a powerful tool for the study of optimization, mathematical programming, game theory, mathematical economics, and other related topics in applied mathematics.

Corollary 8.3.10 (Fan's best approximation in LCS). *Let U be a bounded open convex subset of an LCS E with the zero element $\theta \in U$, and let C be a closed convex subset of E also with zero element $\theta \in C$. Assume that $F : \overline{U} \cap C \to 2^C$ is a QUSC semiclosed 1-set contractive set-valued mapping with nonempty closed convex values and with a closed graph. Then, there exist $x_0 \in \overline{U} \cap X$ and $y_0 \in T(x_0)$ such that $P_U(y_0 - x_0) = d_P(y_0, \overline{U} \cap C) = d_p(y_0, \overline{I_{\overline{U}}(x_0)} \cap C)$, where P_U is the Minkowski p-functional of U in E. More precisely, we have either (I) or (II) holding in the following, where $W_{\overline{U}}(x_0)$ is either the inward set $I_{\overline{U}}(x_0)$ or the outward set $O_{\overline{U}}(x_0)$:*

(I) *F has a fixed point $x_0 \in U \cap C$, $0 = P_U(y_0 - x_0) = P_U(y_0 - x_0) = d_P(y_0, \overline{U} \cap C) = d_p(y_0, \overline{W_{\overline{U}}(x_0)} \cap C))$.*
(II) *There exists $x_0 \in \partial_C(U)$ and $y_0 \in F(x_0) \backslash \overline{U}$ with*

$$P_U(y_0 - x_0) = d_P(y_0, \overline{U} \cap C) = d_p(y_0, \overline{W_{\overline{U}}(x_0)} \cap C)$$
$$= P_U(y_0) - 1 > 0.$$

Proof. When $p = 1$, then it automatically satisfies the inequality $P_U^{\frac{1}{p}}(y) - 1 \leq P_U^{\frac{1}{p}}(y-x)$, and indeed we have that for $x_0 \in \partial_C(U)$, with $y_0 \in F(x_0)$, $P_U(y_0 - x_0) = d_P(y_0, \overline{U} \cap C) = d_p(y_0, \overline{W_{\overline{U}}(x_0)} \cap C) = P_U(y_0) - 1$. The conclusions are given by Theorem 8.3.4. The proof is complete. □

Before ending this section, we would like to point out that similar results on Rothe and Leray–Schauder alternatives have been developed by Isac [118], Park [185], Potter [213], Shahzad [238, 240], Xiao and Zhu [270], and related references therein. These are used as tools for nonlinear analysis in topological vector spaces.

8.4 Principle of Alternative for QUSC Semiclosed 1-Set Contractive Mappings

The goal of this section is to establish the general principles of Birkhoff–Kellogg alternatives for QUSC semiclosed 1-Set contractive mappings and related fixed points for non-self mappings under various boundary conditions.

As applications of the results in Section 8.3 above, we establish the general existence of solutions for Birkhoff–Kellogg problems and the principles of Leray–Schauder alternatives for semiclosed 1-set contractive mappings in locally p-convex spaces, where $p \in (0, 1]$.

Theorem 8.4.1 (Birkhoff–Kellogg alternative in p-vector spaces). *Let U be a bounded open p-convex subset of a locally p-convex space E, where $p \in (0, 1]$ with the zero element $\theta \in U$, and let C be a closed p-convex subset of E also with zero element $\theta \in C$. Also, assume that $F : \overline{U} \cap C \to 2^C$ is a QUSC semiclosed 1-set contractive set-valued mapping with nonempty p-convex values and with a closed graph. In addition, for each $x \in \partial_C(U)$ with $y \in F(x)$, $(P_U^{\frac{1}{p}}(y) - 1)^p \leq P_U^{\frac{1}{p}}(y-x)$ for $0 < p \leq 1$ (which is trivial when $p = 1$), where P_U is the Minkowski p-functional of U. Then, we have either (I) or (II) holding in the following:*

(I) *There exists $x_0 \in \overline{U} \cap C$ such that $x_0 \in F(x_0)$.*
(II) *There exists $x_0 \in \partial_C(U)$ with $y_0 \in F(x_0) \backslash \overline{U}$ and $\lambda > 1$ such that $\lambda x_0 = y_0 \in F(x_0)$, i.e., $F(x_0) \cap \{\lambda x_0 : \lambda > 1\} \neq \emptyset$.*

Proof. By following the argument and notations used in Theorem 8.3.4, we have that either:

(1) F has a fixed point $x_0 \in U \cap C$; or
(2) there exists $x_0 \in \partial_C(U)$ and $y_0 \in F(x_0)$ with $x_0 = f(y_0)$ such that

$$P_U(y_0 - x_0) = d_P(y_0, \overline{U} \cap C) = d_p(y_0, \overline{I_{\overline{U}}(x_0)} \cap C) = P_U(y_0) - 1 > 0,$$

where $\partial_C(U)$ denotes the boundary of U relative to C in E and f is the restriction of the continuous retraction r with respect to the set U in E.

If F has no fixed point, then (2) holds and $x_0 \notin F(x_0)$. As given by the proof of Theorem 8.3.4, we have that $y_0 \in F(x_0)$ and $y_0 \notin \overline{U}$, and thus $P_U(y_0) > 1$ and $x_0 = f(y_0) = \dfrac{y_0}{(P_U(y_0))^{\frac{1}{p}}}$, which means $y_0 = (P_U(y_0))^{\frac{1}{p}} x_0$. Let $\lambda = (P_U(y_0))^{\frac{1}{p}}$. Then, $\lambda > 1$, and we have $\lambda x_0 = y_0 \in F(x_0)$. This completes the proof. $\square$

Theorem 8.4.2 (Birkhoff–Kellogg alternative in LCS). *Let U be a bounded open convex subset of an LCS E with the zero element $\theta \in U$, and let C be a closed convex subset of E also with zero element $\theta \in C$. Assume that $F : \overline{U} \cap C \to 2^C$ is a QUSC semiclosed 1-set contractive set-valued mapping with nonempty convex values and with a closed graph. Then, we have either* (I) *or* (II) *holding in the following, where $W_{\overline{U}}(x_0)$ is either the inward set $I_{\overline{U}}(x_0)$, or the outward set $O_{\overline{U}}(x_0)$:*

(I) *There exists $x_0 \in \overline{U} \cap C$ such that $x_0 \in F(x_0)$.*
(II) *There exists $x_0 \in \partial_C(U)$ with $y_0 \in F(x_0) \backslash \overline{U}$ and $\lambda > 1$ such that $\lambda x_0 = y_0 \in F(x_0)$, i.e., $F(x_0) \cap \{\lambda x_0 : \lambda > 1\} \neq \emptyset$.*

Proof. When $p = 1$, then it automatically satisfies the inequality $P_U^{\frac{1}{p}}(y) - 1 \le P_U^{\frac{1}{p}}(y - x)$, and indeed we have that for $x_0 \in \partial_C(U)$, with $y_0 \in F(x_0)$, $P_U(y_0 - x_0) = d_P(y_0, \overline{U} \cap C) = d_p(y_0, \overline{W_{\overline{U}}(x_0)} \cap C) = P_U(y_0) - 1$. The conclusions are given by Theorems 8.3.5. The proof is complete. $\square$

Indeed, we have the following fixed points for non-self mappings in locally p-convex spaces for $p \in (0, 1]$ under various boundary conditions.

Theorem 8.4.3 (Fixed points of non-self mappings). *Let* U *be a bounded open p-convex subset of a locally p-convex space* E, *where* $p \in (0, 1]$ *with the zero element* $\theta \in U$, *and let* C *be a closed p-convex subset of* E *also with zero element* $\theta \in C$. *Assume that* $F : \overline{U} \cap C \to 2^C$ *is a QUSC semiclosed 1-set contractive set-valued mapping with nonempty p-convex values and with a closed graph. In addition, for each* $x \in \partial_C(U)$ *with* $y \in F(x)$, $(P_U^{\frac{1}{p}}(y)-1)^p \le P_U^{\frac{1}{p}}(y-x)$ *for* $0 < p \le 1$ *(which is trivial when* $p = 1$*), where* P_U *is the Minkowski p-functional of* U. *We consider if* F *satisfies any one of the following conditions for any* $x \in \partial_C(U) \backslash F(x)$:

(i) *for each* $y \in F(x)$, $P_U(y-z) < P_U(y-x)$ *for some* $z \in \overline{I_{\overline{U}}(x)} \cap C$;
(ii) *for each* $y \in F(x)$, *there exists* λ *with* $|\lambda| < 1$ *such that* $\lambda x + (1-\lambda)y \in \overline{I_{\overline{U}}(x)} \cap C$;
(iii) $F(x) \subset \overline{I_{\overline{U}}(x)} \cap C$;
(iv) $F(x) \cap \{\lambda x : \lambda > 1\} = \emptyset$ *(Leray–Schauder boundary condition)*;
(v) $F(\partial U) \subset \overline{U} \cap C$;
(vi) *for each* $y \in F(x)$, $P_U(y - x) \ne ((P_U(y))^{\frac{1}{p}} - 1)^p$.

Then, F *must has a fixed point.*

Proof. By following the argument and symbols used in the proof of Theorem 8.3.4, we have that either:

(1) F has a fixed point $x_0 \in U \cap C$; or
(2) there exists $x_0 \in \partial_C(U)$ and $y_0 \in F(x_0)$ with $x_0 = f(y_0)$ such that

$$P_U(y_0 - x_0) = d_P(y_0, \overline{U} \cap C) = d_p(y_0, \overline{I_{\overline{U}}(x_0)} \cap C)$$

$$= (P_U^{\frac{1}{p}}(y_0) - 1)^p > 0,$$

where $\partial_C(U)$ denotes the boundary of U relative to C in E and f is the restriction of the continuous retraction r with respect to the set U in E.

First, suppose that F satisfies condition (i). If F has no fixed point, then (2) holds and $x_0 \notin F(x_0)$. Then, by condition (i), it follows that $P_U(y_0 - z) < P_U(y_0 - x_0)$ for some $z \in \overline{I_{\overline{U}}(x)} \cap C$, which contradicts the best approximation equations given by (2) above, and thus F must have a fixed point.

Second, suppose that F satisfies condition (ii). If F has no fixed point, then (2) holds and $x_0 \notin F(x_0)$. Then, by condition (ii), there exists $|\lambda| < 1$ such that $\lambda x_0 + (1 - \lambda)y_0 \in \overline{I_U(x)} \cap C$. It follows that

$$P_U(y_0 - x_0) \leq P_U(y_0 - (\lambda x_0 + (1 - \lambda y_0)))$$
$$= P_U(\lambda(y_0 - x_0)) = |\lambda|^p P_U(y_0 - x_0) < P_U(y_0 - x_0).$$

This is impossible, and thus F must have a fixed point in $\overline{U} \cap C$.

Third, suppose that F satisfies condition (iii), i.e., $F(x) \subset \overline{I_U(x)} \cap C$. Then by (2), we have that $P_U(y_0 - x_0)$, and thus $x_0 = y_0 \in F(x_0)$, which means F has a fixed point.

Fourth, suppose that F satisfies condition (iv). If F has no fixed point, then (2) holds and $x_0 \notin F(x_0)$. As given by the proof of Theorem 6.2, we have that $y_0 \notin \overline{U}$, and thus $P_U(y_0) > 1$ and $x_0 = f(y_0) = \dfrac{y_0}{(P_U(y_0))^{\frac{1}{p}}}$, which means $y_0 = (P_U(y_0))^{\frac{1}{p}} x_0$, where $(P_U(y_0))^{\frac{1}{p}} > 1$. This contradicts assumption (iv), and thus F must have a fixed point in $\overline{U} \cap C$.

Fifth, suppose that F satisfies condition (v). Then, $x_0 \notin F(x_0)$. As $x_0 \in \partial_C U$, now by condition (v), we have that $F(\partial U) \subset \overline{U} \cap C$. It follows that for any $y_0 \in F(x_0)$, we have $y_0 \in \overline{U} \cap C$, and thus $y \notin \overline{U} \backslash \cap C$, which implies that $0 < P_U(y_0 - x_0) = d_P(y_0, \overline{U} \cap C) = 0$. This is impossible, and thus F must have a fixed point. Here, as pointed out in Remark 6.2.4, based on condition (v), applying $F(\partial U) \subset \overline{U} \cap C$ is enough to know that the mapping F has a fixed point, making the following general hypothesis unnecessary: "for each $x \in \partial_C(U)$ with $y \in F(x)$, $(P_U^{\frac{1}{p}}(y) - 1)^p \leq P_U^{\frac{1}{p}}(y - x)$ for $0 < p \leq 1$".

Finally, suppose that F satisfies condition (vi). If F has no fixed point, then (2) holds and $x_0 \notin F(x_0)$. Then, condition (v) implies that $P_U(y_0 - x_0) \neq ((P_U(y))^{\frac{1}{p}} - 1)^p$; however, our proof of Theorem 8.3.4 shows that $P_U(y_0 - x_0) = ((P_U(y))^{\frac{1}{p}} - 1)^p$. This is impossible, and thus F must have a fixed point. Then, the proof is complete. □

Now, by taking the set C in Theorem 8.4.1 as the whole p-vector space E itself, we have the following general results for non-self upper semicontinuous set-valued mappings, which include the results of fixed point types given by Rothe, Petryshyn, Altman, and Leray–Schauder as special cases. In particular, taking $p = 1$ and $C = E$

in Theorem 8.4.3, we have the following fixed point theorems for non-self upper semicontinuous set-valued mappings associated with inward or outward sets in LCSs as follows.

Theorem 8.4.4 (Fixed point theorem of non-self mappings with boundary conditions). *Let U be a bounded open convex subset of an LCS E with the zero element $\theta \in U$, and assume that $F : \overline{U} \to 2^E$ is a QUSC semiclosed 1-set contractive set-valued mapping with nonempty convex values and with a closed graph. We consider if F satisfies any one of the following conditions for any $x \in \partial(U) \backslash F(x)$:*

(i) *for each $y \in F(x)$, $P_U(y - z) < P_U(y - x)$ for some $z \in \overline{I_{\overline{U}}(x)}$ (or $z \in \overline{O_{\overline{U}}(x)}$);*
(ii) *for each $y \in F(x)$, there exists λ with $|\lambda| < 1$ such that $\lambda x + (1 - \lambda)y \in \overline{I_{\overline{U}}(x)}$ (or, $\overline{O_{\overline{U}}(x)}$);*
(iii) *$F(x) \subset \overline{I_{\overline{U}}(x)}$ (or $\overline{O_{\overline{U}}(x)}$);*
(iv) *$F(x) \cap \{\lambda x : \lambda > 1\} = \emptyset$;*
(v) *$F(\partial(U) \subset \overline{U}$;*
(vi) *for each $y \in F(x)$, $P_U(y - x) \neq P_U(y) - 1$.*

Then, F must has a fixed point.

In what follows, based on the best approximation theorem in p-seminorm space, we also give some fixed point theorems for non-self set-valued mappings with various boundary conditions, which are related to the study of the existence of solutions for partial differential equations and differential equations with boundary problems (see, Browder [32], Petryshyn [205, 206], and Reich [222]). These play key roles in nonlinear analysis for p-seminorm space, as shown in the following.

First, as discussed in Remark 6.2.4 and the proof of Theorem 8.4.3, with only the strong boundary condition "$F(\partial(U)) \subset \overline{U} \cap C$", we can prove that F has a fixed point, and thus we have the following fixed point theorem of Rothe type in p-vector spaces.

Theorem 8.4.5 (Rothe type). *Let U be a bounded open p-convex subset of a locally p-convex space E, where $P \in (0, 1]$ with the zero element $\theta \in U$. Assume that $F : \overline{U} \to 2^E$ is a QUSC semiclosed 1-set contractive set-valued mapping with nonempty p-convex values* *and with a closed graph and such that $F(\partial(U)) \subset \overline{U}$. Then, F must has a fixed point.*

Now, as applications of Theorem 8.4.5, we give the following Leray–Schauder alternative in p-vector spaces for non-self set-valued mappings associated with boundary conditions, which often appear in applications (see Isac [118] and references therein for the study of complementary problems and related topics in optimization).

Theorem 8.4.6 (Leray–Schauder alternative in locally p-convex spaces). *Let E be a locally p-convex space E, where $p \in (0, 1]$ and $B \subset E$ is a bounded closed p-convex such that zero element $0 \in intB$. Let $F : [0, 1] \times B \to 2^E$ be a QUSC semiclosed 1-set contractive set-valued mapping with nonempty p-convex values and with a closed graph and such that the set $F([0, 1] \times B)$ is relatively compact in E. If the following assumptions are satisfied:*

(1) $x \notin F(t, x)$ for all $x \notin \partial B$ and $t \in [0, 1]$;
(2) $F(\{0\} \times \partial B) \subset B$,

then there is an element $x^* \in B$ such that $x^* \in F(1, x^*)$.

Proof. For any $n \in N$, we consider the mapping

$$
F_n(x) = \begin{cases} F\left(\dfrac{1 - P_B(x)}{\epsilon_n}, \dfrac{x}{P_B(x)}\right) & \text{if } 1 - \epsilon \le P_B(x) \le 1, \\[2ex] F\left(1, \dfrac{X}{1 - \epsilon_n}\right) & \text{if } P_B(x) < 1 - \epsilon_n, \end{cases} \tag{8.1}
$$

where P_B is the Minkowski p-functional of B and $\{\epsilon_n\}_{n \in N}$ is a sequence of real numbers such that $\lim_{n \to \infty} \epsilon_n = 0$ and $0 < \epsilon_n < \frac{1}{2}$ for any $n \in N$. We observe that for each $n \in N$, the mapping F_n is 1-set contractive upper semicontinuous with nonempty closed p-convex values on B. From assumption (2), we have that $F_n(\partial B) \subset B$, and the assumptions of Theorem 8.4.5 are satisfied. Then, for each $n \in N$, there exists an element $u_n \in B$ such that $u_n \in F_n(u_n)$.

We first prove the following statement: "It is impossible to have an infinite number of the elements u_n satisfy the inequality: $1 - \epsilon_n \le P_B(u_n) \le 1$".

If not, we assume to have an infinite number of elements u_n satisfying the following inequality:

$$1 - \epsilon_n \leq P_B(u_n) \leq 1.$$

As $F_n(B)$ is relatively compact and by the definition of mappings F_n, we have that $\{u_n\}_{n \in N}$ is contained in a compact set in E. Without loss of generality (indeed, each compact set is also countably compact), we define the sequence $\{t_n\}_{n \in N}$ by $t_n := \frac{1 - P_B(u_n)}{\epsilon}$ for each $n \in N$. Then, we have that $\{t_n\}_{n \in N} \subset [0, 1]$, and we may assume that $\lim_{n \to \infty} t_n = t \in [0, 1]$. The corresponding subsequence of $\{u_n\}_{n \in N}$ is denoted again by $\{u_n\}_{n \in N}$, and it also satisfies the inequality $1 - \epsilon_n \leq P_B(u_n) \leq 1$, which implies that $\lim_{n \to \infty} P_B(u_n) = 1$.

Now, let u^* be an accumulation point of $\{u_n\}_{n \in N}$ and thus have $\lim_{n \to \infty}(t_n, \frac{u_n}{P_B(u_n)}, u_n) = (t, u^*, u^*)$. By the fact that F is compact, we assume that $u_n \in F(t_n, \frac{u_n}{P_B(u_n)})$ for each $n \in N$. It follows that $u^* \in F(t, u^*)$, which contradicts the assumption (1), as we have $\lim_{n \to \infty} P_B(u_n) = 1$ (which means that $u^* \in \partial B$, but this is impossible).

Thus, it is impossible to have that "to have an infinite number of elements u_n satisfy the inequality: $1 - \epsilon_n \leq P_B(u_n) \leq 1$", which means that there is only a finite number of elements of sequence $\{u_n\}_{n \in N}$ satisfying the inequality $1 - \epsilon_n \leq P_B(u_n) \leq 1$. Now, without loss of generality, for $n \in N$, we have the following inequality:

$$P_B(u_n) < 1 - \epsilon_n.$$

By the fact that $\lim_{n \to}(1 - \epsilon_n) = 1$, $u_n \in F(1, \frac{u_n}{1-\epsilon})$ for all $n \in N$ and assuming that $\lim_{n \to} u_n = u^*$, then the upper semi-continuity of F with nonempty closed values implies that the graph of F is closed, and by the fact $u_n \in F(1, \frac{u_n}{1-\epsilon})$, it implies that $u^* \in F(1, u^*)$. This completes the proof. $\square$

As a special case of Theorem 8.4.6, we have the following principle for the implicit form of Leray–Schauder-type alternative for set-valued mappings in p-vector spaces for $0 < p \leq 1$.

Corollary 8.4.7 (Implicit Leray–Schauder alternative). *Let E be a locally p-convex space E, where $p \in (0, 1]$ and $B \subset E$ a bounded closed p-convex such that zero element $\theta \in intB$. Let*

$F : [0,1] \times B \to 2^E$ *be a QUSC semiclosed 1-set contractive set-valued mapping with nonempty p-convex values and with a closed graph, and the set $F([0,1] \times B)$ is relatively compact in E. If the following assumptions are satisfied:*

(1) $F(\{0\} \times \partial B) \subset B$;
(2) $x \notin F(0,x)$ *for all $x \in \partial B$,*

then at least one of the following properties is satisfied:

(i) *There exists $x^* \in B$ such that $x^* \in F(1, x^*)$.*
(ii) *There exists $(\lambda^*, x^*) \in (0,1) \times \partial B$ such that $x^* \in F(\lambda^*, x^*)$.*

Proof. The result is an immediate consequence of Theorem 8.4.6. This completes the proof. □

We conclude this section by noting that similar results on Rothe and Leray–Schauder alternatives have been developed by Furi and Pera [92], Granas and Dugundji [102], Górniewicz [100], Górniewicz *et al.* [101], Isac [118], Li *et al.* [156], Liu [160], Park [185], Potter [213], Shahzad [238, 240], Song and Huang [252], Xu [278], Xu *et al.* [279], and related references therein. These are used as tools of nonlinear analysis in the Banach space setting and applications to the boundary value problems for ordinary differential equations in noncompact problems, a general class of mappings for nonlinear alternative of Leray–Schauder type in normal topological spaces. Some Birkhoff–Kellogg-type theorems for general class mappings in topological vector spaces are also established by Agarwal *et al.* [1], Agarwal and O'Regan [2, 3], Park [186], and the references therein. In particular, recently, O'Regan [180] used the Leray–Schauder-type coincidence theory to establish some Birkhoff–Kellogg problems and Furi–Pera-type results for a general class of mappings.

8.5 Fixed Points for Non-Self QUSC Semiclosed 1-Set Contractive Set-Valued Mappings

This section focuses on the study of fixed point theorems for non-self QUSC semiclosed 1-set contractive set-valued mappings in locally p-convex spaces for $p \in (0,1]$.

As an application of the approximation result which is Theorem 8.3.4 in Section 8.3 for the 1-set contractive mappings above, we show how it is used as a useful tool for developing fixed point theorems for semiclosed 1-set contractive non-self upper semicontinuous mappings in p-seminorm spaces, where $p \in (0, 1]$, by including seminorm, norm spaces, and uniformly convex Banach spaces as special cases.

We note that when a p-vector space E is with a p-norm, then the convergence of both conditions (H1) and (H) can be described by the weak and strong convergences under the weak and strong topologies, respectively, induced by p-norm for $p \in (0, 1]$. Secondly, if a given p-vector space E has a nonempty open p-convex subset U containing zero, then any mapping satisfying the "(H) condition" is a hemicompact mapping (with respect to P_U for a given bounded open p-convex subset U containing the zero of the p-vector space E), thus satisfying "condition (H)" used in Theorem 6.1.5 in Chapter 6.

By the fact that each semiclosed 1-set mappings satisfies the "(H1) condition", we have the existence of fixed points for the class of semiclosed 1-set mappings. First, as an application of Theorem 8.3.4, we have the following result for non-self mappings in p-seminorm spaces for $p \in (0, 1]$.

Theorem 8.5.1. *Let U be a bounded open p-convex subset of a p-seminorm space E, where $p \in (0, 1]$ and the zero element $\theta \in U$. Assume that $F : \overline{U} \to 2^E$ is a QUSC semiclosed 1-set contractive set-valued mapping with nonempty p-convex values and with a closed graph. In addition, for any $x \in \partial \overline{U}$ and $y \in F(x)$, we have $\lambda x \neq y$ for any $\lambda > 1$ (i.e., the "Leray–Schauder boundary condition"). Then, F has at least one fixed point.*

Proof. By the proof of Theorem 8.3.4 with $C = E$, we actually have either (I) or (II) holding in the following:

(I) F has a fixed point $x_0 \in U$, i.e., $P_U(y_0 - x_0) = 0$.
(II) There exists $x_0 \in \partial(U)$ and $y_0 \in F(x_0)$ with $P_U(y_0 - x_0) = (P_U^{\frac{1}{p}}(y_0) - 1)^p > 0$.

If F has no fixed point, then (II) holds and $x_0 \notin F(x_0)$. By the proof of Theorem 8.3.4 (actually Theorem 6.1.9), we have that $x_0 = f(y_0)$ and $y_0 \notin \overline{U}$. Thus, $P_U(y_0) > 1$ and $x_0 = f(y_0) = \dfrac{y_0}{(P_U(y_0))^{\frac{1}{p}}}$, which means $y_0 = (P_U(y_0))^{\frac{1}{p}} x_0$, where $(P_U(y_0))^{\frac{1}{p}} > 1$, which contradicts the assumption. Thus, F must have a fixed point. The proof is complete. $\qquad\square$

By following the idea used and developed by Browder [32], Li [155], Li *et al.* [156], Goebel and Kirk [96], Petryshyn [205, 206], Tan and Yuan [258], Xu [278], Xu *et al.* [279], and the references therein, we have the following existence theorems for the principle of Leray–Schauder-type alternatives in p-seminorm spaces $(E, \|\cdot\|_p)$ for $p \in (0, 1]$.

Theorem 8.5.2. *Let U be a bounded open p-convex subset of a p-seminorm space $(E, \|\cdot\|_p)$, where $p \in (0, 1]$ and the zero element $\theta \in U$. Assume that $F : \overline{U} \to 2^E$ is a QUSC semiclosed 1-set contractive set-valued mapping with nonempty p-convex values and with a closed graph. In addition, there exist $\alpha > 1$ and $\beta \geq 0$, such that for each $x \in \partial\overline{U}$, we have that for any $y \in F(x)$,*

$$\|y - x\|_p^{\alpha/p} \geq \|y\|_p^{(\alpha+\beta)/p} \|x\|_p^{-\beta/p} - \|x\|_p^{\alpha/p}.$$

Then, F has at least one fixed point.

Proof. By assuming F has no fixed point, we prove the conclusion by showing that the Leray–Schauder boundary condition in Theorem 8.5.1 does not hold. If we assume F has no fixed point, by the boundary condition of Theorem 8.5.1, there exist $x_0 \in \partial\overline{U}$, $y_0 \in F(x_0)$, and $\lambda_0 > 1$ such that $y_0 = \lambda_0 x_0$.

Now, consider the function f defined by

$$f(t) := (t - 1)^\alpha - t^{\alpha+\beta} + 1$$

for each $t \geq 1$. We observe that f is a strictly decreasing function for $t \in [1, \infty)$, as the derivative of $f'(t) = \alpha(t-1)^{\alpha-1} - (\alpha+\beta)t^{\alpha+\beta-1} < 0$ by differentiation. Thus, we have $t^{\alpha+\beta} - 1 > (t - 1)^\alpha$ for $t \in (1, \infty)$. By combining the boundary condition, we have

$$\|y_0 - x_0\|_p^{\alpha/p} = \|\lambda_0 x_0 - x_0\|_p^{\alpha/p} = (\lambda_0 - 1)^\alpha \|x_0\|_p^{\alpha/p}$$

$$< (\lambda_0^{\alpha+\beta} - 1)\|x_0\|_p^{(\alpha+\beta)/p} \|x_0\|_p^{-\beta/p}$$

$$= \|y_0\|_p^{(\alpha+\beta)/p} \|x_0\|_p^{-\beta/p} - \|x_0\|_p^{\alpha/p},$$

which contradicts the boundary condition given by Theorem 8.5.2. Thus, the conclusion follows. $\qquad\square$

Theorem 8.5.3. *Let U be a bounded open p-convex subset of a p-seminorm space $(E, \| \cdot \|_p)$, where $p \in (0,1]$ and the zero element $\theta \in U$. Assume that $F : \overline{U} \to 2^E$ is a QUSC semiclosed 1-set contractive set-valued mapping with nonempty p-convex values and with a closed graph. In addition, there exist $\alpha > 1$ and $\beta \geq 0$ such that for each $x \in \partial \overline{U}$, we have that for any $y \in F(x)$,*

$$\|y + x\|_p^{(\alpha+\beta)/p} \leq \|y\|_p^{\alpha/p}\|x\|_p^{\beta/p} + \|x\|_p^{(\alpha+\beta)/p}.$$

Then, F has at least one fixed point.

Proof. We prove the conclusion by showing the Leray–Schauder boundary condition in Theorem 8.5.1 does not hold. If we assume that F has no fixed point, by the boundary condition of Theorem 8.5.1, there exist $x_0 \in \partial \overline{U}$, $y_0 \in F(x_0)$, and $\lambda_0 > 1$ such that $y_0 = \lambda_0 x_0$.

Now, consider the function f defined by $f(t) := (t+1)^{\alpha+\beta} - t^\alpha - 1$ for $t \geq 1$. We can then show that f is a strictly increasing function for $t \in [1, \infty)$, and thus we have $t^\alpha + 1 < (t+1)^{\alpha+\beta}$ for $t \in (1, \infty)$. By the boundary condition given in Theorem 8.5.3, we have that

$$\|y_0 + x_0\|_p^{(\alpha+\beta)/p} = (\lambda_0 + 1)^{\alpha+\beta}\|x_0\|_p^{(\alpha+\beta)/p}$$
$$> (\lambda_0^\alpha + 1)\|x_0\|_p^{(\alpha+\beta)/p}$$
$$= \|y_0\|_p^{\alpha/p}\|x_0\|_p^{\beta/p} + \|x_0\|_p^{\alpha/p},$$

which contradicts the boundary condition given by Theorem 8.5.3. Thus, the conclusion follows, and the proof is complete. $\square$

Theorem 8.5.4. *Let U be a bounded open p-convex subset of a p-seminorm space $(E, \| \cdot \|_p)$, where $p \in (0,1]$ and the zero element $\theta \in U$. Assume that $F : \overline{U} \to 2^E$ is a QUSC semiclosed 1-set contractive set-valued mapping with nonempty p-convex values and with a closed graph. In addition, there exist $\alpha > 1$ and $\beta \geq 0$ (or alternatively, $\alpha > 1$ and $\beta \geq 0$) such that for each $x \in \partial \overline{U}$, we have that for any $y \in F(x)$,*

$$\|y - x\|_p^{\alpha/p}\|x\|_p^{\beta/p} \geq \|y\|_p^{\alpha/p}\|y + x\|_p^{\beta/p} - \|x\|_p^{(\alpha+\beta)/p}.$$

Then, F has at least one fixed point.

Proof. Similarly as above, we prove the conclusion by showing that the Leray–Schauder boundary condition in Theorem 8.5.1 does not hold. If we assume F has no fixed point, by the boundary condition of Theorem 8.5.1, there exist $x_0 \in \partial \overline{U}$, $y_0 \in F(x_0)$, and $\lambda_0 > 1$ such that $y_0 = \lambda_0 x_0$.

Now, consider the function f defined by $f(t) := (t-1)^\alpha - t^\alpha(t-1)^\beta + 1$ for $t \geq 1$. We can then show that f is a strictly decreasing function for $t \in [1, \infty)$, and thus we have $(t-1)^\alpha < t^\alpha(t+1)^\beta - 1$ for $t \in (1, \infty)$. By the boundary condition given in Theorem 8.5.4, we have that

$$
\begin{aligned}
\|y_0 - x_0\|_p^{\alpha/p}\|x_0\|_p^{\beta/p} &= (\lambda_0 - 1)^\alpha \|x_0\|_p^{(\alpha+\beta)/p} \\
&< (\lambda_0^\alpha(\lambda_0 + 1)^\beta - 1)\|x_0\|_p^{(\alpha+\beta)/p} \\
&= \|y_0\|_p^{\alpha/p}\|y_0 + x_0\|_p^{\beta/p} - \|x_0\|_p^{(\alpha+\beta)/p},
\end{aligned}
$$

which contradicts the boundary condition given by Theorem 8.5.4. Thus, the conclusion follows, and the proof is complete. □

Theorem 8.5.5. *Let U be a bounded open p-convex subset of a p-seminorm space $(E, \|\cdot\|_p)$, where $p \in (0, 1]$ and the zero element $\theta \in U$. Assume that $F : \overline{U} \to 2^E$ is a QUSC semiclosed 1-set contractive set-valued mapping with nonempty p-convex values and with a closed graph. In addition, there exist $\alpha > 1$ and $\beta \geq 0$ such that we have for any $y \in F(x)$,*

$$
\|y + x\|_p^{(\alpha+\beta)/p} \leq \|y - x\|_p^{\alpha/p}\|x\|_p^{\beta/p} + \|y\|_p^{\beta/p}\|x\|^{\alpha/p}.
$$

Then, F has at least one fixed point.

Proof. Similarly as above, we prove the conclusion by showing that the Leray–Schauder boundary condition in Theorem 8.5.1 does not hold. If we assume that F has no fixed point, by the boundary condition of Theorem 8.5.1, there exist $x_0 \in \partial \overline{U}$, $y_0 \in F(x_0)$, and $\lambda_0 > 1$ such that $y_0 = \lambda_0 x_0$.

Now, consider the function f defined by $f(t) := (t+1)^{\alpha+\beta} - (t-1)^\alpha - t^\beta$ for $t \geq 1$. We can then show that f is a strictly increasing function for $t \in [1, \infty)$, and thus we have $(t+1)^{\alpha+\beta} > (t-1)^\alpha + t^\beta$ for $t \in (1, \infty)$.

By the boundary condition given in Theorem 8.5.5, we have

$$\|y_0 + x_0\|_p^{(\alpha+\beta)/p} = (\lambda_0 + 1)^{\alpha+\beta}\|x_0\|_p^{(\alpha+\beta)/p}$$
$$> ((\lambda_0 - 1)^\alpha + \lambda_0^\beta)\|x_0\|_p^{(\alpha+\beta)/p}$$
$$= \|\lambda_0 x_0 - x_0\|_p^{\alpha/p}\|x_0\|_p^{\beta/p} + \|\lambda_0 x_0\|_p^{\beta/p}\|x_0\|_p^{\alpha/p}$$
$$= \|y_0 - x_0\|_p^{\beta/p}\|x_0\|_p^{\alpha/p} + \|y_0\|_p^{\beta/p}\|x_9\|^{\alpha/p},$$

which implies that

$$\|y_0 + x_0\|_p^{(\alpha+\beta)/p} > \|y_0 - x_0\|_p^{\beta/p}\|x_0\|_p^{\alpha/p} + \|y_0\|_p^{\beta/p}\|x_9\|^{\alpha/p}.$$

This contradicts the boundary condition given by Theorem 8.5.5. Thus, the conclusion follows, and the proof is complete. □

As an application of Theorems 8.5.1 by testing the Leray–Schauder boundary condition, we have the following conclusion for each special case, and thus we omit the details of their proofs here.

Corollary 8.5.6. *Let U be a bounded open p-convex subset of a p-seminorm space $(E, \|\cdot\|_p)$, where $p \in (0,1]$ and the zero element $\theta \in U$. Assume that $F : \overline{U} \to 2^E$ is a QUSC semiclosed 1-set contractive set-valued mapping with nonempty p-convex values and with a closed graph. Then, F has at least one fixed point if one of the following (strong) conditions holds for $x \in \partial\overline{U}$ and $y \in F(x)$:*

(i) $\|y\|_p \leq \|x\|_p$,
(ii) $\|y\|_p \leq \|y - x\|_p$,
(iii) $\|y + x\|_p \leq \|y\|_p$,
(iv) $\|y + x\|_p \leq \|x\|_p$,
(v) $\|y + x\|_p \leq \|y - x\|_p$,
(vi) $\|y\|_p \cdot \|y + x\|_p \leq \|x\|_p^2$,
(vii) $\|y\|_p \cdot \|y + x\|_p \leq \|y - x\|_p \cdot \|x\|_p$.

If the p-(semi)norm space E is a uniformly convex Banach space $(E, \|\cdot\|)$ (for p-norm space with $p = 1$), then we have the following general existence result, which can apply to general non-expansive (single-valued or set-valued) mappings, too.

Theorem 8.5.7. *Let U be a bounded open convex subset of a uniformly convex Banach space $(E, \|\cdot\|)$ with zero element $\theta \in U$.*

Assume that $F : \overline{U} \to E$ is a continuous semi-contractive single-valued mapping with nonempty values. In addition, for any $x \in \partial\overline{U}$, we have $\lambda x \neq F(x)$ for any $\lambda > 1$ (i.e., the "Leray–Schauder boundary condition"). Then, F has at least one fixed point.

Proof. By Lemma 7.5.1, F is a continuous semiclosed 1-set contractive mapping. Moreover, by the assumption that E is a uniformly convex Banach space, the mapping $(I - F)$ is closed at zero, and thus F is semiclosed at zero (see Browder [32] or Goebel and Kirk [96]). Thus, all the assumptions of Theorem 8.5.2 are satisfied. The conclusion follows by Theorem 8.5.2. The proof is complete. □

By considering the p-seminorm space $(E, \|\cdot\|)$ with a seminorm for $p = 1$, the following result is a special case of the corresponding results from Theorems 8.5.2–8.5.5, and thus we omit its proof.

Corollary 8.5.8. *Let U be a bounded open convex subset of a norm space $(E, \|\cdot\|)$. Assume that $F : \overline{U} \to 2^E$ is a QUSC semiclosed 1-set contractive set-valued mapping with nonempty p-convex values and with a closed graph. Then, F has at least one fixed point if there exist $\alpha > 1$ and $\beta \geq 0$ such that any one of the following conditions satisfied:*

(i) *for each $x \in \partial\overline{U}$ and $y \in F(x)$, $\|y - x\|^\alpha \geq \|y\|^{(\alpha+\beta)}\|x\|^{-\beta} - \|x\|^\alpha$,*

(ii) *for each $x \in \partial\overline{U}$ and $y \in F(x)$, $\|y + x\|^{(\alpha+\beta)} \leq \|y\|^\alpha\|x\|^\beta + \|x\|^{(\alpha+\beta)}$,*

(iii) *for each $x \in \partial\overline{U}$ and $y \in F(x)$, $\|y - x\|^\alpha\|x\|^\beta \geq \|y\|^\alpha\|y + x\|^\beta - \|x\|^{(\alpha+\beta)}$,*

(iv) *for each $x \in \partial\overline{U}$ and $y \in F(x)$, $\|y + x\|^{(\alpha+\beta)} \leq \|y - x\|^\alpha\|x\|^\beta + \|y\|^\beta\|x\|^\alpha$.*

Remark 8.5.9. As discussed in Lemma 7.5.1 and the proof of Theorem 8.5.7, when p-vector spaces are uniformly convex Banach spaces, the semi-contractive or nonexpansive mappings automatically satisfy "condition (H1)" required by Theorem 8.5.1, and thus the mappings are indeed semiclosed. In addition, Theorems 8.5.1–8.5.6 improve or unify the corresponding results given by Browder [32], Li [155], Li *et al.* [156], Goebel and Kirk [96], Petryshyn [205, 206], Reich [222], Tan and Yuan [258], Xu [272], Xu [278], Xu *et al.* [279], and results from the references therein by extending non-self mappings to the classes of semiclosed 1-set contractive set-valued mappings in p-seminorm spaces with $p \in (0, 1]$, which include norm spaces or Banach spaces when $p = 1$ as a special class.

Before concluding, we would like to share with readers that the main goal of this section is to develop some new results and tools in nonlinear analysis for 1-set contractive mappings under the general framework of p-vector spaces, where $p \in (0, 1]$. We expect that these new results would become useful tools for the study of optimization, nonlinear programming, variational inequality, complementarity, game theory, mathematical economics, and other related social science areas. Finally, we note that the results established in this book not only unify or improve the corresponding results in the existing literature for nonlinear analysis, but they can also be regarded as a continuation of related work established recently by Yuan [286–291] and the references therein.

8.6 Fixed Points for Non-Self Nonexpansive Mappings

The goal of this section is to explore corresponding fixed point theorems for non-self set-valued nonexpansive mappings under the general locally complete convex spaces by using the fixed point theorem established by Caristi [39] in 1976 and its generalization as a fundamental and powerful tool. In this section, we focus on the study of general fixed point theorems for non-self contraction and non-expansive set-valued mappings under the framework of locally complete convex spaces, which indeed unify or improve the corresponding results in the existing literature.

For the convenience of our discussion, we now recall some notions introduced in Section 7.5 in Chapter 7. Similarly to above, for a given space E, we denote by $K(E)$ the family of nonempty compact subsets E.

Now, let E denote a Hausdorff LCS and $\mathfrak{F}$ denote the family of continuous seminorms generating the topology of E. For each $p \in \mathfrak{F}$ and $A, B \in K(E)$, we can define the metric $\delta(A, B)$ between the subsets A and B by

$$\delta(A, B) := \sup\{p(a - b) : a \in A, b \in B\},$$

and the Hausdorff metric $d_{Hp}(A, B)$ by

$$d_{Hp}(A, B) := \max\{\sup_{a \in A} \inf_{b \in B} p(a - b), \sup_{b \in B} \inf_{a \in A} p(a - b)\}.$$

It is clear that when p is only a seminorm in E (and thus E is a normed space), d_{Hp} is a traditional Hausdorff metric on $K(E)$ (e.g., see Ko and Tsai [143]). Secondly, by the definition of Hausdorff metric d_{Hp} for each $p \in \mathfrak{F}$, we know that if $A, B \in K(E)$, then for each $a \in A$, there exists $b \in B$ such that $p(a - b) \leq d_{Hp}(A, B)$ (e.g., by the definition of the Hausdorff metric and the compactness of subsets, similar to the idea used by Nadler [173]).

Let E denote a Hausdorff locally convex topological vector space and $\mathfrak{F}$ denote the family of continuous seminorms generating the topology of E, with K being a nonempty subset of E. We recall that a mapping $T : K \to K(E)$ is said to be a $\mathfrak{F}$ **contraction** set-valued mapping if for each $p \in \mathfrak{F}$, there exists a constant $k_p \in (0, 1)$ such that $d_{Hp}(T(x), T(y)) \leq k_p p(x - y)$. When $k_p = 1$ for all $p \in \mathfrak{F}$, T is said to be a non-expansive set-valued mapping if for any $x, y \in K$, we have $d_{Hp}(T(x), T(y))) \leq p(x - y)$ for all $x, y \in K$.

By following Chen and Singh [57], we recall that for a given LCS E, which is said to satisfy the "P-**Opial condition**" if for each $x \in E$ and every net (x_α) converging weakly to x, then for each $P \in \mathfrak{F}$, we have

$$\liminf P(x_\alpha - y) > \liminf P(x_\alpha - x)$$

for any $y \neq x$.

Here, we would like to remark that the notion of the "P-**Opial condition**" (denoted by the uppercase letter P) was originally introduced by Chen and Singh [57] in 1992 given for the LCS E, which is different from Definition 3.3.9 for the "p-**Opial condition**" (denoted by the lowercase letter p) under the framework of p-normed spaces. There, $p \in (0, 1]$ is a positive real number less than or equal to 1, not representing a family member of seminorms which generate the topological structure for the LCS E.

Let K be a nonempty bounded closed convex subset of a normed space $(E, \|\cdot\|)$. We say that K has the fixed point property (FPP) for nonexpansive mapping T if for every nonexpansive mapping $T : K \to K$ (i.e., $\|Tx - Ty\| \leq \|x - y\|$ for all $x, y \in K$), T has a fixed point $x \in K$ such that $Tx = x$. The space E is said to have the FPP if any nonempty bounded closed convex subset of E has the FPP for nonexpansive mapping; the space E has the weak fixed point property (WFPP) if any weakly compact convex subset of E has the FPP for nonexpansive mapping. For a reflexive Banach space, both properties are obviously the same.

The famous question of whether a Banach space has the WFPP had remained open for a long time. It has been answered in the negative by Alspach [9] and Sadovski [236] who constructed following examples, respectively.

Example 8.6.1 (Alspach [9]). Let $X = L_1[0, 1]$, and let

$$K = \left\{ f \in L_1[0, 1], \int_0^1 f = 1, 0 \le f \le 2, a.e. \right\}.$$

It is easy to see that K is a weakly closed, convex subset of the order interval $\{f : 0 \le f \le 2\}$, and thus K is weakly compact because order intervals in $L_1[0, 1]$ are weakly compact. This is a direct consequence of uniform integrability by Dunford and Schwartz [72, p. 3]. Now, we define a (nonexpansive) mapping $T : K \to K$ by

$$T(f(t)) := \begin{cases} 2f(2t) \wedge 2, & \text{if } 0 \le t \le \frac{1}{2}, \\ [2f(2t-1) - 2] \vee 0, & \text{if } \frac{1}{2} < t \le 1. \end{cases} \tag{8.2}$$

Then, we check that T is an isometry on K, but T has no fixed point.

Example 8.6.2 (Sadovski [236]). Let $E = c_0$ and $K = \{x \in c_0 : \|x\| \le 1\}$. Define a mapping $T : K \to k$ by $T(x) := (1, x_1, x_2, x_3, \ldots)$ for each $x = (x_{1,2}, x_3, \ldots) \in K$. Then, K is bounded, closed, and convex; T is an isometry, i.e., $\|Tx - Ty\| = \|x - y\|$ for any $x, y \in K$, but T is fixed point free. Thus, c_0 does not have the FPP.

The above two simple examples suggest that to obtain positive results in the problem of the existence of fixed points for nonexpansive mappings, it is necessary to impose some restrictions either on the mapping T or on the underlying space E.

Now, it is time for us to establish fixed point theorems for non-self contraction and non-self expansive set-valued mappings by using the Caristi fixed point theorem [39], which is a powerful tool under the framework of locally complete convex spaces.

If B is a balanced and convex subset of a Hausdorff locally convex real vector space E, by following Pérez Carreras and Bonet [204] and Horváth [109], the subset B is called a disk. Let E_B denote the linear span of B endowed with the topology generated by the Minkowski functional (gauge) p_B of B. In addition, when B is bounded, p_B is a norm, and the norm topology is finer than the topology inherited from E.

Here, we recall that if (E_B, p_B) is a Banach space, then B is said to be a Banach disk. We say that E is a locally complete space if each closed, bounded disk is a Banach disk. By the definition, each complete LCS is locally complete, but the converse is not true; see more details provided by Pérez Carreras and Bonet [204] and the references therein.

Hereinafter in this section, all LCSs are assumed to be locally complete unless specified otherwise.

Let M be a subset of a given LCS E, and assume that $T : M \to K(E)$ is a set-valued mapping with nonempty values. Then, we say that:

(1) T satisfies the boundary condition (α): if for each $x \in M$ and any $y \in T(x)$, we have $(x, y] \cap M \neq \emptyset$, where $(x, y] := \{(1 - \lambda)x + \lambda y : \lambda \in (0, 1]\}$ (and also, we define $[x, y] := \{(1 - \lambda)x + \lambda y : \lambda \in [0, 1]\}$); and

(2) T is called weakly inward if for each $x \in M$, $T(x) \subset \overline{I_M(x)}$, where $I_M(x) := \{z \in E : x = x + \lambda(y - x), y \in M, \text{ and } \lambda \in [1, +\infty)\}$.

Now, we recall the following result, which is the famous Caristi fixed point theorem in the setting of Hausdorff locally complete convex spaces, which is actually a special case of Theorem 3.4.14 given in locally complete spaces (see also more from Caristi [39], Kirk and Saliga [136], Cammaroto *et al.* [37], Fang [87], Latif *et al.* [149], Qiu [215–220], Qiu and Rolewicz [221], Zhong [300], and Zhong *et al.* [302], as well as Feng and Liu [88], Lau and Yao [152], and references therein for some new results in another direction not covered by the current version of this book).

Lemma 8.6.3. *Let E be a locally complete convex space and $T : E \to E$ be any arbitrary (single-valued) mapping with nonempty values. Suppose there exists a lower semicontinuous function $f : E \to [0, +\infty)$ such that for each for each given $p \in \mathfrak{F}$, we have* *for any $x \in E$, $p(x - T(x)) \leq f(x) - f(T(x))$. Then, T has a fixed point.*

Proof. This is a special case of Theorem 3.4.14. The proof is complete. □

Here, we would like point out that Lemma 8.6.3 is actually Theorem 2.1 of Latif *et al.* [149], which is a special form of Theorem 3.2 given by Cammaroto *et al.* [37], with the original result of Theorem 3.1 established by Fang [87]. In addition, the general form of Lemma 8.6.3 is also given by Theorem 4.1 or 4.2 of Qiu [220], for which we state its another special form as Lemma 8.6.4 in the following, which includes the above Lemma 8.6.3 as its special case.

We now have the following result, which is a generalization of Lemma 8.6.3 above and is also a variant of Lemma 1.2 given by Zhong *et al.* [302]) (indeed, it is a special case of Theorem 4.1 established by Qiu [220]).

Lemma 8.6.4. *Let E be a locally complete convex space, $f : E \to [0, +\infty)$ be a bounded below, lower semicontinuous function, and $h : [0, +\infty)$ be a nonnegative continuous nondecreasing function such that $\int_0^{+\infty} \frac{dr}{1+h(r)} = +\infty$. Let $T : E \to E$ be a (single-valued) mapping with nonempty values such that for any given $x_0 \in E$ and $p \in \mathfrak{F}$ and for any $x \in E$, we have $\frac{p(x-f(x))}{1+h(p(x_0-x))} \leq f(x) - f(T(x))$. Then, T has a fixed point.*

Proof. This is a special case of Theorem 4.1 of Qiu [220]. Indeed, assume that $\psi(s) := \frac{1}{1+h(s)}$, and $\phi(s) := s$ for each $s \in [0, +\infty)$ in Theorem 4.1 of Qiu [220]. Then, for each given $p_\lambda \in \mathfrak{F}$, with the properties of h and f defined by Lemma 8.6.4 above, it follows that $\phi_\lambda(p_\lambda(x - T(x))) = p_\lambda(x - T(x))$. Let $\psi_\lambda(f(x)) := \psi_\lambda (p_\lambda(x_0 - x))$ for each $x \in E$. Then, we have $\psi_\lambda(f(x)) = \psi_\lambda(p_\lambda (x_0 - x)) = \frac{1}{1+h(p_\lambda(x_0-x))}$ (by definitions of ψ and ϕ) for each $x \in X$. It is clear that all the hypotheses of Theorem 4.1 of Qiu [220] are satisfied, and thus T has a fixed point by Theorem 4.1 of Qiu [220]. This completes the proof. □

Here, we also note that Lemma 8.6.4 is actually Theorem 2.2 of Latif *et al.* [149] (which is actually a special case of Theorem 4.1 or 4.2 by Qiu [220]). Indeed, it is a version of Lemma 1.2 given by Zhong *et al.* [302] under the framework of locally complete convex spaces, which can be proved by using the same idea used for proving Theorem 2.1 by Zhong [300]. In addition, the corresponding results on the Ekeland principle and the Caristi fixed point theorem in locally p-convex spaces for $p \in (0,1]$ have also been established by Qiu and Rolewicz [221] (but we omit their discussion here due to space considerations).

Now, by Lemma 8.6.3, we first have the following fixed point theorem for non-self $\mathfrak{F}$ contract set-valued mapping in locally complete convex spaces.

Theorem 8.6.5. *Let M be a nonempty closed subset of locally complete convex space E and $T : M \to K(E)$ be a $\mathfrak{F}$ contraction set-valued mapping satisfying the boundary condition (α). Then, T has a fixed point.*

Proof.   Let $p \in \mathfrak{F}$ be any given one seminorm in E. For each $x \in M$, as $T(x)$ is compact, we choose $y \in T(x)$ such that

$$p(x - y) = d_p(x, T(x))(:= \inf\{p(x - z) : z \in T(x)\}).$$

We denote by z_{xp} the farthest point from x in $[x, y] \cap M$, which means z_{xp} satisfies the following equation:

$$p(x - z_{xp}) = \max\{p(x - w) : w \in [x, y] \cap M \neq \emptyset\}.$$

Then, it is easy to verify that the following equality holds by using the definition of z_{xp} and the triangle inequality property of the seminorm p:

$$p(x - y) = p(x - z_{xp}) + p(z_{xp} - y).$$

Since T is a $\mathfrak{F}$ contraction set-valued mapping with nonempty compact values, it follows that for each $p \in \mathfrak{F}$, there exists $k_p \in (0,1)$, and we have

$$
\begin{aligned}
d_p(z_{xp}, T(z_{xp})) &\leq p(z_{xp} - y) + d_p(y, T(z_{xp})) \\
&\leq p(z_{xp} - y) + d_p(T(x), T(z_{xp})) \\
&\leq p(z_{xp} - y) + k_p p(x - z_{xp}) \\
&= p(x - y) - p(x - z_{xp}) + k_p p(x - z_{xp}) \\
&= d_p(x, T(x)) - (1 - k_p)p(x - z_{xp}).
\end{aligned}
$$

Thus, we have

$$p(x - z_{xp}) \leq (1 - k_p)^{-1}\{d_p(x, T(x)) - d_p(z_{xp}, T(z_{xp}))\}.$$

Now, we define the mapping $f : M \to M$ by $f_p(x) := z_{xp}$ for each $x \in M$, and also define a non-negative function $\phi_p : M \to [0, +\infty)$ by $\phi(x) := (1 - k_p)^{-1}d_p(x, T(x))$ for each $x \in M$. Then, by the definitions of f and ϕ, it follows that

$$p(x - f_p(x)) \leq \phi_p(x) - \phi_p(f_p(x)),$$

for each $x \in M$. By the fact that M is a closed subset of a complete space E, it is complete, and hence by Lemma 8.6.3, f has a fixed point $u \in M$ such that $u = f_p(u)$. Note that $f_p(u) = u = z_{up}$, and by the definition of z_{up}, it is the farthest point from u in $[u, y] \cap M$, which means $p(u - z_{up}) = \max\{p(u - w) : w \in [u, y] \cap M \neq \emptyset\}$. This will help us prove $d(u, T(u)) = 0$, which implies that u is a fixed point of T, as will be shown in the following in detail.

Now, by the boundary condition (α) for the $\mathfrak{F}$ mapping T, we have $(u, y] \cap M \neq \emptyset$. Without loss of generality, there exists a non-zero positive number $\lambda_0 \in (0, 1]$ such that $w_0 = (1 - \lambda_0)u + \lambda_0 y$, and $w_0 \in (u, y] \cap M \neq \emptyset$. By the definition of z_{up}, we have $p(u - z_{up}) = \max\{p(u - w) : w \in [u, y] \cap M \neq \emptyset\}$, and it follows that $p(u - z_{up}) \geq p(u - w_0) = \lambda_0 p(u - y) = \lambda_0 d(u, T(u))$ (as given in the beginning of the proof above). Since $u = z_{up}$, it follows that

$$\lambda_0 d(u, T(u)) = p(u - z_{up}) = p(z_{up} - z_{up}) = 0,$$

which implies that $d(u, T(u)) = 0$ as $\lambda_0 > 0$. This means u is a fixed point of T. The proof is complete. $\qquad\square$

We note that Theorem 8.6.5 extends Theorem 2 of Massa [165] (see also Massa *et al.* [166]) for non-self $\mathfrak{F}$ contraction set-valued mappings in Hausdorff LCSs.

Now, as an application of Lemma 8.6.3, we have the following fixed point result for non-self $\mathfrak{F}$ contraction set-valued mappings with different boundary conditions.

Theorem 8.6.6. *Let M be a nonempty closed subset of locally complete convex space E and $T : M \to K(E)$ be a $\mathfrak{F}$ contraction set-valued mapping such that for each given $p \in \mathfrak{F}$ and for any $x \in M$, we have $\{z \in T(x) : p(x - z) = d_p(x, T(x))\} \cap \overline{I_M(x)} \neq \emptyset$ (also called the "intersection boundary condition"). Then, T has a fixed in M.*

Proof. Suppose that T has no fixed point. Then, for any given $p \in \mathfrak{F}$ and for any $x \in M$, we have $d_p(x, T(x)) > 0$. As T is a $\mathfrak{F}$ contractive set-valued mapping, there exists $k_p \in (0, 1)$ such that $d_{Hp}(T(x), T(y)) \leq k_p p(x - y)$. Now, choose a constant $q \in (0, 1)$, denoted by $k := k_p < \frac{1-q}{1+q}$, where we know that $k_p \in (0, 1)$. In addition, by the assumption for the boundary condition of the $\mathfrak{F}$ mapping F, the following set is not empty, i.e.,

$$\{z \in T(x) : p(x - z) = d_p(x, T(x))\} \cap \overline{I_M(x)} \neq \emptyset.$$

Since $T(x)$ is also nonempty compact for $x \in M$, there exists $z \in T(x) \cap \overline{I_M(x)}$ such that $d_p(x, T(x)) = p(x - z) > 0$. In addition, by the definition of $I_M(x)$ and $z \in T(x) \cap \overline{I_M(x)}$, it follows that we can choose some $t \in (0, 1)$ (see also the proof in detail in Theorem 1.2 by Caristi [39]) such that

$$t^{-1} d_p(1 - t)x + tz, M) < qp(x - z).$$

Now, let $w := (1 - t) + tz$. Then, there exists some $y \in M$ such that

$$p(w - y) < qtp(x - z) = qp((1 - t)x + tz - x) = qp(w - x).$$

Since $p(y - x) \leq p(y - w) + p(w - x)$, it follows that

$$p(y - x) - p(w - x) \leq p(w - y) < qp(w - x),$$

which implies

$$p(y - x) < (1 + q)p(w - x).$$

Therefore, the following inequality holds:

$$(q - 1)p(w - x) < \frac{q - 1}{q + 1}p(x - y).$$

On the other hand, as $w = (1 - t) + tz$, where $t \in (0, 1)$, it is easy to verify that the following equality is true:

$$p(w - x) + p(w - z) = p(x - z).$$

Since $T(x)$ is a $\mathfrak{F}$ contraction set-valued mapping with nonempty compact values and by the definition of the Hausdorff metric d_{Hp}, we can choose $u \in T(x)$ and $v \in T(y)$ such that $p(w - u) = d_p(w, T(x))$ and

$$p(u - v) \leq d_{Hp}(T(x), T(y)) \leq k_p p(x - y).$$

Now, by putting all the above inequalities and the equality (i.e., $p(w - x) + p(w - z) = p(x - z)$) together, we have the following inequality holding:

$$
\begin{aligned}
d_p(y, T(y)) &\leq p(y - v) \leq p(y - w) + p(w - u) + p(u - v) \\
&\leq qp(w - x) + d_p(w, T(x)) + kp(x - y) \\
&< qp(w - x) + p(w - z) + kp(x - y) \\
&= qp(w - x) + p(x - z) - p(w - x) + kp(x - y) \\
&= (q - 1)p(w - x) + p(x - z) + kp(x - y) \\
&< \frac{q - 1}{q + 1}p(x - y) + p(x - z) + kp(x - y) \\
&= d_p(x, T(x)) - \left(\frac{1 - q}{1 + q} - k\right)p(x - y).
\end{aligned}
$$

Now, let $c = \frac{1-q}{1+q} - k$, and we have $d_p(y, T(y)) < d_p(x, T(x)) - cp(x - y)$. Therefore, we have the following inequality:

$$p(x - y) < \frac{d_p(x, T(x))}{c} - \frac{d_p(y, T(y))}{c}.$$

Now, we define two mappings $f : M \to M$ and $\phi : M \to [0, +\infty)$ by $f(x) := y$ and $\phi(x) := \frac{d_p(x, T(x))}{c}$ for each $x \in M$. It is clear that we have

$$p(x - f(x)) < \phi(x) - \phi(f(x)).$$

Now, by Lemma 8.6.3, f has a fixed point $x_0 \in M$, and thus $f(x_0) = x_0$. On the other hand, we have

$$0 = p(x_0 - f(x_0)) < \phi(x_0) - \phi(f(x_0)) = \phi(x_0) - \phi(x_0) = 0,$$

which is impossible. Hence, T must have a fixed point. This completes the proof. $\qquad\square$

As a special case of Theorem 8.6.6, we have the following result.

Corollary 8.6.7. *Let M be a closed subset of locally complete convex space E, and let $T : M \to K(E)$ be a weakly inward $\mathfrak{F}$ contraction mapping, i.e., $T(x) \subset \overline{I_M(x)}$ for each $x \in M$. Then, T has a fixed point.*

Remark 8.6.8. We would like to point out that our Theorem 8.6.6 extends Theorem 3.3 given by Zhang [298]. Secondly, it is worth mentioning that the intersection condition of Theorem 8.6.6 cannot be replaced by the condition $T(x) \cap \overline{I_M(x)} \neq \emptyset$ for each $x \in M$ even in the setting of Banach spaces, as shown by Zhang [298], Xu [274], and the references therein.

Next, by applying Lemma 8.6.4 above and combining a contractive condition, basically due to Zhong [300], we have the following fixed point theorem for non-self $\mathfrak{F}$ contractive weakly inward set-valued mappings in the setting of locally convex topological vector spaces.

Here, we assume that the mapping $h : [0, +\infty) \to [0, +\infty)$ is a continuous nondecreasing function satisfying $\int_0^{+\infty} \frac{dr}{1+h(r)} = +\infty$. Now, we have the following fixed point theorem for non-self contractive set-valued mappings for spaces satisfying the p-Opial condition.

Theorem 8.6.9. *Let M be a closed subset of a locally complete convex space E and $T : M \to K(E)$ be a weakly inward mapping, for a given $x_0 \in M$ and a given constant number $\sigma \in (0, 1]$. For each given $p \in \mathfrak{F}$, the following condition holds:*

$$d_{Hp}(T(x), T(y)) \leq \left(1 - \frac{\sigma}{1 + h(p(x_0 - x))}\right) p(x - y),$$

for any $x, y \in M$. Then, T has a fixed point.

Proof. Suppose that T has no fixed point. Then, $d_p(x, T(x)) > 0$ for all $x \in M$. By the assumption, the function h is a continuous nondecreasing function satisfying $\int_0^{+\infty} \frac{dr}{1+h(r)} = +\infty$, we can choose $c \in (0, \sigma)$, where $\sigma \in (0, 1]$, and let $q(x) := \frac{\sigma - c}{2(1 + h(p(x_0 - x)))}$ such that $q(x) < 1$. Since T is with nonempty values, there exists $z \in T(x)$ such that $d_p(x, T(x)) = p(x - z) > 0$. By the assumption, T is weakly inward, and we have $T(x) \subset \overline{I_M(x)}$ for $x \in M$. Thus, there exist

$y \in M$ and $\lambda \geq 1$ such that $z' := x + \lambda(y - x)$ with the following (approximation) inequality property:

$$P(z - z') = p(z - (x + \lambda(y - x))) < q(x)p(x - z).$$

Let $t := \frac{1}{\lambda}$ and $w := (1-t)x+tz$. Note that $w-y = (1-t)x+tz-y = t(z - (x + \frac{1}{t}(y - x)) = t(z - z')$, and it follows that

$$p(w - y) = t(p(z - (x + \lambda(y - x)))) = tp(z - z').$$

Thus, we have: $p(w - y) = tp(z - z') < tq(x)p(x - z)$, $p(w - x) = tp(x - z)$, and $p(w - z) = (1 - t)p(x - z)$.

Secondly, as $w = (1-t)x+tz$, it is easy to verify that the following equality holds: $p(w - x) + p(w - z) = p(x - z)$.

As $p(y - x) \leq p(y - w) + p(w - x)$, it implies that $p(y - x) - p(w - x) \leq p(w - y)$, and thus we have

$$p(x - y) < p(w - x) + tq(x)p(x - z) = (1 + q(x))p(w - x).$$

Therefore, we obtain the following inequality:

$$(q(x) - 1)p(w - x) < \frac{q(x) - 1}{q(x) + 1}p(x - y),$$

where $q(x) \in (0, 1)$. Since T is with compact values and by the definition of Hausdorff metric d_{Hp}, we can choose $u \in T(x)$ and $v \in T(y)$ such that $p(w - u) = d_p(w, T(x))$ and

$$p(u - v) \leq d_{Hp}(T(x), T(y)).$$

Now, putting all the above inequalities and the equality (i.e., $p(w - x) + p(w - z) = p(x - z)$) together and also considering the assumption on the mapping T, we have the following inequalities:

$$\begin{aligned}
d_p(y, T(y)) &\leq p(y - v \leq p(y - w) + p(w - u) + p(u - v) \\
&< q(x)p(w - x) + p(w - z) + d_{Hp}(T(x), T(y)) \\
&= q(x)p(w - x) - p(w - x) + p(x - z) + d_{Hp}(T(x), T(y)) \\
&= (q(x) - 1)p(w - x) + p(x - z) + d_{Hp}(T(x), T(y)) \\
&< \left(\frac{q(x) - 1}{q(x) + 1} + 1\right) p(x - y) + p(x - z)
\end{aligned}$$

$$- \frac{\sigma}{1 + h(p(x_0 - x))p(x_0 - x))} p(x - y)$$

$$< 2 \left(\frac{\sigma - c}{2(1 + h(p(x_0 - x)))} \right) p(x - y) + p(x - z)$$

$$- \frac{\sigma}{1 + h(p(x_0 - x))} p(x - y)$$

$$< d_p(x, T(x)) - \frac{c}{1 + h(p(x_0 - x))} p(x - y).$$

Thus, it follows that

$$\frac{p(x - y)}{1 + h(p(x_0 - x))} < \frac{d_p(x, T(x))}{c} - \frac{d_p(y, T(y))}{c}.$$

Now, we define two mappings for $f : M \to M$ and $\phi : M \to [0, +\infty)$ by $f(x) = y$ and $\phi(x) = \frac{d_p(x, T(x))}{c}$ for each $x \in M$. Note that ϕ is continuous and non-negative, and

$$\frac{p(x - f(x))}{1 + h(p(x_0 - x))} < \phi(x) - \phi(f(x)).$$

Now, applying Lemma 8.6.4, f has a fixed point. However, the above inequality shows that $0 < 0$, which is impossible. This contradiction shows that T must have a fixed point. The proof is complete.   □

Remark 8.6.10. We note that Theorem 8.6.9 extends Theorem 2.5 by Zhong *et al.* [302] to Hausdorff LCSs. Now, as an application of Theorem 8.6.8, we have the following fixed point theorem for non-self expansive set-valued mappings with boundary condition (α) in locally complete convex spaces, which satisfy the P-Opial condition introduced by Chen and Singh [57] in 1992.

Theorem 8.6.11. *Let M be a nonempty weakly compact convex subset of a locally complete convex space E satisfying the P-Opial condition. Let $T : M \to K(E)$ be a non-self nonexpansive set-valued mapping with the boundary condition (α) (and thus $T(M)$ is bounded). Then, mapping $(I - T)$ is demiclosed, i.e., the set $(I-T)(M)$ is closed in $E_w \times E$, where E_w is E with its weak topology and I is the identify mapping; and T has at least one fixed point.*

Proof. Since E satisfies the P-Opial condition, we have that if $x_n \to x$ weakly and $y \neq x$, then $\limsup \|x_n - x\| < \limsup \|x_n - y\|$. By following the same argument used in Lemma 7.5.2, we can prove that the set $(I - T)(M)$ is closed in $E_w \times E$, where E_w is E with its weak topology and I is the identify mapping. This will help us prove that T has a fixed point. In order to do so, we now show that there is a sequence $\{x_n\} \subset M$ and a sequence $\{y_n\} \subset E$ with $y_n \in T(x_n)$ such that $\lim_{n\to\infty}(x_n - y_n) = 0$, and then $0 \in (I - T)(M)$.

Now, let q be a given point in M. For each $n \in \mathbb{N}$, we define a set-valued mapping $T_n : M \to K(E)$ by

$$T_n(x) = \left(1 - \frac{1}{n}\right)Tx + \frac{1}{n}q,$$

for each $x \in M$. Then it is clear that $T_n(x) \subset E$ and T_n satisfies the boundary condition (α), since so does T. As each T_n is a $\mathfrak{F}$ contraction set-valued mapping for $n \in \mathbb{N}$, by Theorem 8.6.5, there exists $x_n \in M$ such that $x_n \in T_n(x_n)$. This means that there exists $y_n \in T(x_n)$ such that $x_n - y_n = \frac{1}{n}q - \frac{1}{n}y_n$ for $n \in \mathbb{N}$. As $T(M)$ is bounded, $\lim_{n\to\infty}(x_n - y_n) = 0$. Now, by following the argument used in Lemma 7.5.2 or by Theorem 1 of Chen and Singh [57], we have that the set $(I - T)(M)$ is closed in $E_w \times E$. Then, it follows that $0 \in (I - T)K$. This means that there exists $x \in K$ such that $x \in T(x)$, which means T has at least one fixed point in M. This completes the proof. $\square$

Now, we have the following fixed point result for non-self nonexpansive mappings which are weakly inward in LCSs.

Theorem 8.6.12. *Let M be a nonempty weakly compact convex subset of a locally complete convex space E, and let $T : M \to K(E)$ be a nonexpansive and weakly inward mapping. If $(I - T)$ is demiclosed, then T has a fixed point.*

Proof. Similarly to the above, Let q be a point in M. For each $n \in \mathbb{N}$, we define a set-valued mapping $T_n : M \to K(X)$ by

$$T_n(x) = \left(1 - \frac{1}{n}\right)Tx + \frac{1}{n}q,$$

for each $x \in M$. Then, it is clear that $T_n(x) \subset E$ and T_n is also weakly inward, since so does T. As each T_n is a $\mathfrak{F}$ contraction set-valued mapping for $n \in \mathbb{N}$, by Corollary 8.6.7, there exists $x_n \in M$

such that $x_n \in T_n(x_n)$. This means that there exists $y_n \in T(x_n)$ such that $x_n = (1 - \frac{1}{n})y_n + \frac{1}{n}q$ for $n \in \mathbb{N}$, which means we have $(x_n - y_n) = \frac{1}{n-1}(q - x_n)$. As M is weakly compact, the sequence $\{q - x_n\}$ is also bounded. In addition. without loss of generality, assume that x_n converges weakly to z in M. Now, for any $p \in \mathfrak{F}$, we have

$$\lim_{n \to +\infty} p(x_n - y_n) = \lim_{n \to +\infty} \frac{1}{n-1} p(q - x_n) = 0.$$

By the assumption that $(I - T)$ is demiclosed, it implies that $0 \in (I - T)z$, which means $z \in T(z)$. This completes the proof. □

As an application of Theorem 8.6.12, we have the following general fixed point theorem for nonexpansive set-valued mappings that are weakly inward in locally complete convex spaces which satisfy the P-Opial condition; see more results by Yanagi [280] and related references therein.

Theorem 8.6.13. *Let M be a nonempty weakly compact convex subset of a locally complete convex space E which satisfies the P-Opial condition. Let $T : M \to K(X)$ be a nonexpansive and weakly inward mapping. Then, T has a fixed point.*

Proof. As E satisfies the P-Opial condition and T is nonexpansive, it implies that the mapping $(I - T)$ is demiclosed by Theorem 8.6.11 above or by following the same proof of Lemma 7.5.2. Then, the conclusion follows by Theorem 8.6.12, and this completes the proof. □

Remark 8.6.14. We first note that the intersection condition used in Theorem 8.6.6 and the weakly inward condition used in Theorems 8.6.12 and 8.6.13 cannot not be replaced by the assumption that $T(x) \cap I_M(x) \neq \emptyset$ for each $x \in M$, as shown by the example provided by Xu [274, p. 705] (see also Zhang [298] and the references therein).

Before concluding this section, it is natural to establish Leray–Schauder principle for nonexpansive mappings which are either single-valued or set-valued mappings. Here, we have the following general condition for nonexpansive single-valued mappings in Banach spaces which are either uniformly convex or satisfying the

Opial condition (see also a similar result given by Theorem 7.2.11 in Chapter 7).

Theorem 8.6.15. *Let X be a Banach space which is either uniformly convex or satisfies the Opial condition, and let G be an open, convex, and weakly compact (thus bounded) subset of X and containing zero element, i.e., $\theta \in G$. Assume that $T : \overline{G} \to X$ is a nonexpansive mapping with nonempty values and satisfying the Leray–Schauder condition, i.e., for each $x \in \partial G$, $\lambda x \neq T(x)$ for any $\lambda \in (1, +\infty)$. Then, T must have a fixed point in G.*

Proof. Let $p = 1$ in Theorem 6.1.9. By the assumption, X is either uniformly convex or satisfying the Opial condition. It follows that the nonexpansive mapping T satisfies condition (H1) in Theorem 6.1.9 by Lemmas 7.2.3 and 7.5.9. Thus, the conclusion follows by Theorem 6.1.9. The proof is complete. $\square$

By Theorem 8.6.15, we have the following Leray–Schauder principle for nonexpansive mappings.

Theorem 8.6.16. *Let X be a Banach space which is either uniformly convex or satisfies the Opial condition, and let G be an open, convex, and weakly compact (thus bounded) subset of X and containing zero element $\theta \in G$. Assume that $T : \overline{G} \to X$ is a nonexpansive mapping with nonempty values. Then, the following conclusion holds:*

(1) *T has a fixed point $x \in G$ such that $x = T(x)$; or*
(2) *for each $x \in \partial G$, there exists $\lambda \in (1, +\infty)$ such that $\lambda x = T(x)$ (the Leray–Schauder condition).*

Proof. This is the consequence of Theorem 8.6.15. The proof is complete. $\square$

Remark 8.6.17. There exist comprehensive references on the study of the Leray–Schauder principle for nonexpansive singe- or set-valued mappings. Some of them are the works of Browder [31–36], Morales [171], Park [185–189], and related reference. In particular, see also Park [190–200] for a list of comprehensive references for new developments in fixed point theory and related applications to ordered spaces and associated discussion there.

Before closing this chapter, we would like to point out that, as shown by the discussion given in Section 3.4 in Chapter 3, the Caristi-type fixed point theorem [39] can be used as a fundamental tool to establish the majority of the principles of nonlinear functional analysis. In addition, as mentioned in the proof of Lemma 8.6.3, the Ekeland variational principle and its equivalent Caristi fixed point theorem under locally compete p-convex spaces for $p \in (0, 1]$ or in abstract spaces have been established by a number of scholars. For example, see Fang [87], Qiu [215–220], Qiu and Rolewicz [221], Park [187–200], Zhong and Zhao [301], and the references therein. Thus, the corresponding fixed point theorems for non-self nonexpansive mappings of single-values or set-valued values and related nonlinear alternative principle can be established for p-vector spaces and locally p-convex spaces. We leave those exciting topics to interested readers and young scholars in the area of nonlinear functional analysis and applications. In particular, we do not touch upon any studies related to nonlinear analysis based on the fixed point theorem established in Chapter 5 for USC set-valued mappings, in general, and p-vector spaces for $p \in (0, 1]$, which is a huge topic as well.

In addition, this book does not touch upon many other topics related to fixed point theory, in particular, on the study of various iteration algorithms for fixed points and related applications. This has been an active area since Mann's work [164] in 1953 and Ishikawa's works [114, 115] in the 1970s on the iteration algorithm of nonexpansive mappings in Banach spaces. The interested readers can find more recent studies in this area from Takahashi *et al.* [256], Zhu and Yao [304] for the developments in different iterative methods for generalized split feasibility problems and split variational inclusions related to fixed points and applications, and also Dong *et al.* [69] for the recent progress and applications on Krasnosel'skiĭ-Mann iterative methods. Moreover, we also refer Cobzaş [61], Jachymski *et al.* [117], Park [195], Park and Rhoades [201], and related references therein to the interested readers for a comprehensive review of the discussion on the various circumstances in which fixed point results imply completeness for metric spaces involving the Ekeland variational principle and its equivalent, the Caristi fixed point theorem, as well as other fixed point results having this property in metric spaces, including quasi-metric spaces and partial metric spaces for the discussion on topology and order and on fixed points in ordered structures and their completeness properties.

In the following chapter, we provide a short summary of the references for information and material that we used and supplementary reference materials which we did not cover in detail in the current version of this book with a brief explanation. Of course, for comprehensive references and recent developments in the general framework of fixed point theory and its study and applications, we refer the interested readers to Li [157, 158], Park [189–200], Xu [272–275], Xu and Muglia [277], Yuan [286–291], Yuan and Xiao [294, 295], and related references therein.

Chapter 9

Notes and Remarks

This chapter provides a summary of references for information and material used or referred in this book, as well as supplementary reference materials which are not covered in detail in the current version of this book, with a brief explanation.

Chapter 1:

The main material and information in this chapter follow a new writing style inspired by books authored by Berge [19] and Yuan [286–291], together with some material from Birkhoff–Kellogg [23], Ennassik et al. [78], Ennassik and Taoudi [79], Schauder [237], Goebel and Kirk [96], Granas and Dugundji [102], Kirk [130–134], Mauldin [167], Park [189], Takahashi [255], Xiao and Zhu [270], Zeidler [297], and the references therein.

Secondly, information from Poincare [211], Bernstein [20], Leray and Schauder [154], Schauder [237], Granas and Dugundji [102], Isac [118], Rothe [230, 231], and Zeidler [297] were referred.

In addition, recent developments on the fixed point theory in p-vector and locally p-convex spaces, mainly from Ennassik et al. [78], Ennassik and Taoudi [79], Yuan [286–291], and Yuan and Xiao [294, 295] were also referred.

Chapter 2:

The main material and information in this chapter were extracted from the following works: Axler [11], Balachandran [13], Bayoumi [15], Bayoumi et al. [16], Ding [65], Ennassik and Taoudi [79], Ennassik et al. [78], Jarchow [116], Kalton [119–121],

Kalton *et al.* [122], Qiu and Rolewicz [221], Rolewicz [228], Wang [263], Xiao and Lu [268], Xiao and Zhu [270], and Yuan [286–291].

Chapter 3:

For this chapter, the primary references were as follows: Ansari [7], Browder [30–36], Cauty [42, 43], Chang *et al.* [48], Du [70], [71], Ennassik *et al.* [78], Fan [81], Goebel and Kirk [96], Granas and Dugundji [102], Kirk [130–134], Mauldin [167], Michael [169], Park [184–190], Takahashi [255], Xiao and Zhu [270], Zeidler [297], and the references therein. Lemma 3.1.4 for the Dungundji-type extension theorem is an original idea and appears for the first time in this book. It is expected to serve as a useful tool for the nonlinear analyses of p-vector spaces in functional analysis.

In addition, for fixed point theorems in p-vector and locally p-convex spaces, we also referred to Ennassik *et al.* [78], Ennassik and Taoudi [79], Yuan [286–291], and Yuan and Xiao [294, 295]. The recent results from these works were also referred.

Moreover, we would like to point out that as an application of Lemma 3.2.2, Theorem 3.2.3 states that a s-convex compact subset D in a complete p-normed space $(X, \| \cdot \|_p)$ has a linear (continuous) mapping, $F : D \to F(D) (\subset l^p)$, which is a homeomorphism for $0 < s \leq p \leq 1$ due to the universal property of l^p. On the other hand, by the fact that the space $L^p[0, 1]$ for $p \in (0, 1)$ has no separability property, i.e., the dual space $L^p[0, 1])^* = \{0\}$, but, Yu (see [282]) unfortunately uses Roberts' example in [224] for a compact convex set with no extreme points in the (non-separable) space $L^p[0, 1]$ when applying Theorem 2.1 of Bessaga and Pełczyński (see [22]) to prove the non-existence of the homeomorphism $F : D \to F(D)$ given above in Theorem 3.13. Thus, the Corollary 2.5 of Yu [282] for the counterexample is not true for $0 < s < p < 1$.

Chapter 4:

The contents in this chapter were adapted from Yuan [284–291], plus some from Granas and Dugundji [102], Bonsall [24], Isac [118], Guo *et al.* [104], Machrafi and Oubbi [162], and Park [189–200].

Chapter 5:

The main material and information in this chapter were adapted from Ageev and Repovš [4], Alghamdi *et al.* [5], KKM [142], Park [189–200], Yuan [284–291].

In addition, for some material and information, the following were referred: Dugungji [73] and Kelly [126], Fan [82, 85, 86], Shih and Tan [243, 244], Tarafdar and Yuan [259], Granas and Dugundji [102], Goebel and Kirk [96], Repovš *et al.* [223].

Moreover, material from Kuratowski [148], Darbo [63], Alghamdi et al. [5], Machrafi and Oubbi [162], Nussbaum [176], Sadovskii [235], Silva et al. [249], Xiao, and Lu [268] are also cited.

For Proposition 5.6.1 and discussion on the equivalence of p-norms of finite n-dimensional spaces for $p \in (0, 1]$, it is mainly derived from the work of Tychonoff [260] in 1935 and the discussion provided in Theorem 1.21 of Rudin [232, p. 16]; also, see Theorem 3.5.6 of Jarchow [116, p. 66].

In addition, as pointed out in Remark 5.4.2, one of the most significant results in Chapter 5 is Theorem 5.6.6, which is the first fixed point theorem for USC set-valued mappings in topological vector spaces and p-vector spaces, which not only unifies the fixed point theorems of both single-valued continuous and USC mappings in Hausdorff topological vector spaces (and p-vector spaces) but also provides a unified positive answer to the Schauder conjecture. The key idea in the proof of Theorem 5.6.6 is partly due to the (classic) fact that all p-norms in finite-dimensional spaces are equivalent, combining the proof method originally developed by Fan [81] in 1952, with the application of Kelton's embedding theorem [119] in 1977, which states that a compact convex subset in topological vector spaces can be linearly embedded into locally p-convex spaces for $p \in (0, 1)$. This method should prove useful for the development of nonlinear analyses for p-vector spaces, in general.

Chapter 6:
The major references for this chapter were Yuan [284–291], plus Park [189–200] and the references therein.

Chapter 7:
The main material and information were adapted from Agarwal *et al.* [1], Ben-El-Mechaiekh and Mechaiekh [17], Ben-El-Mechaiekh and Saidi [18], Browder [33], Cellina [45], Chang [47], Chang *et al.* [49], Ennassik *et al.* [78], Fan [81, 82], Górniewicz [100], Granas and Dugundji [102], Guo *et al.* [104], Nhu [175], Park [189], Reich [222], Smart [250], Tychonoff [260], Weber [264, 265], Xiao and Lu

[268], Xiao and Zhu [270], Xu [273], Zeidler [297], and related references therein.

We also like to point out that the results given in this chapter are new and a continuation of related work discussed in Chapter 7; see also the recent works by Yuan [286–291] in locally p-convex spaces, for $p \in (0, 1]$.

Chapter 8:

The primary references for this chapter were as follows: Agarwal *et al.* [1], Ben-El-Mechaiekh and Mechaiekh [17], Ben-El-Mechaiekh and Saidi [18], Browder [33], Cauty [42, 43], Cellina [45], Chang [47], Chang *et al.* chang1993, Cobzaş [61], Ennassik *et al.* [78], Fan [81, 82], Górniewicz [100], Granas and Dugundji [102], Guo *et al.* [104], Jachymski *et al.* [117], Nhu [175], Park [189–195], Reich [222], Smart [250], Tychonoff [260], Weber [264, 265], Xiao and Lu [268], Xiao and Zhu [270], Xu [273], Yuan [283, 285], Zeidler [297], and related references therein.

In addition, Chen and Singh [57], Carbone and Conti [38], Chang and Yen [55], Chang *et al.* [53, 54], Kim *et al.* [129], Shahzad [238, 240], Singh [248], O'Regan [181], Tan and Yuan [258], and the references therein were also refereed.

Additional Note and Remark as the Final Takeaways for This Book

We would first like to point out that the major results given in this book are new and a continuation of related results provided and discussed by Yuan [286–291] recently in locally p-convex spaces, for $p \in (0, 1]$.

In addition, some important material and principles for nonlinear analysis were adapted from Browder [31–36], Morales [171], Park [185, 189], Fang [87], Qiu [215–220], Qiu and Rolewicz [221], Park [187–200], Zhong and Zhao [301], and the references therein.

Moreover, this book does not cover any content on the various iterative algorithms of fixed points and related applications, which has been a very active area since Mann's work [164] in 1953 and Ishikawa's work [114, 115] in the 1970s for the iteration algorithm of nonexpansive mappings in Banach spaces. Interested readers can find more recent studies in this area by referring to Takahashi *et al.* [256], Zhu and Yao [304] for the development on different iterative methods for generalized split feasibility problems and split variational inclusions related to fixed points and applications, and

Dong *et al.* [69] for the recent progress and applications related to the Krasnosel'skiĭ–Mann iterative methods.

As we know, the study of fixed point theory and its relation to nonlinear functional analysis is a large and active field, involving many scholars worldwide and covering numerous topics that cannot be listed fully in this monograph. Here, we only mention a few of them. For example, Ben-El-Mechaiekh and Mechaiekh [17] for the development of topological fixed point theory; Cauty [44], Dobrowolski [67], Ennassik and Taoudi [79], Ennassik *et al.* [78], Yuan [287], and Yuan [290] for the study of fixed point theory related to the Schauder conjecture in (metric) topological vector spaces; Li [157, 158] for a discussion on the extension of Tychonoff's fixed point theorem to pseudonorm adjoint topological vector spaces and also for fixed point theorems without continuity in metric vector spaces; Qin *et al.* [214] for a discussion on efficient extragradient methods for bilevel pseudomonotone variational inequalities with non-Lipschitz operators and their applications; Popescu [212] for a the new generalization of Ćirić's multi-valued operators, which were developed by Ćirić [59] in the early 2010s; review papers and handbooks or monographs by Kirk [134], Kirk and Shahzad [137], Kirk and Simes [138], and Rus *et al.* [233] for comprehensive studies; Petruşel *et al.* [207] for a recent discussion on the existence of common fixed points for a general class of operators; Granas and Dugundji [102] for the general fixed point theory; Guo [103] and Guo *et al.* [104] for some recent developments in random functional analysis and fixed point theory; Cho [58] for a survey on the fixed point theory, including topological fixed point and metric fixed point theories; Takahashi *et al.* [256] for a study on different iterative methods for generalized split feasibility problems related to fixed points and applications; Chang *et al.* [48] for the graph approximation method used in the study of the fixed point theory in locally p-convex spaces; Almezel *et al.* [6], O'Regan [181], and Ruzhansky *et al.* [234] for discussions on developments in fixed point theory and applications in general; and Qiu [220], Zhong *et al.* [302], and Zhou *et al.* [303] for a study of the Ekeland variational principle and its relation to fixed point theory and applications.

Finally, for a list of comprehensive references, recent developments in the general framework of fixed point theory, and its study and applications, we refer interested readers to Jachymski *et al.* [117]

on the development of fixed point theory with Banach fixed point theorem and selected topics from its hundred-year history; secondly, we suggest Chang *et al.* [50–52], Park [189–200], Xu [272–276], and Xu and Muglia [277] for refinements of the convergence results related to iteration for fixed points and gradient-projection algorithms under CAT(0) spaces or in fuzzy environments; and finally, the readers can refer to Yuan [286–291], Yuan and Xiao [294, 295], and the related references therein.

Of course, the biggest shortcoming of this book is the lack of discussion on topics, such as partial differential equations, optimization, and game theory, and disciplines, such as social sciences and economics, as well as other related areas that use the fixed point approach for nonlinear analysis. We refer interested readers to some of the classical monographs, such as Agarwal *et al.* [1], Agarwal and O'Regan [3], Alghamdi *et al.* [5], Aubin and Ekeland [10], Bauschke and Combettes [14], Borwein and Lewis [25], Brézis [26], Browder [35, 36], Chang [46], Cobzaş [61], Ekeland [77], Feng and Liu [88], Granas and Dugundji [102], Isac [118], Jachymski *et al.* [117], Kalton *et al.* [122], Lau and Yao [152], Lax [153], Park [189–195], Rockafellar [226], Rockafellar and Wets [227], Rothe [231], Takahashi [255], Yuan [283, 284], Zălinescu [296], Zeidler [297], Zhang [299], and the comprehensive references therein.

Last but not least, in particular, we would like to point out that the Dugundji-type extension theorem (i.e., Lemma 3.1.4) in Section 3.1 of Chapter 3, and the selection theorem (i.e., Lemma 5.3.4) established in Section 5.5 of Chapter 5 are new tools developed in this book; in addition, the method developed in Section 5.6 of Chapter 5 to prove fixed point theorems for upper semicontinuous set-valued mappings under the framework of general p-vector spaces would be a useful tool/approach for the development of a new theory in nonlinear analysis and topology for p-vector spaces and related applications for $p \in (0, 1]$.

Bibliography

[1] Agarwal RP, Meehan M, O'Regan D. Fixed point theory and applications. Cambridge Tracts in Mathematics. Vol 141. Cambridge: Cambridge University Press; 2001.

[2] Agarwal RP, O'Regan D. Birkhoff-Kellogg theorems on invariant directions for multimaps. Abstr Appl Anal. 2003;2003(7):435–448.

[3] Agarwal RP, O'Regan D. Essential U_c^k-type maps and Birkhoff-Kellogg theorems. J Appl Math Stoch Anal. 2004;2004(1):1–8.

[4] Ageev SM, Repovš D. A selection theorem for strongly regular multivalued mappings. Set-Valued Anal. 1998;6(4):345–362.

[5] Alghamdi MA, O'Regan D, Shahzad N. Krasnosel'skii type fixed point theorems for mappings on nonconvex sets. Abstr Appl Anal. 2012;2020:1–23. Article ID 267531.

[6] Almezel S, Ansari QH, Khamsi MA. Topics in fixed point theory. Cham: Springer; 2014.

[7] Ansari QH. Ekeland's variational principle and its extensions with applications. In: Almezel S, Ansari Q, Khamsi M, editors. Topics in fixed point theory. Cham: Springer; 2014. p. 65–100. Available from: https://doi.org/10.1007/978-3-319-01586-6_3.

[8] Askoura Y, Godet-Thobie C. Fixed points in contractible spaces and convex subsets of topological vector spaces. J Convex Anal. 2006;13(2):193–205.

[9] Alspach DE. A fixed point free nonexpansive map. Proc Am Math Soc. 1981;82:423–424.

[10] Aubin JP, Ekeland I. Applied nonlinear analysis. Mineola: Dover Publications, Inc. 2006.

[11] Axler S. Linear algebra done right. Undergraduate Texts in Mathematics. Cham: Springer; 2015.

[12] Ayerbe Toledano, JM. Domínguez Benavides T, López Acedo G. Measures of noncompactness in metric fixed point theory. Basel: Springer Basel AG; Birkhauser; 1997.

[13] Balachandran VK. Topological algebras. Vol 185. Amsterdam: Elsevier; 2000.

[14] Bauschke HH, Combettes PL. Convex analysis and monotone operator theory in Hilbert spaces. 2nd ed. Cham: Springer; 2017.

[15] Bayoumi A. Foundations of complex analysis in non locally convex spaces. Function theory without convexity condition. North-Holland mathematics studies. Vol 193. Amsterdam: Elsevier Science B.V.; 2003.

[16] Bayoumi A, Faried N, Mostafa R. Regularity properties of p-distance transformations in image analysis. Int J Contemp Math Sci. 2015;10:143–157.

[17] Ben-El-Mechaiekh H, Mechaiekh YA. (2022): An elementary proof of the Brouwer's fixed point theorem. Arab. J. Math. 11, 179–188 (2022). https://doi.org/10.1007/s40065-022-00366-0.

[18] Ben-El-Mechaiekh H, Saidi FB. On the continuous approximation of upper semicontinuous set-valued maps. Quest Answ Gen Topol. 2013;31(2):71–78.

[19] Berge C. Espaces Topologiques, Fonctions Multivoques. Paris: Dunod; 1959. First English Translation: Patterson EM. Topological spaces including a treatment of multi-valued functions, vector spaces and convexity. Edinburgh: Oliver & Boyd Ltd.; 1963.

[20] Bernstein S. Sur les equations de calcul des variations. Ann Sci Ecole Normale Sup. 1912;29:431–485.

[21] Bernuées J, Pena A. On the shape of p-convex hulls $0 < p < 1$. Acta Math Hungar. 1997;74(4):345–353.

[22] Bessaga C, Pełczyński A. Selected Topics in Infinite-Dimensional Topology, Vol. 58, Warsaw, Poland: PWN-Polish Scientific Publishers; 1975.

[23] Birkhoff GD, Kellogg OD. Invariant points in function space. Trans Am Math Soc. 1922;23(1):96–115.

[24] Bonsall FF. Lectures on some fixed point theorems of functional analysis. Notes by Vedak KB. Bombay: Tata Institute of Fundamental Research; 1962.

[25] Borwein JM, Lewis AS. Convex analysis and nonlinear optimization. CMS Books in Mathematics/Ouvrages de Mathematiques de la SMC. Vol 3. New York: Springer-Verlag; 2000.

[26] Brézis H. Functional analysis, Sobolev spaces and partial differential equations. Universitext. New York: Springer; 2011.

[27] Brézis H, Browder F. A general principle on ordered sets in nonlinear functional analysis. Adv Math. 1976;21:355–364.

[28] Brodskii MS, Milman DP. On the center of a convex set. Dokl Akad Nauk SSSR (NS). 1945;59:837–840 (in Russian).

[29] Brouwer LEJ. Beweis des ebenen Translationssatzes. Math Ann. 1912; 72(1):37–54.

[30] Browder FE. Nonexpansive nonlinear operators in a Banach space. Proc Natl Acad Sci USA. 1965;54:1041–1044.

[31] Browder FE. Convergence of approximants to fixed points of non-expansive nonlinear mappings in Banach spaces. Arch Ration Mech Anal. 1967;24(1):82–90.

[32] Browder FE. Semicontractive and semiaccretive nonlinear mappings in Banach spaces. Bull Am Math Soc. 1968;74:660–665.

[33] Browder FE. The fixed point theory of multi-valued mappings in topological vector spaces. Math Ann. 1968;177:283–301.

[34] Browder FE. On a theorem of Caristi and Kirk. Fixed Point Theory and Its Applications. Proceedings of Seminar. Halifax: Dalhousie University; 1975. p. 23–27. New York: Academic Press, Inc. [Harcourt Brace Jovanovich, Publishers]; 1976.

[35] Browder FE. Nonlinear functional analysis. Proceedings of Symposia in Pure Mathematics. Vol 18, Part 2. Providence: American Mathematical Society; 1976.

[36] Browder FE. Fixed point theory and nonlinear problems. Bull Am Math Soc (NS). 1983;9(1):1–39.

[37] Cammaroto F, Chinni A, Sturiale G. A remark on Ekeland's principle in locally convex topological vector spaces. Math Comput Model. 1999;30:75–79.

[38] Carbone A, Conti G. Multivalued maps and existence of best approximations. J Approx Theory. 1991;64:203–208.

[39] Caristi J. Fixed point theorems for mappings satisfying inwardness conditions. Trans Am Math Soc. 1976;215:241–251.

[40] Caristi J, Kirk WA. Geometric fixed point theory and inwardness conditions. The Geometry of Metric and Linear Spaces. Proceedings of a Conference Held at Michigan State University, East Lansing, Michigan, 1974. Lecture Notes in Mathematics. Vol. 490. Berlin: Springer-Verlag; 1975. pp. 74–83.

[41] Cauty R. Solution du problème de point xe de Schauder. Fund Math. 2001;170(3):231–246. DOI: 10.4064/fm170-3-2.

[42] Cauty R. Rétractès absolus de voisinage algébriques (French) [Algebraic absolute neighborhood retracts]. Serdica Math J. 2005;31(4):309–354.

[43] Cauty R. Le théorème de Lefschetz — Hopf pour les applications compactes des espaces ULC. (French) [The Lefschetz — Hopf theorem for compact maps of uniformly locally contractible spaces]. J Fixed Point Theory Appl. 2007;1(1):123–134.

[44] Cauty R. Points xes des applications compactes dans les espaces ULC. 2010. arXiv:1010.2401v1.

[45] Cellina A. Approximation of set valued functions and fixed point theorems. Ann Mat Pura Appl. 1969;82(4):17–24.

[46] Chang KQ. Infinite-dimensional Morse theory and its applications. Seminaire de Mathématiques Supérieures [Seminar on Higher Mathematics]. Vol. 97. Montreal: Presses de l'Université de Montréal; 1985.

[47] Chang SS. Some problems and results in the study of nonlinear analysis. Proceedings of the Second World Congress of Nonlinear Analysts, Part 7. Athens, 1996. Nonlinear Anal. 1997;30(7):4197–4208.

[48] Chang SS, Cho YJ, Park S, Yuan GX. Chapter 5: Fixed point theorems for quasi upper semicontinuous set-valued mappings in p-vector and locally p-convex spaces. In: Debnath P, Torres DFM, Cho YJ, editors. Advanced mathematical ananlysis and its applications. 1st ed. Boca Raton: Chapman and Hall/CRC; 2024. pp. 57–72. Available from: https://doi.org/10.1201/9781003388678.

[49] Chang SS, Cho YJ, Zhang Y. The topological versions of KKM theorem and Fan's matching theorem with applications. Topol Methods Nonlinear Anal. 1993;1(2):231–245.

[50] Chang SS, Salahuddin, Ahmad MK, Wang XR. Generalized vector variational like inequalities in fuzzy environment. Sets Syst. 2015;265:110–120.

[51] Chang SS, Tang JF, Wen CF. A new algorithm for monotone inclusion problems and fixed points on Hadamard manifolds with applications. Acta Math Sci Ser B (Engl Ed). 2021;41(4):1250–1262.

[52] Chang SS, Wang L, Zhao LC, Liu XD. A proximal point algorithm for finding minimizers and fixed points of quasi-pseudo-contractive mappings in CAT(0) spaces. J Fixed Point Theory Appl. 2021;23(1):18 p. Paper No. 5.

[53] Chang TH, Huang YY, Jeng JC. Fixed point theorems for multifunctions in S-KKM class. Nonlinear Anal. 2001;44:1007–1017.

[54] Chang TH, Huang YY, Jeng JC, Kuo KH. On S-KKM property and related topics. J Math Anal Appl. 1999;229:212–227.

[55] Chang TH, Yen CL. KKM property and fixed point theorems. J Math Anal Appl. 1996;203:224–235.

[56] Chen YQ. Fixed points for convex continuous mappings in topological vector spaces. Proc Am Math Soc. 2001;129(7):2157–2162.

[57] Chen YK, Singh KL. Fixed points for nonexpansive multivalued mapping and the Opial's condition. Jñānābha. 1992;22:107–110.

[58] Cho YJ. Survey on metric fixed point theory and applications. In: Ruzhansky M, Cho Y, Agarwal P, Area I, editors. Advances in

real and complex analysis with applications. Trends in Mathematics. Singapore: Birkhäuser; 2017. p. 183–241. Available from: https://doi.org/10.1007/978-981-10-4337-6_9.

[59] Ćirić L. Multi-valued nonlinear contraction mappings. Nonlinear Anal. 2009;71(7–8):2716–2723.

[60] Clarkson JA. Uniformly convex spaces. Trans Am Math Soc. 1926;40:396–414.

[61] Cobzaş S. Fixed points and completeness in metric and generalized metric spaces. J Math Sci (NY). 2020;250(3):475–535.

[62] Daneš J. A geometric theorem useful in nonlinear functional analysis. Boll Un Mat Ital. 1972;6:369–372.

[63] Darbo G. Punti uniti in trasformazioni a condominio non compatto. Rend Sem Mat Univ Padova. 1955;24:84–92.

[64] Diestel J. Geometry of Banach spaces-selected topics. Lecture Notes in Mathematics. Vol 485. Berlin: Springer-Verlag; 1975.

[65] Ding GG. New theory in functional analysis. Beijing: Academic Press; 2007.

[66] Djebali S. Algebraic Topology: Constructions, Retractions, and Fixed Point Theory. Walter de Gruyter GmbH & Co.KG, Berlin; 2024.

[67] Dobrowolski T. Revisiting Cauty's proof of the Schauder conjecture. Abstr Appl Anal. 2003;7:407–433.

[68] Dobrowolski T, Marciszewski W. Rays and the fixed point property in noncompact spaces. Tsukuba J Math. 1997;21(1):97–112. Available from: https://doi.org/10.21099/tkbjm/1496163163.

[69] Dong QL, Cho YJ, He SN, Pardalos PM, Rassias TM. The Krasnosel'skiǐ-Mann iterative method — recent progress and applications. SpringerBriefs in Optimization. Cham: Springer; 2022.

[70] Du WS. A simple proof of Caristi's fixed point theorem without using Zorn's lemma and transfinite induction. Thai J Math. 2016;14(2):259–264.

[71] Du WS. A direct proof of Caristi's fixed point theorem. Appl Math Sci. 2016; 10(46):2289–2294.

[72] Dunford N, Schwartz JT. Linear operators: General theory. Pure and Applied Mathematics. Vol 7. New York: Interscience; 1958.

[73] Dugundji J. Topology. Allyn and Bacon, Inc., Boston, USA; 1978.

[74] Edelstein M. The construction of an asymptotic center with a fixed point property. Bull Am Math Soc. 1972;78:206–208.

[75] Ekeland I. Sur les problémes variationnels. C R Acad Sci Paris Sér A-B. 1972;275(1972):A1057–A1059.

[76] Ekeland I. On the variational principle. J Math Anal Appl. 1974;47:324–353.

[77] Ekeland I. On convex minimization problems. Bull Am Math Soc. 1979;1:445–474.

[78] Ennassik M, Maniar L, Taoudi MA. Fixed point theorems in r-normed and locally r-convex spaces and applications. Fixed Point Theory. 2021;22(2):625–644.

[79] Ennassik M, Taoudi MA. On the conjecture of Schauder. J Fixed Point Theory Appl. 2021;23(4):15 p. Paper No. 52.

[80] Ewert J, Neubrunn T. On quasi-continuous multivalued maps. Demonstr Math. 1988;21(3):697–711.

[81] Fan K. Fixed-point and minimax theorems in locally convex topological linear spaces. Proc Natl Acad Sci USA. 1952;38:121–126.

[82] Fan K. A generalization of Tychonoff's fixed point theorem. Math Ann. 1960/61;142:305–310.

[83] Fan K. Sur un théorème minimax. C R Acad Sci Paris. 1964;259:3925–3928.

[84] Fan K. Extensions of two fixed point theorems of F E Browder. Math Z. 1969;112:234–240.

[85] Fan K. A minimax inequality and applications. Inequalities, III. Proceedings of Third Symposium, University of California, Los Angeles, California, 1969; Dedicated to the memory of Theodore S. Motzkin. New York: Academic Press; 1972. pp. 103–113.

[86] Fan K. Some properties of convex sets related to fixed point theorems. Math Ann. 1984;266(4):519–537.

[87] Fang J. The variational principle and fixed point theorems in certain topological spaces. J Math Anal Appl. 1996;202:398–412. DOI: 10.1006/jmaa.1996.0323.

[88] Feng YQ, Liu SY. Fixed point theorems for multi-valued contractive mappings and multi-valued Caristi type mappings. J Math Anal Appl. 2006;317(1):103–112.

[89] Fuchssteiner B. Verallgemeinerte Konvexilatsbcgrifle und der Satz von Krein-Milman. Math Ann. 1970;186:149–154.

[90] Fuchssteiner B. Verallgemeinerte KonvexitStsbegriffe und Lp-Raume. Math Ann. 1970;186:171–176.

[91] Fuchssteiner B. Lattices and Choquet's theorem. Funct Anal. 1974;17:377–387.

[92] Furi M, Pera MP. A continuation method on locally convex spaces and applications to ordinary differential equations on noncompact intervals. Ann Polon Math. 1987;47(3):331–346.

[93] Gal SG, Goldstein JA. Semigroups of linear operators on p-Fréchet spaces $0 < p < 1$. Acta Math Hungar. 2007;114(1–2):13–36.

[94] Gholizadeh L, Karapinar E, Roohi M. Some fixed point theorems in locally p-convex spaces. Fixed Point Theory Appl. 2013; 2013(312):10 p.

[95] Goebel K. On a fixed point theorem for multivalued nonexpansive mappings. Ann Univ Mariae Curie-Sklodowska Sect A. 1975;29: 69–72.

[96] Goebel K, Kirk WA. Topics in metric fixed point theory. Cambridge Studies in Advanced Mathematics. Vol 28. Cambridge: Cambridge University Press; 1990.

[97] Goebel K, Kirk WA. Some problems in metric fixed point theory. J Fixed Point Theory Appl. 2008;4(1):13–25.

[98] Goebel K, Reich S. Uniform convexity, hyperbolic geometry and non-expansive mappings. New York: Marcel Dekker; 1984.

[99] Göhde D. Zum Prinzip der kontraktiven Abbildung (in German). Math Nachr. 1965;30:251–258.

[100] Górniewicz L. Topological fixed point theory of multivalued mappings. Mathematics and Its Applications. Vol. 495. Dordrecht: Kluwer Academic Publishers; 1999.

[101] Górniewicz L, Granas A, Kryszewski W. On the homotopy method in the fixed point index theory of multivalued mappings of compact absolute neighborhood retracts. J Math Anal Appl. 1991;161(2): 457–473.

[102] Granas A, Dugundji J. Fixed point theory. Springer Monographs in Mathematics. New York: Springer-Verlag; 2003.

[103] Guo TX. Survey of recent developments of random metric theory and its applications in China. I. Acta Anal Funct Appl. 2001;3(2): 129–158.

[104] Guo TX, Wang YC, Xu HK, Yuan GX, Chen G. The noncompact Schauder xed point theorem in random normed modules and its applications. Math Ann. 2024;391:3863–3911.

[105] Halpern BR, Bergman GH. A fixed-point theorem for inward and outward maps. Trans Am Math Soc. 1965;130:353–358.

[106] He W, Yannelis NC. Equilibria with discontinuous preferences: new fixed point theorems. J Math Anal Appl. 2017;450(2):1421–1433.

[107] He W, Yannelis NC. Existence of equilibria in discontinuous Bayesian games. J Econom Theory. 2016;162:181–194.

[108] Holá L, Mirmostafaee AK. On continuity of set-valued mappings, Topol Appl. 2022;320(2022):11 p. Paper No. 108200. Available from: http://doi.org.zzulib.vpn358.com/10.1016/j.topol.2022.108200.

[109] Horváth J. Topological vector spaces and distributions. Vol 1. Reading: Addison–Wesley; 1966.

[110] Huang NJ, Lee BS, Kang MK. Fixed point theorems for compatible mappings with applications to the solutions of functional equations arising in dynamic programmings. Int J Math Math Sci. 1997;20(4):673–680.

[111] Husain T, Latif A. Fixed points of multivalued nonexpansive maps. Math Jpn. 1988;33: 385–391.

[112] Husain T, Tarafdar E. Fixed point theorems for multivalued mappings of nonexpansive type. Yokohama Math J. 1980;28(1–2):1–6.

[113] Hussain N, Khan AR. Applications of the best approximation operator to *-nonexpansive maps in Hilbert spaces. Numer Funct Anal Optim. 2003;24(3–4):327–338.

[114] Ishikawa S. Fixed points by a new iteration method. Proc Am Math Soc. 1974;44:147–150.

[115] Ishikawa S. Fixed points and iteration of a nonexpansive mapping in a Banach space. Proc Am Math Soc. 1976;59(1):65–71.

[116] Jarchow H. Locally convex spaces. Stuttgart: B.G. Teubner; 1981.

[117] Jachymski J, Jóźwik I, Terepeta M. The Banach Fixed Point Theorem: selected topics from its hundred-year history. Rev Real Acad Cienc Exactas Fis Nat Ser A-Mat. 2024;118:140. Available from: https://doi.org/10.1007/s13398-024-01636-6.

[118] Isac G. Leray-Schauder type alternatives, complementarity problems and variational inequalities. Nonconvex Optimization and Its Applications. Vol 87. New York: Springer; 2006.

[119] Kalton NJ. Compact p-convex sets. Q J Math Oxf Ser. 1977;28(2): 301–308.

[120] Kalton NJ. Universal spaces and universal bases in metric linear spaces. Stud Math. 1977;61:161–191.

[121] Kalton NJ. Compact and strictly singular operators on Orlicz spaces. Isr J Math. 1977;26:126–136.

[122] Kalton NJ, Peck NT, Roberts JW. An F-space sampler. London Mathematical Society Lecture Note Series. Vol 89. Cambridge: Cambridge University Press; 1984.

[123] Karlovitz LA. Existence of fixed points of nonexpansive mappings in a space without normal structure. Pacific J Math. 1976;66(1):153–159.

[124] Karlovitz LA. On nonexpansive mappings. Proc Am Math Soc. 1976;55(2):321–325.

[125] Kaniok L. On measures of non compactness in general topological vector spaces. Comment Math Univ Carol. 1990;31(3):479–487.

[126] Kelley JL. General topology. Princeton: Van Nostrand; 1957.

[127] Khan MA, McLean RP, Uyanik M. On equilibria in constrained generalized games with the weak continuous inclusion property. J Math Anal Appl. 2024;537(1):19 p. Paper No. 128258.

[128] Khan MA, McLean RP, Uyanik M. On constrained generalized games with action sets in non-locally-convex and non-Hausdorff topological vector spaces. J Math Econom. 2024;111:8 p. Paper No. 102964.

[129] Kim IS, Kim K, Park S. Leray-Schauder alternatives for approximable maps in topological vector spaces. Math Comput Model. 2002;35: 385–391.

[130] Kirk WA. A fixed point theorem for mappings which do not increase distances. Am Math Mon. 1965;72:1004–1006.

[131] Kirk WA. Caristi's fixed point theorem and the theory of normal solvability. Fixed Point Theory and Its Applications. Proceedings of Seminar, Dalhousie University, Halifax, N.S., 1975. New York: Academic Press, Inc. [Harcourt Brace Jovanovich, Publishers]; 1976. p. 109–120.

[132] Kirk WA. Caristi's fixed point theorem and metric convexity. Colloq Math. 1976;36(1):81–86.

[133] Kirk WA. An abstract fixed point theorem for nonexpansive mappings. Proc Am Math Soc. 1981;82(4):640–642.

[134] Kirk WA. Metric fixed point theory: a brief retrospective. Fixed Point Theory Appl. 2015;215(2015). Available from: https://doi.org/10.11 86/s13663-015-0464-5.

[135] Kirk WA, Caristi J. Mapping theorems in metric and Banach spaces. Bull Acad Polon Sci Sér Sci Math Astron Phys. 1975;23(8):891–894.

[136] Kirk WA, Saliga LM. The Brézis-Browder order principle and extensions of Caristi's theorem. Nonlinear Anal. 2001;47(4):2765–2778.

[137] Kirk WA, Shahzad N. Fixed point theory in distance spaces. Cham: Springer; 2014.

[138] Kirk WA, Sims B. Handbook of metric fixed point theory. Dordrecht: Springer; 2001. Available from: https://doi.org/10.1007/978-94-017-1748-9_12.

[139] Klee VL. Some topological properties of convex sets. Trans Am Math Soc. 1955;78(1):3045. Available from: https://doi.org/10.1090/S0002-9947-1955-0069388-5.

[140] Klee VL. Leray-Schauder theory without local convexity. Math Ann. 1960:141:286–296.

[141] Klee VL. Convexity of Chevyshev sets. Math Ann. 1960/61;142: 292–304.

[142] Knaster H, Kuratowski C, Mazurkiwiecz S. Ein beweis des fixpunktsatzes für n-dimensional simplexe. Fund Math. 1929;63:132–137.

[143] Ko HM, Tsai YH. Fixed point theorems for point to set mappings in locally convex spaces and a characterization of complete metric spaces. Bull Acad Sin. 1979;6(4):461–470.

[144] Kozlov V, Thim J, Turesson B. A fixed point theorem in locally convex spaces. Collect Math. 2010;61(2):223–239.

[145] Kozlowski WM. A purely metric proof of the Caristi fixed point theorem. Bull Aust Math Soc. 2017;95(2):333–337.

[146] Kreyszig E. Introductory functional analysis with applications. Wiley Classics Library. New York: John Wiley & Sons, Inc.; 1989.

[147] Kryszewski W. Graph-approximation of set-valued maps on noncompact domains. Topol Appl. 1998;83(1):1–21.

[148] Kuratowski K. Sur les espaces complets. Fund Math. 1930;15: 301–309.

[149] Latif A, Hussain N, Kutbi MA. Applications of Caristi's fixed point results. J Inequal Appl. 2012;40(2012). Available from: https://doi. org/10.1186/1029-242X-2012-40.

[150] Lami Dozo E. Multivalued nonexpansive mappings and Opial's condition. Proc Am Math Soc. 1973;38:286–292.

[151] Lassonde M. Sur un principe géométrique en analyse convexe (French. English summary) [On a geometric principle in convex analysis]. Stud Math. 1991;101(1991):1–18.

[152] Lau AT-M, Yao LJ. Common fixed point properties for a family of set-valued mappings. J Math Anal Appl. 2018;459(1):203–216.

[153] Lax PD. Functional analysis. Pure and Applied Mathematics (New York). New York: Wiley-Interscience [John Wiley & Sons]; 2002.

[154] Leray J, Schauder J. Topologie et equations fonctionnelles. Ann Sci Ecole Normale Sup. 1934;51:45–78.

[155] Li GZ. The fixed point index and the fixed point theorems of 1-set-contraction mappings. Proc Am Math Soc. 1988;104:1163–1170.

[156] Li GZ, Xu SY, Duan HG. Fixed point theorems of 1-set-contractive operators in Banach spaces. Appl Math Lett. 2006;19(5):403–412.

[157] Li JL. An extension of Tychonoff's fixed point theorem to pseudonorm adjoint topological vector spaces. Optimization. 2021;70(5–6):1217–1229.

[158] Li JL. Fixed point theorems without continuity in metric vector spaces. Fixed Point Theory. 2024;25(2):621–634.

[159] Lim TG. A fixed point theorem for multivalued nonexpansive mappings in a uniformly convex Banach space. Bull Am Math Soc. 1974;80:1123–1126.

[160] Liu LS. Approximation theorems and fixed point theorems for various classes of 1-set-contractive mappings in Banach spaces. Acta Math Sin (Engl Ser). 2001;17(1):103–112.

[161] López-Acedo G, Xu HK. Remarks on multivalued nonexpansive mappings. Soochow J Math. 1995;21(1):107–115.

[162] Machrafi N, Oubbi L. Real-valued non compactness measures in topological vector spaces and applications. [Corrected title: Real-valued non compactness measures in topological vector spaces and applications]. Banach J Math Anal. 2020;14(4):1305–1325.

[163] Mańka R. The topological fixed point property - an elementary continuum-theoretic approach. Fixed Point Theory and Its Applications. Vol. 77. Warsaw: Banach Center Publications, Polish Academy of Sciences Institute of Mathematics; 2007. pp. 183–200.

[164] Mann WR. Mean value methods in iteration. Proc Am Math Soc. 1953;4:506–510.

[165] Massa S. Some remarks on Opial spaces. Boll Un Mat Ital. 1983;6: 65–70.

[166] Massa S, Roux D, Singh SP. Fixed point theorems for multifunctions. J Pure Appl Math. 1987;18(6):512–514.

[167] Mauldin RD. The Scottish book. Mathematics from the Scottish Café with selected problems from the new Scottish book. 2nd ed. Birkhauser, Basel, Switzerland; 2015.

[168] McLennan A. Advanced fixed point theory for economics. Singapore: Springer; 2018.

[169] Michael, E. A note on a set-valued extension property. Topology Appl. 2011;158(13):1526–1528.

[170] Milman D. On some criteria for the regularity of spaces of type (B). C.R.(Doklady) Acad. Sci. U.R.S.S. 1938;20:243–246.

[171] Morales CH. The Leray-Schauder condition for continuous pseudo-contraction mappings. Proc Am Math Soc. 2009;137(3):1013–1020.

[172] Muglia L, Marino G. Some results on the approximation of solutions of variational inequalities for multivalued maps on Banach spaces. Mediterr J Math. 2021;18(4):19 p. Paper No. 157.

[173] Nadler SB. Multi-valued contraction mappings. Pacific J Math. 1969;30:475–488.

[174] Neubrunn T. quasi-continuity. Real Anal Exch. 1988/89;14(2): 259–306. DOI: 10.2307/44151947.

[175] Nhu NT. The fixed point property for weakly admissible compact convex sets: searching for a solution to Schauder's conjecture. Topol Appl. 1996;68(1):1–12.

[176] Nussbaum RD. The fixed point index and asymptotic fixed point theorems for k-set-contractions. Bull Am Math Soc. 1969;75:490–495.

[177] Oettli W, Théra M. Equivalents of Ekeland's principle. Bull Austral Math Soc. 1993;48(3):385–392.

[178] Okon T. The Kakutani fixed point theorem for Robert spaces. Topol Appl. 2002;123(3):461–470.

[179] Opial Z. Weak convergence of the sequence of successive approximations for nonexpansive mappings. Bull Am Math Soc. 1967;73: 595–597.

[180] O'Regan D. Abstract Leray-Schauder type alternatives and extensions. An Stiin Univ "Ovidius" Constana Ser Mat. 2019;27(1): 233–243.

[181] O'Regan D. Continuation theorems for Monch countable compactness-type set-valued maps. Appl Anal. 2021;100(7): 1432–1439.

[182] O'Regan D, Precup R. Theorems of Leray-Schauder type and applications. Gordon and Breach Science Publishers, London, UK; 2001.

[183] Oubbi L. Algebras of Gelfand-continuous functions into Arens-Michael algebras. Commun Korean Math Soc. 2019;34(2):585–602.

[184] Park S. Some coincidence theorems on acyclic multifunctions and applications to KKM theory. Fixed Point Theory and Applications (Halifax, NS, 1991). River Edge: World Scientific Publishing; 1992. pp. 248–277.

[185] Park S. Generalized Leray-Schauder principles for compact admissible multifunctions. Topol Methods Nonlinear Anal. 1995;5(2): 271–277.

[186] Park S. Generalized Leray-Schauder principles for condensing admissible multifunctions. Ann Mat Pura Appl. 1997;172(4):65–85.

[187] Park S. The KKM principle in abstract convex spaces: Equivalent formulations and applications. Nonlinear Anal. 2010;73(4):1028–1042.

[188] Park S. On the KKM theory of locally p-convex spaces. Nonlinear Analysis and Convex Analysis. Vol 2011. Japan: Institute of Mathematical Research (Kyoto University); 2016. p. 70–77. Available from: http://hdl.handle.net/2433/231597.

[189] Park S. One hundred years of the Brouwer fixed point theorem. J Natl Acad Sci ROK Nat Sci Ser. 2021;60(1):1–77.

[190] Park S. Some new equivalents of the Brouwer fixed point theorem. Adv Theory Nonlinear Anal Appl. 2022;6(3):300–309. Available from: https://doi.org/10.31197/atnaa.1086232.

[191] Park S. Equivalents of various maximum principles. Results Nonlinear Anal. 2022;5:169–174.

[192] Park S. Equivalents of ordered fixed point theorems of Kirk, Caristi, Nadler, Banach, and others. Adv Theory Nonlinear Anal Appl. 2022;6:420–439.

[193] Park S. Foundations of ordered fixed point theory. J Natl Acad Sci ROK Nat Sci Ser. 2022;61:247–287.

[194] Park S. Equivalents of some ordered fixed point theorems. J Adv Math Comput Sci. 2023;38:52–67.

[195] Park S. Historical remarks on the Caristi related theorems. J Nonlinear Convex Anal. 2023;24:2531–2542.

[196] Park S. Remarks on the metatheorem in ordered fixed point theory. In: Debnath P, Torres DFM, Cho YJ, editors. Advanced mathematical analysis and its applications. CRC Press, Boca Raton, FL, USA; 2023. p. 11–27.

[197] Park S. History of the metatheorem in ordered fixed point theory. J Natl Acad Sci ROK Natl Sci Ser. 2023;62:373–410.

[198] Park S. The use of quasi-metric in the metric fixed point theory. J Nonlinear Convex Anal. 2024;25(7):1553–1564.

[199] Park S. The realm of the Rus-Hicks-Rhoades maps in the metric fixed point theory. J Natl Acad Sci ROK Natl Sci Ser. 2024;63(1):1–45.

[200] Park S. Improving many metric fixed point theorems. Lett Nonlinear Anal Its Appl. 2024;2(2):35–61. Available from: https://lettersinnon linearanalysis.com/index.php/lnaa.

[201] Park S, Rhoades BE. Comments on characterizations for metric completeness. Math Jpn. 1986;31(1):95–97.

[202] Penot JP. Fixed point theorems without convexity. Bull Soc Math Fr Mém. 1979;60:129–152.

[203] Penot JP. The drop theorem, the petal theorem and Ekeland's variational principle. Nonlinear Anal. 1986;10(9):813–822.

[204] Pérez Carreras P, Bonet J. Barrelled locally convex spaces. North-Holland Mathematics Studies. Vol 131. Amsterdam: North-Holland Publishing Co.; 1987.

[205] Petryshyn WV. Construction of fixed points of demicompact mappings in Hilbert spaces. J Math Anal Appl. 1966;14:276–284.

[206] Petryshyn WV. Fixed point theorems for various classes of 1-set-contractive and 1-ball-contractive mappings in Banach spaces. Trans Am Math Soc. 1973;182:323–352.

[207] Petruşel A, Petruşel G, Yao JC. Common fixed point results for a general class of operators. J Nonlinear Convex Anal. 2022;23(11): 2687–2694.

[208] Petruşel A, Rus IA, Serban MA. Basic problems of the metric fixed point theory and the relevance of a metric fixed point theorem for a multivalued operator. J Nonlinear Convex Anal. 2014;15(3):493–513.

[209] Pettis BJ. A proof that every uniformly convex space is reflexive. Duke Math J. 1939;5:249–253.

[210] Pietramala P. Convergence of approximating fixed points sets for multivalued nonexpansive mappings. Comment Math Univ Carol. 1991;32(4):697–701.

[211] Poincare H. Sur un theoreme de geometric. Rend Circ Mat Palermo. 1912;33:357–407.

[212] Popescu O. A new generalization of Ćirić's multi-valued operators. Fixed Point Theory. 2024;25(2):705–722.

[213] Potter AJB. An elementary version of the Leray-Schauder theorem. J Lond Math Soc. 1972;5(2):414–416.

[214] Qin X, Petruşel A, Tan B, Yao JC. Efficient extragradient methods for bilevel pseudomonotone variational inequalities with non-Lipschitz

operators and their applications. Fixed Point Theory. 2024;25(1): 309–331.

[215] Qiu JH. Local completeness and dual local quasi-completeness. Proc Am Math Soc. 2001;129:1419–1425.

[216] Qiu JH. Local completeness and drop theorem. J Math Anal Appl. 2002;266:288–297.

[217] Qiu JH. Ekeland's variational principle in locally complete spaces. Math Nachr. 2003;257:55–58.

[218] Qiu JH. Local completeness, drop theorem and Ekelands variational principle. J Math Anal Appl. 2005;311:23–39.

[219] Qiu JH. The density of extremal points in Ekeland's variational principle. J Math Anal Appl. 2007;328:946–957.

[220] Qiu JH. Ekeland's variational principle in locally convex spaces and the density of extremal points. J Math Anal Appl. 2009;360(1): 317–327.

[221] Qiu JH, Rolewicz S. Ekeland's variational principle in locally p-convex spaces and related results. Stud Math. 2008;186(3):219–235.

[222] Reich S. Fixed points in locally convex spaces. Math Z. 1972;125: 17–31.

[223] Repovš D, Semenov PV, Ščepin EV. Approximation of upper semicontinuous maps on paracompact spaces. Rocky Mt J Math. 1998;28(3):1089–1101.

[224] Roberts JW. A compact convex set with no extreme points. Stud Math. 1977;60(3):255–266.

[225] Robertson LB. Topological vector spaces. Publ Inst Math. 1971;12(26):19–21.

[226] Rockafellar RT. Convex analysis. Princeton: Princeton University Press; 1970.

[227] Rockafellar RT, Wets RJ-B. Variational analysis. Grundlehren der Mathematischen Wissenschaften. Vol. 317. Heidelberg: Springer; 1998.

[228] Rolewicz S. Metric linear spaces. Warszawa: PWN-Polish Scientific Publishers; 1985.

[229] Rolewicz S. On drop property. Stud Math. 1986;85(1):27–35.

[230] Rothe EH. Some homotopy theorems concerning Leray-Schauder maps. Dynamical systems, II (Gainesville, Fla., 1981). New York: Academic Press; 1982. pp. 327–348.

[231] Rothe EH. Introduction to various aspects of degree theory in Banach spaces. Mathematical Surveys and Monographs. Vol. 23. Providence: American Mathematical Society; 1986.

[232] Rudin W. Functional analysis. 2nd ed. New York: McGraw-Hill; 1991.

[233] Rus IA, Petruşel A, Petruşel G. Fixed point theory. Cluj University Press, Cluj-Napoca, Romania; 2008.

[234] Ruzhansky M, Cho YJ, Agarwal P, Iván A. Advances in real and complex analysis with applications. Trends in Mathematics. Singapore: Birkhäuser/Springer; 2017.

[235] Sadovskii BN. On a fixed point principle [in Russian]. Funkt Analiz Prilozh. 1967;1(2):74–76.

[236] Sadovskii BN. Application of topological methods in the theory of periodic solutions of nonlinear differential-operator equations of neutral type. Dokl Akad Nauk SSSR. 1971;200:1037–1040 (in Russian). Sov Phys Dokl. 1971;12.

[237] Schauder J. Der Fixpunktsatz in Funktionalraumen. Stud Math. 1930;2:171–180.

[238] Shahzad N. Approximation and Leray-Schauder type results for $\mathfrak{U}_c^k$ maps. Topol Methods Nonlinear Anal. 2004;24(2):337–346.

[239] Shahzad N. Fixed point and approximation results for multimaps in $S - KKM$ class. Nonlinear Anal. 2004;56(6):905–918.

[240] Shahzad N. Approximation and Leray-Schauder type results for multimaps in the S-KKM class. Bull Belg Math Soc. 2006;13(1):113–121.

[241] Shapiro JH. Examples of proper, closed, weakly dense subspaces in non locally convex F-spaces. Israel J Math. 1969;7:369–380.

[242] Shapiro JH. A fixed-point Farrago. Universitext. Cham: Springer; 2016.

[243] Shih MH, Tan KK. Shapley selections and covering theorems of simplexes. Lecture Notes in Pure and Applied Mathematics. Vol. 107. New York: Marcel Dekker, Inc.; 1987. pp. 245–251.

[244] Shih MH, Tan KK. Browder — Hartman — Stampacchia variational inequalities for multi-valued monotone operators. J Math Anal Appl. 1988;134(2):431–440.

[245] Sezer S, Eken Z, Tinaztepe G, Adilov G. p-convex functions and some of their properties. Numer Funct Anal Optim. 2021;42(4):443–459.

[246] Siegel J. New proof of Caristi's fixed point theorem. Proc Am Math Soc. 1977;66(1):54–56.

[247] Simons S. Boundness in linear topological spaces. Trans Am Math Soc. 1964;113:169–180.

[248] Singh SP, Watson B, Srivastava F. Fixed point theory and best approximation: the KKM-map principle. Mathematics and Its Applications. Vol. 424. Dordrecht: Kluwer Academic Publishers; 1997.

[249] Silva EB, Fernandez DL, Nikolova L. Generalized quasi-Banach sequence spaces and measures of noncompactness. An Acad Bras Cie. 2013;85(2):443–456.

[250] Smart DR. Fixed point theorems. Cambridge: Cambridge University Press; 1980.

[251] Subrahmanyam PV. Elementary fixed point theorems. Forum for Interdisciplinary Mathematics. Singapore: Springer; 2018.

[252] Song YS, Huang Y. Fixed point property and approximation of a class of nonexpansive mappings. Fixed Point Theory Appl. 2014;2014(01):11 p.

[253] Tabor JA, Tabor JO, Idak M. Stability of isometries in p-Banach spaces. Funct Approx. 2008;38:109–119.

[254] Takahashi W. Existence theorems generalizing fixed point theorems for multivalued mappings. In: Baillon J-B, Théra M, editors. Fixed point theory and applications. Pitman Research Notes in Mathematics. Vol. 252. Harlow: Longman; 1991. pp. 397–406.

[255] Takahashi W. Nonlinear functional analysis. Yokohama: Yokohama Publishers; 2000.

[256] Takahashi W, Xu HK, Yao JC. Iterative methods for generalized split feasibility problems in Hilbert spaces. Set-Valued Var Anal. 2015;23(2):205–221.

[257] Tan DN. On extension of isometries on the unit spheres of L^p - spaces for $0 < p \leq 1$. Nonlinear Anal. 2011;74:6981–6987.

[258] Tan KK, Yuan GX. Random fixed-point theorems and approximation in cones. J Math Anal Appl. 1994;185:378–390.

[259] Tarafdar E, Yuan GX. A remark on coincidence theorems. Proc Am Math Soc. 1994;122(3):957–959.

[260] Tychonoff A. Ein Fixpunktsatz. Math Ann. 1935;111:767–776.

[261] van Dulst D. Equivalent norms and the fixed point property for nonexpansive mappings. J Lond Math Soc (2). 1982;25(1):139–144.

[262] Waelbroeck L. Topological vector spaces and algebras. Springer Lecture Notes, Springer, New York, USA. Vol 230. 1970.

[263] Wang JY. An introduction to locally p-convex spaces. Beijing: Academic Press; 2013. pp. 26–64.

[264] Weber H. Compact convex sets in non-locally convex linear spaces. Dedicated to the memory of Professor Gottfried Köthe. Note Mat. 1992;12:271–289.

[265] Weber H. Compact convex sets in non-locally convex linear spaces, Schauder-Tychonoff fixed point theorem. Topology, measures, and fractals (Warnemunde, 1991). Mathematical Research. Vol 66. Akademie-Verlag: Berlin; 1992. p. 37–40.

[266] Wiweger A. Linear spaces with mixed topology. Stud Math. 1961;20:47–68.

[267] Wong CS. On a fixed point theorem of contractive type. Proc Am Math Soc. 1976;57(2):283–285.

[268] Xiao JZ, Lu Y. Some fixed point theorems for s-convex subsets in p-normed spaces based on measures ofnoncompactness. J Fixed Point Theory Appl. 2018;20(2):22 p. Paper No. 83.

[269] Xiao JZ, Wang ZY. Foundations of abstract analysis. 2nd ed. (in Chinese). Beijing: Tsinghua University Press; 2022.

[270] Xiao JZ, Zhu XH. Some fixed point theorems for s-convex subsets in p-normed spaces. Nonlinear Anal. 2011;74(5):1738–1748.

[271] Xiao JZ, Zhu XH. Fixed points of nonexpansive operators and normal structure concerning s-Orlicz convex sets. J Math Anal Appl. 2024;128620. Available from: https://doi.org/10.1016/j.jmaa.2024.128620.

[272] Xu HK. Inequalities in Banach spaces with applications. Nonlinear Anal. 1991;16(12):1127–1138.

[273] Xu HK. Metric fixed point theory for multivalued mappings. Diss Math (Rozpr Mat). 2000;389:39 p.

[274] Xu HK. Multivalued nonexpansive mappings in Banach spaces. Nonlinear Anal. 2001;43(6):693–706.

[275] Xu HK. On weakly nonexpansive and ∗-nonexpansive multivalued mappings. Math Jpn. 2001;36(3):441–445.

[276] Xu HK. Refinements of some convergence results of the gradient-projection algorithm. Ann Math Sci Appl. 2024;8(2):347–363.

[277] Xu HK, Muglia L. On solving variational inequalities defined on fixed point sets of multivalued mappings in Banach spaces. J Fixed Point Theory Appl. 2020;22(4):17 p. Paper No.79.

[278] Xu SY. New fixed point theorems for 1-set-contractive operators in Banach spaces. Nonlinear Anal. 2007;67(3):938–944.

[279] Xu SY, Jia BG, Li GZ. Fixed points for weakly inward mappings in Banach spaces. J Math Anal Appl. 2006;319(2):863–873.

[280] Yanagi K. On some fixed point theorems for multivalued mappings. Pacific J Math. 1980;87(1):233–240.

[281] Yi HW, Zhao YC. Fixed point theorems for weakly inward multivalued mappings and their randomizations. J Math Anal Appl. 1994;183:613–619. DOI: 10.1006/jmaa.1994.1167.

[282] Yu L. Remarks on "Some fixed point theorems for s-convex subsets in p-normed spaces". J. Anal. 2024;32(2):1139–1143.

[283] Yuan GX. The study of minimax inequalities and applications to economies and variational inequalities. Mem Am Math Soc. 1998;132(625).

[284] Yuan GX. KKM theory and applications in nonlinear analysis. Monographs and Textbooks in Pure and Applied Mathematics. Vol 218. New York: Marcel Dekker, Inc.; 1999.

[285] Yuan GX. Nonlinear analysis in p-vector spaces for single-valued 1-set contractive mappings. Fixed Point Theory Algorithms Sci Eng. 2022;26. Available from: https://doi.org/10.1186/s13663-022-00735-6.

[286] Yuan GX. Nonlinear analysis by applying best approximation method in *p*-vector spaces. Fixed Point Theory Algorithms Sci Eng. 2022;20. Available from: https://doi.org/10.1186/s13663-022-00730-x.

[287] Yuan GX. Fixed point theorems and applications in *p*-vector spaces. Fixed Point Theory Algorithms Sci Eng. 2023;10. Available from: https://doi.org/10.1186/s13663-023-00747-w.

[288] Yuan GX. Corrigendum: fixed point theorem and related nonlinear analysis by the best approximation method in *p*-vector spaces. Numer Funct Anal Optim. 2023;44(10):1094–1096.

[289] Yuan GX. Fixed point theorem and related nonlinear analysis by the best approximation method in *p*-vector spaces. Numer Funct Anal Optim. 2023;44(4):221–295. DOI: 10.1080/01630563.2023.2167088.

[290] Yuan GX. Fixed point theorems for continuous single-valued and upper semicontinuous set-valued mappings in *p*-vector and locally *P*-convex spaces. Topol Methods Nonlinear Anal. 2024;63(1): 209–225. DOI: 10.12775/TMNA.2023.027.

[291] Yuan GX. Fixed point theorems and principles of nonlinear alternatives in *p*-vector spaces. Acta Math Sin (in Chinese). 2024;67(5):962– 986. Available from: https://link.cnki.net/urlid/11.2038.O1.2024031 4.0941.004.

[292] Yuan GX. Fixed point theorems of compact USC set-valued mappings in Hausdorff topological vector spaces. Preprint on Researchgate.net. 2025. Available from: https://www.researchgate.net/publication/39 0165768. DOI: 10.13140/RG.2.2.23629.55527.

[293] Yuan GX. Fixed point theorems of set-valued mappings in *p*-vector spaces with applications to generalized games. Preprint on Researchgate.net. 2025. Available from: https://www.researchgate.net/publi cation/390167767. DOI: 10.13140/RG.2.2.30969.58727.

[294] Yuan GX, Xiao JZ. Some fixed point theorems in *p*-normed spaces. J of Convex and Nonlinear Anal (forthcoming), 2025.

[295] Yuan GX, Xiao JZ. Browder-Göhde-Kirk type fixed point theorem for nonexpansive mappins in *p*-normed spaces with normal structure. Appl Math Opt (forthcoming), 2025.

[296] Zălinescu C. Convex analysis in general vector spaces. River Edge: World Scientific; 2002. p. 33.

[297] Zeidler E. Nonlinear functional analysis and its applications. Vol I. Fixed-point theorems. New York: Springer Verlag; 1986.

[298] Zhang S. Star-shaped sets and fixed points of multivalued mappings. Math Jpn. 1991;36(2):327–334.

[299] Zhang SQ. Functional analysis and its applications (in Chinese). Beijing: China Science Press; 2018.

[300] Zhong CK. A generalization of Ekeland's variational principle and application to the study of the relation between the weak P.S. condition and coercivity. Nonlinear Anal. 1997;29(12):1421–1431.

[301] Zhong CK, Zhao PH. Locally Ekeland's variational principle and some surjective mapping theorems. Chin Ann Math Ser B. 1998;19(3):273–280.

[302] Zhong CK, Zhu J, Zhao PH. An extension of multi-valued contraction mappings and fixed points. Proc Am Math Soc. 1999;128(8):2439–2444.

[303] Zhou ZA, Tang ZY, Zhao KQ. Ekeland's variational principle for the set-valued map with variable ordering structures and applications. J Ind Manag Optim. 2024;20(1):260–269.

[304] Zhu LJ, Yao YH. Modified splitting algorithms for approximating solutions of split variational inclusions in Hilbert spaces. Fixed Point Theory. 2024;25(2):773–784.

Glossary

$(E, D; \Gamma)$ denotes an abstract convex space.

A_0, A and $\partial(A)$ denote its interior, closure and boundary, respectively for a given subset A if there is no confusion or specification.

D_c denotes the Chebyshev center of a subset D.

$I_C^p(x)$ denotes the p-inward set of a subset C at the point x.

$O_C^p(x)$ denotes the p-outward set of a subset C at the point x.

$U \mathfrak{R} V$ denotes the equivalence relation between U and V.

X^* denotes the dual space of space X.

Δ_n denotes an n-dimensional simplex.

$\Delta_p(C)$ denotes the smallest absolutely p-convex sets containing set C.

$\Gamma_A = \Gamma(A)$ denotes the convex hull of a subset A.

$\Gamma_p(C)$ denotes the smallest p-convex sets containing C.

Ω denotes $\{1, 2, \ldots, n\}$ or $\{1, 2, \ldots, \}$, respectively, if there is no confusion, or unless specified.

$\alpha_U(A)$ denotes a subset A's U-measure of noncompactness.

$\beta_H(D)$ denotes the Hausdorff measure of noncompactness for a subset D.

$\beta_K(D)$ denotes the Kurotowskii measure of noncompactness for a subset D.

$\beta_U(A)$ denotes a subset A's U-measure of noncompactness.

$\mathcal{C}(K)$ denotes the Banach space of all real-valued continuous functions defined on the set (space) K.

$\mathcal{M}(K)$ denotes the dual space of $\mathcal{C}(K)$.

$\mathbb{C}$ (or $\mathbb{K}$) denotes the complex field (or real line (field)).

$\mathbb{R}$ (or $\mathbb{K}$) also denotes the real line (field) if no confusion.

$\mathbb{N}$ denotes the set of all positive integers.

θ denotes the origin element of a p-vector space if there is no confusion.

$\mathbb{R}^n$ denotes a finite n-dimensional Euclidean space.

$\mathcal{U}$ denotes the base of p-vector spaces X's topology.

$\mathcal{B}_\theta$ denotes the family of all balanced zero element (θ) neighborhoods in space E.

$\mathcal{B}_\theta$ denotes the family of all shrinkable zero element (θ) neighborhoods in E if no confusion.

$\overline{AC}_p(K)$ denotes the closure of the absolutely p-convex hull of a given subset K in X.

$\overline{C}_p(K)$ denotes the closure of p-convex hull of a subset K in X.

$\partial(C)$ denotes the set of all frontal points of C with respect to the origin element θ in space X.

$\partial_a(C)$ denotes the set of all frontal points of C with respect to the point a in the space X.

$\partial_p(C)$ denotes the set of all p-extreme points of C.

$\mathrm{Aff}(S)$ denotes the affine hull of a subset S.

$\mathrm{lin}(A)$ (or $\mathrm{span}(A)$) denotes the linear hull (span) of a subset A.

$aint_M A$ denotes the algebraic interior of a subset A in M.

$bd(S)$ or $\partial(S)$ denotes the boundary of a subset S if no confusion or unless specified.

$core(C)$ denotes the set of all internal points (i.e., core) of a subset C.

$core(A)$ or A^i also denotes the core of a subset A if no confusion.

$d_H(A_1, A_2)$ denotes the Hausdorff metric between two subsets A_1 and A_2.

$d_{H_p}(A, B)$ denotes the Hausdorff metric between subsets A and B for a given seminorm p in a locally convex space E.

$d_{P_U}(x, C) := \inf\{P_U(x, y) : y \in C\}$, where P_U is a Minkowski p-functional of an open subset U containing the zero element θ.

$diam(D)$ or $\delta(D)$ denotes the diameter of a subset D.

$dim(E)$ denotes the dimension of a vector space E.

$dom(J_c)$ denotes the effective domain of the Minkowski functional J_c.

$int(A)$ denotes the interior of a subset A.

$\overline{co}_p(A)$ also denotes the closed p-convex hull of a subset A.

$co_p(A)$ also denotes the p-convex hull of a subset A.

q_c denotes the Minkowski p-functional of C in the space X.

$r(D)$ denotes the Chebyshev radius of a subset D.

$ri(A)$ denotes the relative interior of a subset A in E.

$rint(A)$ denotes the topological interior of a subset A in $\text{Aff}(A)$ (or $\text{aff}(A)$).

$x \to y$ denotes x (strongly) converges to y.

$x \stackrel{*}{\rightharpoonup} y$ denotes x weakly converges to y.

PCWD denotes proper, closed and weakly dense (subspace).

span[B] also denotes the linear subspace spanned by a subset B in X.

QUSC denotes quasi upper semicontinuous.

LSC denotes lower semicontinuous.

KKM principle denotes Knaster, Kuratowski and Mazurkiewicz lemma (theorem).

BCZN denotes the basic collections of zero neighborhoods.

SCZN denotes the sufficient collections of zero neighborhoods.

LS denotes the Leray–Schauder boundary condition.

FPP denotes the fixed point property.

WFPP denotes weak fixed point property.

Index